Fundamentals of Modern Digital Systems

B. R. Bannister

and

D. G. Whitehead

University of Hull

Fundamentals of Modern Digital Systems

Second Edition

Springer-Verlag New York Inc.

Softcover reprint of the hardcover 1st edition 1987

First edition 1983
Second edition 1987

This edition published 1987 by
MACMILLAN EDUCATION LTD
London and Basingstoke

Sole distributors in the USA and its dependencies
Springer-Verlag New York Inc.
175 Fifth Avenue,
New York, NY 10010
USA.

Springer-Verlag New York

Library of Congress Cataloging-in-Publication Data
Bannister, B. R. (Brian Roy)
Fundamentals of modern digital systems.
Includes bibliographies and index.
1. Digital electronics. 2. Electronic digital
computers — Circuits. I. Whitehead, D. G. (Donald
Gill) II. Title.
TK7868.D5B42 1987 621.3815 87-20515

ISBN-13: 978-1-4684-6296-8 e-ISBN-13: 978-1-4684-6294-4
DOI: 10.1007/ 978-1-4684-6294-4

Contents

Preface to Second Edition

"One should teach things which will still be valid
in twenty years' time: the fundamental concepts
and underlying principles. Anything else is dishonest."
Professor G. Strachey, 1969

Professor Strachey's basic requirement of any course is just as important today as it was twenty years ago, but the difficulty is always in deciding exactly what constitutes the core material. In a field developing as rapidly as digital electronics it is even more difficult to determine how many of the current technques should be included to allow the student to appreciate the significance of that core material. Twenty years ago the integrated logic circuit itself had yet to make an impact.

This book, therefore, presents the fundamental principles common to all modern digital electronic systems, and is intended to provide a solid foundation for the study of such systems at first degree or equivalent level. Some of the topics covered in the later chapters are also relevant to post-graduate studies. This second edition reflects the continuing rapid pace of developments in the digital field over recent years, and is a completely revised, updated, and in some cases rearranged book, re-emphasizing the underlying principles in a modern framework.

The first three chapters introduce the ideas basic to all digital systems; the two-state algebra used in combinational logic (chapter 1), the two-valued representation of data (chapter 2), and the methods of building two-state electronic circuits used in the different families of logic (chapter 3). Chapter 1 now includes an introduction to symmetric functions, and bar codes have been introduced in chapter 2 to illustrate some of the special coding techniques. Chapter 3 has been reshaped to cater for the shift in emphasis over recent years from bipolar to unipolar circuitry, with an outline of VLSI principles. The need to consider the testing of circuits at all stages in the design process is stressed throughout the book.

The next three chapters, 4, 5, and 6, cover design methods involving the use of large scale integrated devices; chapter 4 considers logic design techniques for combinational circuits and includes a new section on custom design. The material in chapter 5 of the first edition has been reworked and considerably extended so that sequential logic is now split into two chapters: 5 covers sequential logic components, and 6 deals with the more formal analysis and design of sequential circuits. This allows an introduction to linear feedback shift registers and timing circuits to be included in chapter 5, and the algorithmic state machine chart representation in chapter 6.

Chapter 7 explains the principles of memory systems, and chapter 8 is devoted to programmable devices. Again, the fundamental concepts are emphasized rather than specific device details, though typical usage is quoted freely to illustrate the ideas. Finally, chapter 9 deals with some of the techniques associated with the operation of digital systems, especially those involving data transmission and conversion between the analog and digital forms.

One of the major decisions in any book of this type is what logic symbols to use. Although international standards have now moved to rectangular symbol shapes, we have decided to use the well-known Mil. Spec. symbols which indicate the logic function by their shape and have the advantage of both simplicity and being readily understood. We have, however, included in appendix A an introduction to the current international standard, which is designed to allow meaningful symbols to be developed for even very complex circuits such as those found in VLSI devices.

Worked examples are included throughout the text, and each chapter concludes with problems for the reader and references for further work.

We are indebted to our many students, past and present, who have knowingly and unknowingly contributed to the formation of our ideas on the presentation of the material, and who have made this new edition both possible and worth while.

Hull, BRB
1987 DGW

1 Combinational Logic

To express any body of ideas in a meaningful form, some sort of logical framework is required. This book is concerned with digital systems, and the framework used is that based on a mathematical logic developed largely by the English mathematician George Boole. Boole published the basic axioms and rules for a two-valued algebra in 1854, but for the rest of the nineteenth century his work remained firmly in the province of mathematics (see Boole, 1953). Huntington (1904) published a set of postulates for a two-state algebra which forms the basis of our modern approach to boolean algebra. However, it was not until 1938 that this algebra was shown to be a useful tool for the engineer. Shannon (1938) introduced a switching algebra, adapting boolean algebra for use in the analysis of relay switching networks used in telephone systems. The development of digital systems since the 1940s, initially restricted to the digital computer, now extends over a seemingly unlimited range of applications. A grasp of the structure of Boole's two-state logic is essential to the understanding of switching theory, which is itself fundamental to the design of all digital systems.

Many books have been written on the mathematical aspects of switching theory but here we concentrate on its engineering applications. And in addition to our two-state algebra we make use of two further two-state features: firstly, the representation of numerical data in binary form and the extension of coding ideas to include other non-numeric data; and secondly, the ability to represent logical values in terms of voltage levels existing in circuits switching between two well-defined voltage levels. These circuits range from the simple switch to complex integrated circuits containing thousands of transistors. In the first few chapters, then, we shall concentrate on these fundamental engineering features, with the intention of providing the tools necessary for the analysis and manipulation of digital switching circuits. At the present time, switching theory cannot provide all the answers to the problems of the logic designer, especially when large complex systems are considered. It is also true that when real-life problems are encountered, the number of variables present can make the manipulation of the logical expressions almost impossibly complicated. However, it is often possible to decompose such systems into smaller subsystems, more amenable to analysis by the designer, or to employ suitably programmed computers to handle the unwieldy expressions.

Digital switching circuits can be divided, very broadly, into two types known as *combinational logic* circuits and *sequential logic* circuits. A combinational logic circuit is one in which the response is determined wholly by the present state of its inputs. A sequential logic circuit involves the concept of system memory in which the immediate response depends in part on the previous response. As we shall see in later chapters, such circuits can be broken down into a combinational circuit plus memory and feedback units. This first chapter, therefore, will consider the simpler problem of specifying and analysing combinational logic circuits, and we must initially introduce some of the basic axioms of boolean algebra.

1.1 Sets and boolean algebra

It is common usage to call a group of objects with similar properties a *set*. For example, we speak of a set of cutlery, a set of wheels or a set of chess pieces. The term set can be extended to define a group of objects or ideas that have one or more common properties. Thus all people who live in New York have at least one thing in common, they live in New York. They are the *elements* of a set. We could also have a set of sets and this would be termed a *class*.

The general convention is to use capital letters to denote the sets and lower case letters to denote the elements, so, if the set A has a finite number of elements, we write A = (a, b, c, . . .).

A set can often be broken down into subsets by grouping according to some other common feature. For example, the chess set, C = (32 chessmen), but the pieces can be grouped into several subsets where the elements could be all white pawns, all court pieces, and so on. Then we would write C = (a, b, c) where a is the subset (all white pawns), b is the subset (all black pawns) and c is the subset (all court pieces).

Two sets may be said to be equal iff (if and only if) they contain the same elements but not necessarily the same number of each element. For example, a car needs a set of four wheels but an equal set for a tricycle will contain only three wheels.

It is often helpful to represent sets graphically, so that interrelationships show up more clearly. The diagrams we use are termed Venn or Veitch diagrams (Veitch, 1952). In figure 1.1 all elements of set A are contained within the rectangle A, and all elements of set B are included within the rectangle B. Any

Figure 1.1

element outside a rectangle is not a member of that set. In this example, rectangle A could represent the set of all Frenchmen, and B the set of all Germans. A third set, C, of Europeans contains both A and B as *proper subsets* in that each is totally contained in C, as in figure 1.2.

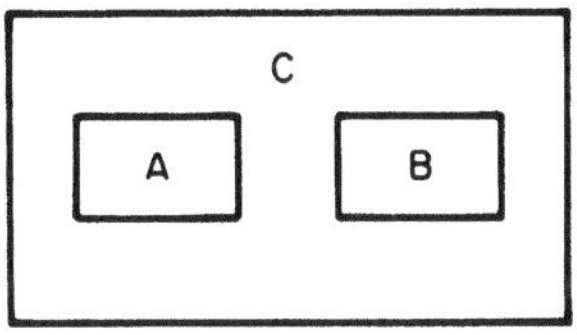

Figure 1.2

If now we extend set A to include all French-speaking people, and set B to include all German-speaking people, then the diagram must be revised since some Frenchmen speak German and vice versa. The area common to A and B in figure 1.3 is the *intersection* of A and B, and represents those Europeans who speak both French *and* German. We shall represent the intersection of A and B by A • B. The area covered by either A or B (or both) is the *union* of sets A and B, and represents those Europeans who speak French *or* German. We describe the union of A and B by A + B (A • B is read as 'A and B'; A + B is read as 'A or B')*.

If a collection of the subsets of a set A has a union which is equal to A, the collection is said to be a *cover* of the set A. If all the elements of the cover are mutually *disjoint* (that is, they do not intersect), then the cover is a *partition* of the set A. Figure 1.3 represents a covering of the set of Europeans in that it

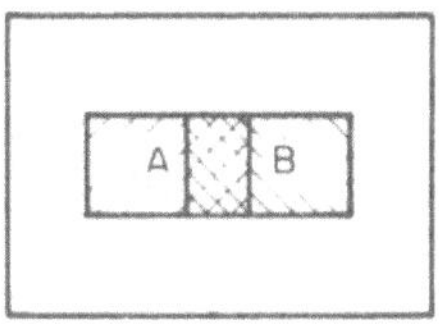

Figure 1.3

includes the set of French-speaking people, German-speaking people, and those who speak both French and German. It is not a partition, however, since there is an intersection. If we could ensure that no European speaks both French and German, the subsets would become disjoint and the set partitioned. This could be represented by figure 1.2. Finally we note there is a fourth subset in figure 1.3, and that is the group of Europeans who speak neither French nor German.

The Venn diagram can be used as a convenient way of demonstrating the validity of any algebraic representation of a set.

*This convention ties in with general engineering usage: mathematical texts use $A \cap B$ rather than A • B and $A \cup B$ rather than A + B.

Example 1.1: Demonstrate the validity of the expression

$$A + (A \cdot B) = A \cdot (A + B) = A$$

Figure 1.4 shows the Venn diagram for $A + (A \cdot B)$ and for $A \cdot (A + B)$.

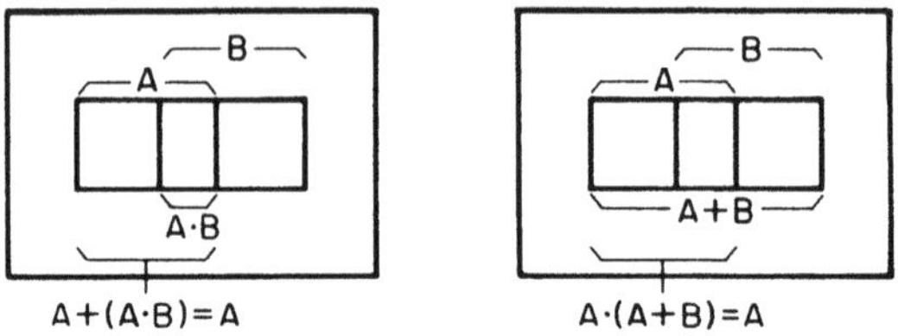

Figure 1.4 Example 1.1

In both cases the diagram reduces to A, as shown.

The method can be extended to show the truth of expressions involving more than two variables, though for more than four variables it becomes too unwieldy.

Example 1.2: Use Venn diagrams to prove that

$$(A \cdot B) + (B \cdot C) + (C \cdot A) = (A + B) \cdot (B + C) \cdot (C + A)$$

The technique is to draw diagrams for both sides of the equality, to show that they result in identical patterns. The three-variable diagram is of the form shown in figure 1.5a, so, considering the left-hand side of the expression, we get the pattern of figure 1.5b and the right-hand side gives the pattern of figure 1.5c. The two are seen to be identical.

We can also use the method with numerical data.

Example 1.3: In the electronic engineering degree course, all final-year students take mathematics and at least two of the following options: Communications Theory, Logic Systems, Advanced Circuit Design. Last year 9 took Communications Theory, 14 chose Logic Systems, and 13 studied Advanced Circuit Design. Only 2 students took all three options. How many final-year students were there, and how many took Logic Systems with Advanced Circuit Design only?

There are three intersecting sets. Let set A be those taking Advanced Circuit Design, set B those taking Logic Systems, and set C those taking Communications Theory. The Venn diagram for three intersecting sets is as shown in figure 1.6.

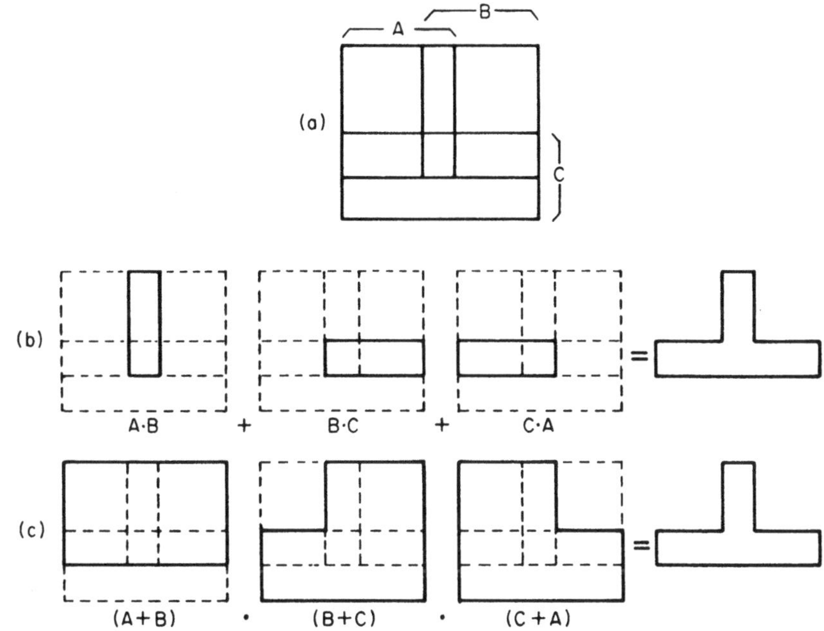

Figure 1.5 Example 1.2

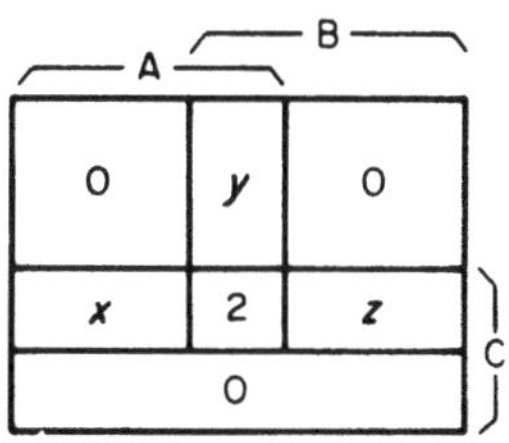

Figure 1.6 Example 1.3

Considering the students who take Advanced Circuit Design, it can be seen from the Venn diagram that these are represented by

$$x + y + 2 = 13$$

Similarly, we obtain for the other disciplines

$$y + z + 2 = 14$$

$$x + z + 2 = 9$$

Solving these simultaneous equations yields

$$x = 3; z = 4; y = 8$$

The number that took Logic Systems with Advanced Circuit Design only is represented by the subset y, and $y = 8$. The total number of students is the sum of $x + y + z + 2 = 17$.

We have so far represented the elements of a set as being contained within a rectangle. All elements outside the rectangle must form another set which is known as the *complementary* set. The complement of set A is the set $\overline{A}$ (read as 'not A'). It follows that any set and its complement must contain *all* elements, which form the *universal* set, denoted by *I*. Thus we can write $A + \overline{A} = I$.

The complement of the universal set must contain no elements, and is termed the *null* or *empty* set, denoted by *0*. Since there can be no intersection of a set with its complement, it must also follow that $A \cdot \overline{A} = 0$.

The laws of boolean algebra, as applicable to switching theory, are similarly defined in terms of the two simple *connectives*, AND and OR, and the *inversion*, NOT. The connective AND corresponds to the intersection in set theory, and has the same symbol (•); the connective OR corresponds to the union, and has the symbol +. For convenience, where no ambiguity is likely to arise, the AND symbol is normally implied and we shall follow the convention from now on. Any apparent ambiguity can be resolved by remembering that AND takes precedence over OR. Thus

$$A + BC \equiv A + (B \cdot C)$$
$$(A + B)C \equiv (A + B) \cdot C$$

Certain fundamental laws of boolean algebra can be defined

$$\left.\begin{aligned} A(A + B) &= A \\ A + AB &= A \end{aligned}\right\} \text{Absorption laws for AND and OR}$$

$$\left.\begin{aligned} A(BC) &= (AB)C \\ A + (B + C) &= (A + B) + C \end{aligned}\right\} \text{Associative laws for AND and OR}$$

$$\left.\begin{aligned} AB &= BA \\ A + B &= B + A \end{aligned}\right\} \text{Commutative laws for AND and OR}$$

$$\left.\begin{aligned} A(B + C) &= AB + AC \\ A + BC &= (A + B)(A + C) \end{aligned}\right\} \text{Distributive laws for AND and OR}$$

In a two-state system the complement of A, $\overline{A}$, is also available. Switching theory is concerned with the presence or absence of signals such as voltage, current or fluid flow. In set theory the universal set *I* contains all elements and the null set *0* is empty. The universal set leads to the idea of a signal being present at all times, or of a logical statement being valid irrespective of other conditions. The symbol used for such a signal or statement is 1. When a signal is not present, or a statement is invalid, the symbol is 0. Thus $\overline{1} = 0$ and $\overline{0} = 1$: Also $A + \overline{A} = 1$ and $A\overline{A} = 0$.

Additional relationships can be shown using these ideas.

Example 1.4: Show that $A \cdot 1 = A$.

Since $(A + \overline{A}) = 1$

$$\begin{aligned} A \cdot 1 &= A(A + \overline{A}) \\ &= AA + A\overline{A} \\ &= A + 0 \\ &= A \end{aligned}$$

Example 1.5: Show that $A + 1 = 1$.

Again using $(A + \overline{A}) = 1$

$$\begin{aligned} A + 1 &= A + (A + \overline{A}) \\ &= (A + A) + \overline{A} \\ &= A + \overline{A} \\ &= 1 \end{aligned}$$

1.2 The truth table

A logical function is made up of boolean variables which can have only the value zero or one, and the validity or truth value of a function is dependent on the values of the variables involved. Just as the Venn diagram enabled us to show the interrelationships between sets, so the truth value of the function is more clearly shown in a *truth table*. A truth table is a list of the truth values of the function, f, for all possible values, zero and one, of its variables. Thus n variables, each having two states, give 2^n combinations leading to 2^n rows in the truth table. Figure 1.7 shows the truth tables (a) for f = AB and (b) for f = A + B.

A	B	f = AB
0	0	0
0	1	0
1	0	0
1	1	1

(a)

A	B	f = A+B
0	0	0
0	1	1
1	0	1
1	1	1

(b)

Figure 1.7 Truth tables: (a) for f = AB; (b) for f = A + B

The truth table allows the proof of identities by *perfect induction*, but becomes unwieldy above three or four variables.

Example 1.6: Show that $\overline{XY} = \overline{X} + \overline{Y}$.

The process involves showing that each side of the identity gives the same function column, in much the same way as Venn diagrams are used. Figure 1.8 gives the two tables which are seen to be identical in the final column.

X	Y	XY	$\overline{XY}$
0	0	0	1
0	1	0	1
1	0	0	1
1	1	1	0

X	Y	$\overline{X}$	$\overline{Y}$	$\overline{X}+\overline{Y}$
0	0	1	1	1
0	1	1	0	1
1	0	0	1	1
1	1	0	0	0

Figure 1.8 Example 1.6

The example illustrates a very important relationship which is known as de Morgan's law for AND, and allows us to introduce a new concept at this point. Any boolean function has a *dual* function which can be obtained by replacing each connective by its dual, that is AND by OR and OR by AND. By taking the dual of each side of de Morgan's law for AND, for example

$$\overline{XY} = \overline{X} + \overline{Y}$$

becomes

$$\overline{X + Y} = \overline{X}\,\overline{Y}$$

This is de Morgan's law for OR, and can readily be verified by truth table. If a theorem is valid its dual is also valid, and this can be of great assistance in problem solving.

Example 1.7: Prove the distributive laws for AND and OR.

The proof by perfect induction of the first distributive law, A(B + C) = AB + AC, is shown in figure 1.9. By taking the dual of the identity, we obtain A + BC = (A + B) (A + C) which is the second distributive law, and must also be valid.

A	B	C	B+C	A(B+C)	AB	AC	AB+AC
0	0	0	0	0	0	0	0
0	0	1	1	0	0	0	0
0	1	0	1	0	0	0	0
0	1	1	1	0	0	0	0
1	0	0	0	0	0	0	0
1	0	1	1	1	0	1	1
1	1	0	1	1	1	0	1
1	1	1	1	1	1	1	1

Figure 1.9 Example 1.7

Duality also allows us to arrive at an alternative form of a given boolean relationship, by complementing variables as well as replacing each connective by its dual. In fact, the dual of a variable is its complement, and it follows that the dual of zero is one, and vice versa. Given the function f = A(B + C), for example, we can use duality to give the alternative form $f = \overline{\overline{A} + \overline{B}\,\overline{C}}$, since $\overline{f} = \overline{A} + \overline{B}\,\overline{C}$.

1.3 The Karnaugh map

The *Karnaugh map* (Karnaugh, 1953) can be considered as a development of the Venn diagram. As we shall see the Karnaugh map (or K-map) is an invaluable aid to all aspects of logic design and we shall encounter it frequently.

The K-map for a single variable is shown in figure 1.10a. In terms of set theory we have divided the universal set into two complementary subsets A and $\overline{A}$, whereas the Venn diagram allows the area for $\overline{A}$, representing those elements

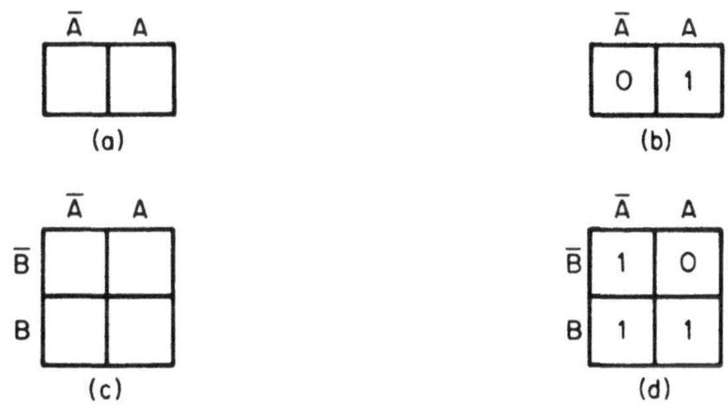

Figure 1.10 Karnaugh maps: (a) for single variable; (b) for f = A; (c) for two variables; (d) for $f = \overline{A} + B$

outside the set A, to extend to infinity. As before, the variable A can exist only in one subset or the other, at any time. The map of the function f = A is shown in figure 1.10b. The two-variable map, figure 1.10c, is formed from two overlapping single-variable maps, giving as many cells to the map as there are combinations of the variables. Using the two-variable map we can, for example, represent the function $f = \overline{A} + B$ by writing a one on each cell covered by $\overline{A}$ or B, as in figure 1.10d. The resulting pattern indicates which of the four possible combinations of the two variables, A and B, are included in the function $f = \overline{A} + B$.

In algebraic form, $f = \overline{A} + B$

$$\begin{aligned} \text{becomes } f &= \overline{A}(\overline{B} + B) + (\overline{A} + A)B \\ &= \overline{A}\overline{B} + \overline{A}B + AB \\ &= \overline{A}\overline{B}(1) + \overline{A}B(1) + A\overline{B}(0) + AB(1) \end{aligned}$$

The entries in brackets indicate what value the corresponding combination of A and B should take on the map.

Three or four variables are dealt with by extending the mapping principle, ensuring that all possible intersections are represented, as in figure 1.11. Always, n variables require 2^n cells on the map.

A different method of labelling the maps is usually more convenient. We refer to any cell by its matrix 'address' where each column or row is labelled as zero or one depending on whether it relates to the variable or its complement respec-

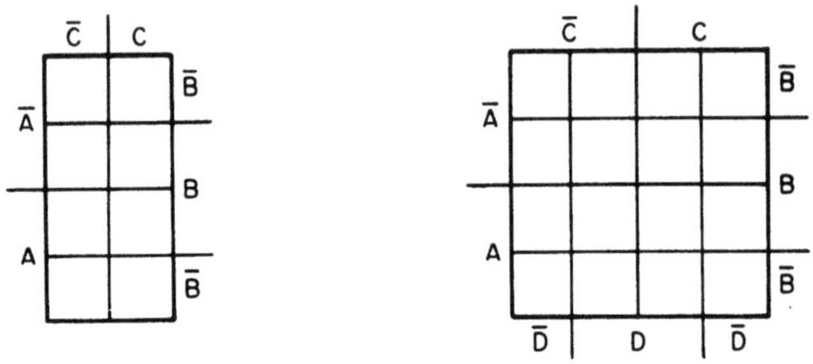

(a) (b)

Figure 1.11 (a) A three-variable Karnaugh map. (b) A four-variable Karnaugh map

tively. Thus, 0101 indicates the cell defined by $\bar{A}B\bar{C}D$ and the function $f = \bar{A}B\bar{C}D$ gives the map shown in figure 1.12a. The mapping of any other function is similar.

AB \ CD	00	01	11	10
00	0	0	0	0
01	0	1	0	0
11	0	0	0	0
10	0	0	0	0

(a)

AB \ CD	00	01	11	10
00	0	0	0	0
01	0	0	1	0
11	0	0	1	0
10	1	1	1	1

(b)

Figure 1.12 (a) Labelling maps by matrix address. (b) Example 1.8

Example 1.8: Map the function $f = A\bar{B} + BCD$.

The term $A\bar{B}$ gives the 'address' 10 indicating a one entry on each of four cells with the address 10XX, since the values of C and D are not specified and all possible combinations of C and D are included. Similarly the term BCD gives the address X111 indicating one entries on two cells, since A is not specified. The resulting map is shown in figure 1.12b.

A five-variable map can be drawn (figure 1.13) but any further increase in the number of variables leads to an unwieldy map, since the intersection requirements must be satisfied. The five-variable map can be considered as a single-variable map for E, with each cell containing a four-variable map for A, B, C and D. In theory, any number of variables can be represented by drawing suf-

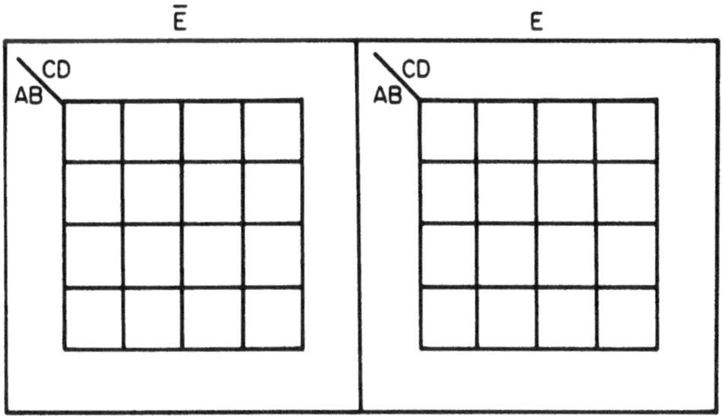

Figure 1.13 A five-variable Karnaugh map

ficient maps within maps, though the value as a visual aid falls so rapidly as to limit its practicability to five variables in most cases.

However, a technique known as *variable-entered mapping* can be helpful in reducing the size of map required, especially where a function contains a large number of infrequently occurring variables. We will illustrate the technique by means of an example, but for convenience we restrict ourselves to four variables, and consider again the function $f = A\bar{B} + BCD$.

In fully expanded form

$$\begin{aligned} f &= A\bar{B}\bar{C}\bar{D} + A\bar{B}\bar{C}D + A\bar{B}C\bar{D} + A\bar{B}CD + \bar{A}BCD + ABCD \\ &= A\bar{B}(\bar{C}\bar{D} + \bar{C}D + C\bar{D} + CD) + \bar{A}B(CD) + AB(CD) \\ &= A\bar{B}(1) + \bar{A}B(CD) + AB(CD) \end{aligned}$$

Rearranging the order of terms, and introducing the missing combination of A and B we get $f = \bar{A}\bar{B}(0) + \bar{A}B(CD) + A\bar{B}(1) + AB(CD)$ so that the map is as shown in figure 1.14a. The method may be more clearly seen from the truth table of the function (figure 1.14b) where we have divided the table into four sections defined by A and B. Within each section the subfunction is defined in terms of C and D. The first section has the output function at zero regardless of the values of C and D, so we enter 0. The second section is dependent on CD being one, so we enter CD, and so on.

The choice of map variables is not limited to A and B. The method applies to any variables and any size map.

One use of the Karnaugh map is in proving identities.

Example 1.9: Show that $\overline{(A + B)(C + D)} = \bar{A}\bar{B} + \bar{C}\bar{D}$.

The first thing to do is to draw the map for the left-hand side of the function. This is easily achieved by considering the AND of the two terms in brackets. Figure 1.15 shows the individual maps for these terms, and the

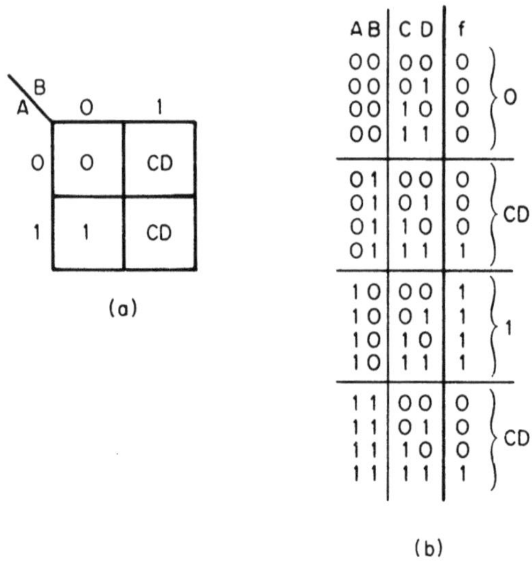

Figure 1.14 A variable-entered map

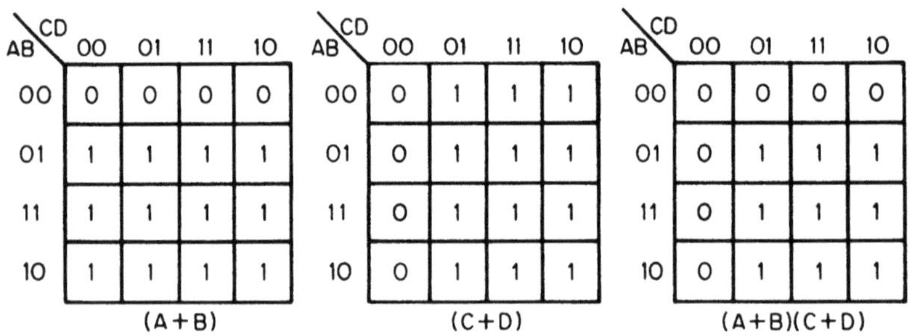

Figure 1.15 Example 1.9

result of ANDing them, but note that it is usually easier to sketch the function on a single map by including in each cell an entry for each of the bracketed terms. An AND function then gives a final one only in those cells containing all ones whereas an OR function gives a one in those cells containing at least one one. This is shown for our present function in figure 1.16a. The NOT of the function is arrived at by changing each zero to one and one to zero, as in figure 1.16b, and it now remains to show that the right-hand side of the identity gives the same pattern. This is left as an exercise for the reader.

1.4 Canonical forms

Any boolean function can be expanded into either of two standard forms: *sum of products*, S of P, which is AND terms ORed together, and *product of sums*,

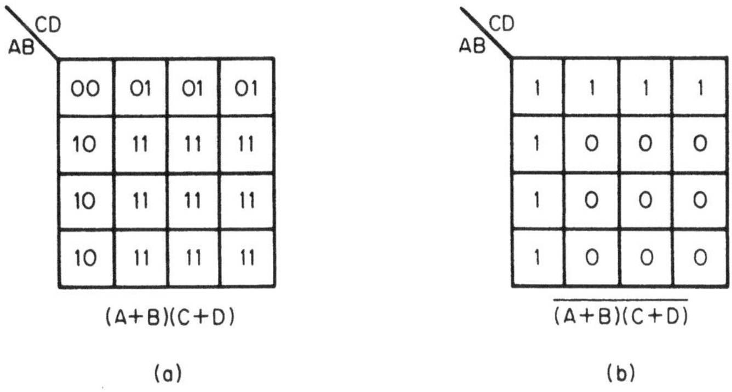

Figure 1.16 Example 1.9

P of S, which is OR terms ANDed together. For example, the function $f = (AB + C)(BC + \overline{D})$ can be expanded to $f = BC + AB\overline{D} + C\overline{D}$, which is the sum of products form, or $f = (A + C)(B + C)(B + \overline{D})(C + \overline{D})$, the product of sums form.

A *canonical term* of a function is defined as a term containing exactly one occurrence of each of the variables of the function in either the true or the complement form. The terms in the previous example are not canonical since they do not contain each variable. We define the *canonical form* of a function as the expanded form which contains only canonical terms.

Recalling that $f + 0 = f$ and that $A\overline{A} = 0$, we can write $f + A\overline{A} = f$. Similarly, since $f \cdot 1 = f$, and $(A + \overline{A}) = 1$, $f \cdot (A + \overline{A}) = f$. These relationships are used in *Shannon's Canonical Expansion Theorems*, which can be stated in a general form as:

$$f(A, B, C \ldots) = f(A, B, C \ldots) + A\overline{A} + B\overline{B} + C\overline{C} + \ldots$$

and $$f(A, B, C \ldots) = f(A, B, C \ldots)(A + \overline{A})(B + \overline{B})(C + \overline{C}) \ldots$$

Let us now expand the function $f = (AB + C)(BC + \overline{D})$ to the canonical sum of products form.

$$\begin{aligned}
f &= (AB + C)(BC + \overline{D}) \\
&= BC + AB\overline{D} + C\overline{D} \\
&= (BC + AB\overline{D} + C\overline{D})(A + \overline{A})(B + \overline{B})(C + \overline{C})(D + \overline{D}) \\
&= BC(A + \overline{A})(D + \overline{D}) + AB\overline{D}(C + \overline{C}) + C\overline{D}(A + \overline{A})(B + \overline{B}) \\
&= ABCD + \overline{A}BCD + ABC\overline{D} + \overline{A}BC\overline{D} + ABC\overline{D} + AB\overline{C}\overline{D} + ABC\overline{D} + \overline{A}BC\overline{D} \\
&\qquad + A\overline{B}C\overline{D} + \overline{A}\overline{B}C\overline{D}
\end{aligned}$$

and $$f = ABCD + \overline{A}BCD + ABC\overline{D} + \overline{A}BC\overline{D} + AB\overline{C}\overline{D} + A\overline{B}C\overline{D} + \overline{A}\overline{B}C\overline{D}$$

when duplicated terms are absorbed.

Note that in the early stages of expansion some of the additional terms, $(A + \overline{A})(B + \overline{B})$ etc., are absorbed. Thus, in practice, it is only necessary to include with each product term the additional terms containing the missing variables. In the expansion above this technique allows us to go almost directly to the fourth row.

Plotting our expanded function on a K-map, figure 1.17, gives us the same pattern as the original function, since we have not altered the truth values. But it can now be seen that each canonical term defines a single cell, giving as many ones on the map as there are terms in the S of P canonical form. Since each term

AB \ CD	00	01	11	10
00	0	0	0	0010 1
01	0	0	0111 1	0110 1
11	1100 1	0	1111 1	1110 1
10	0	0	0	1010 1

Figure 1.17

specifies a minimum area on the map this has become known as the *minterm* expansion of the expression. It is often convenient to represent a function in its minterm form using a series of decimal numbers, where each number corresponds to the binary address of the cell of the map containing the one. Thus

$$\begin{aligned} f &= ABCD + \bar{A}BCD + ABC\bar{D} + \bar{A}BC\bar{D} + AB\bar{C}\bar{D} + A\bar{B}C\bar{D} + \bar{A}\bar{B}C\bar{D} \\ &\equiv 1111 \quad 0111 \quad 1110 \quad 0110 \quad 1100 \quad 1010 \quad 0010 \\ &\equiv 15 \quad 7 \quad 14 \quad 6 \quad 12 \quad 10 \quad 2 \end{aligned}$$

which is expressed as

$$f(ABCD) = \Sigma m(2, 6, 7, 10, 12, 14, 15)$$

Since the K-map is an extension of the truth table, it follows that each term in the S of P canonical form enters a one on the row of the table defined by that term, as in figure 1.18.

Having shown that a function can be expanded to the S of P canonical form, let us now expand the same function to its P of S canonical form. This time we use the alternative form of Shannon's Expansion Theorem, giving

$$\begin{aligned} f &= (AB + C)(BC + \bar{D}) + A\bar{A} + B\bar{B} + C\bar{C} + D\bar{D} \\ &= (A + C)(B + C)(B + \bar{D})(C + \bar{D}) + A\bar{A} + B\bar{B} + C\bar{C} + D\bar{D} \\ &= [(A + C) + B\bar{B} + D\bar{D}]\ [(B + C) + A\bar{A} + D\bar{D}]\ [(B + \bar{D}) + A\bar{A} + C\bar{C}] \\ &\quad [(C + \bar{D}) + A\bar{A} + B\bar{B}] \end{aligned}$$

At this point, let us just have a look at the expression we have developed. Having expanded the function into the P of S form using the absorption law in reverse, we find that each bracket has the missing terms added in $X\bar{X}$ form, in

	A	B	C	D	f
0	0	0	0	0	0
1	0	0	0	1	0
→ 2	0	0	1	0	1
3	0	0	1	1	0
4	0	1	0	0	0
5	0	1	0	1	0
→ 6	0	1	1	0	1
→ 7	0	1	1	1	1
8	1	0	0	0	0
9	1	0	0	1	0
→ 10	1	0	1	0	1
11	1	0	1	1	0
→ 12	1	1	0	0	1
13	1	1	0	1	0
→ 14	1	1	1	0	1
→ 15	1	1	1	1	1

Figure 1.18

much the same way as before. With a little practice, therefore, we can go straight to this stage. Now continuing our expansion gives

$$\begin{aligned} f = {} & (A + C + B + D\bar{D})(A + C + \bar{B} + D\bar{D})(B + C + A + D\bar{D})(B + C + \bar{A} + D\bar{D}) \\ & (B + \bar{D} + A + C\bar{C})(B + \bar{D} + \bar{A} + C\bar{C})(C + \bar{D} + A + B\bar{B})(C + \bar{D} + \bar{A} + B\bar{B}) \end{aligned}$$

$$\begin{aligned} = {} & (A + C + B + D)(A + C + B + \bar{D})(A + C + \bar{B} + D)(A + C + \bar{B} + \bar{D}) \\ & (B + C + A + D)(B + C + A + \bar{D})(B + C + \bar{A} + D)(B + C + \bar{A} + \bar{D}) \\ & (B + \bar{D} + A + C)(B + \bar{D} + A + \bar{C})(B + \bar{D} + \bar{A} + C)(B + \bar{D} + \bar{A} + \bar{C}) \\ & (C + \bar{D} + A + B)(C + \bar{D} + A + \bar{B})(C + \bar{D} + \bar{A} + B)(C + \bar{D} + \bar{A} + \bar{B}) \end{aligned}$$

Omitting repeated terms and rearranging each bracket content to alphabetical order

$$\begin{aligned} f = {} & (A + B + C + D)(A + B + C + \bar{D})(A + \bar{B} + C + D)(A + \bar{B} + C + \bar{D}) \\ & (\bar{A} + B + C + D)(\bar{A} + B + C + \bar{D})(A + B + \bar{C} + \bar{D})(\bar{A} + B + \bar{C} + \bar{D}) \\ & (\bar{A} + \bar{B} + C + \bar{D}) \end{aligned}$$

Each term of the P of S canonical form specifies a maximum area of ones (or single zero) on the map so is known as a *maxterm*. Again we can express the terms in a decimal coded form.

Expansion of our function gave us

$$\begin{array}{lcccc} f = & (A + B + C + D) & (A + B + C + \bar{D}) & (A + \bar{B} + C + D) & (A + \bar{B} + C + \bar{D}) \\ \equiv & (1111) & (1110) & (1011) & (1010) \\ \equiv & 15 & 14 & 11 & 10 \\ & & & & \\ & (\bar{A} + B + C + D) & (\bar{A} + B + C + \bar{D}) & (A + B + \bar{C} + \bar{D}) & (\bar{A} + B + \bar{C} + \bar{D}) \\ & (0111) & (0110) & (1100) & (0100) \\ & 7 & 6 & 12 & 4 \\ & & & & \\ & (\bar{A} + \bar{B} + C + \bar{D}) & & & \\ & (0010) & & & \\ & 2 & & & \end{array}$$

which is expressed as

$$f(ABCD) = \Pi M(2, 4, 6, 7, 10, 11, 12, 14, 15)$$

We know that the map for this function, figure 1.17, contains nine zeros, and we note that the P of S canonical form of the function has nine terms. It is possible to show a relationship between each term and a zero cell by defining a new function g, the dual of the function f. By duality

$$g = \overline{f} = \overline{A}\,\overline{B}\,\overline{C}\,\overline{D} + \overline{A}\,\overline{B}\,\overline{C}D + \overline{A}B\overline{C}\,\overline{D} + \overline{A}B\overline{C}D + A\overline{B}\,\overline{C}\,\overline{D} + A\overline{B}\,\overline{C}D$$
$$+ \overline{A}\,\overline{B}CD + A\overline{B}CD + AB\overline{C}D$$

This function, g, being the inverse of the original function, f, must define the positions of those cells containing zero. The minterms for function g are 0000, 0001, 0100, 0101, 1000, 1001, 0011, 1011 and 1101, and each term gives the address of one of the zeros on the map as in figure 1.19.

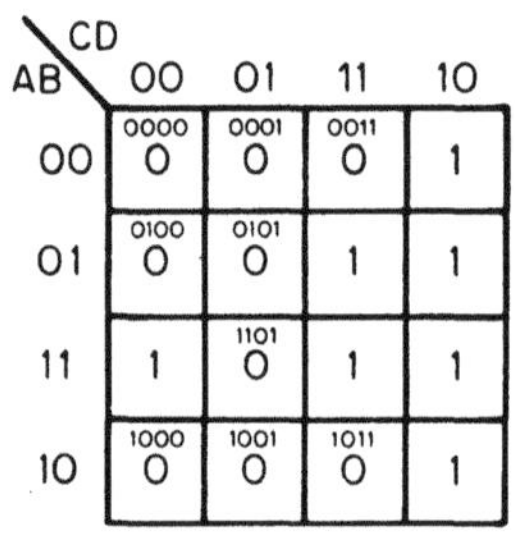

Figure 1.19

Now, each canonical term in the dual function has been obtained by complementing the variables of the P of S terms. Thus these dual minterms could have been found by subtracting each maxterm from $2^n - 1$, in this case 15. Equally, it follows that maxterms can be derived directly from the K-map by subtracting the addresses of the zero cells from $(2^n - 1)$.

Example 1.10: Derive the maxterm form of the function

$$f(ABCD) = \Sigma m(2, 6, 7, 10, 12, 14, 15)$$

The zero cells are defined by the missing terms in the function,

giving $\overline{f} = \Sigma m(0, 1, 3, 4, 5, 8, 9, 11, 13)$

Hence, by subtracting each from 15, the maxterm form of the original function will be

$$f(ABCD) = \Pi M(15, 14, 12, 11, 10, 7, 6, 4, 2)$$

1.5 Symmetric functions

Certain functions have symmetrical properties which can be used to advantage when simplifying logic circuits, and are particularly useful in VLSI design [Lewin (1985), Mukherjee (1986)]. Note that in some earlier texts [Mead and Conway (1980)] symmetric functions are referred to as *tally functions*.

If the interchanging of any two or more variables of a function f(ABC . . .) leaves the function unchanged, then the function is said to be *partially symmetric*, and the variables which may be interchanged are the *variables of symmetry*.

Example 1.11: Show that $f(ABCD) = (A + \overline{B})CD$ is partially symmetric about A and B, and also about C and D.

(a) Symmetry exists about A and $\overline{B}$ if we can replace A with $\overline{B}$, $\overline{A}$ with B, B with $\overline{A}$, and $\overline{B}$ with A. Interchanging in this way, the function becomes $f(ABCD) = (\overline{B} + A)CD$, which is identical to the original since the commutative law applies.

(b) Similarly, symmetry exists about C and D if we can replace C with D, $\overline{C}$ with $\overline{D}$, D with C, and $\overline{D}$ with $\overline{C}$. The function now becomes $f(ABCD) = (A + \overline{B})DC$, which is again identical to the original.

Example 1.11 illustrates two variants of symmetry. The first variables of symmetry, A and B, are each interchangeable with the inverse of the other (A with $\overline{B}$, and $\overline{A}$ with B) and this form is called *equivalence symmetry*, ES. The other variables of symmetry, C and D, are each interchangeable with the same form of the other (C with D, $\overline{C}$ with $\overline{D}$) and this form is called *non-equivalence symmetry*, NES.

The presence of symmetry in a function may be detected by inspection of the expansion of the function. When equivalence symmetry about A and B is present, the expansion contains identical terms for $\overline{A}\overline{B}$ and AB. Similarly, when non-equivalence symmetry about A and B is present, the expansion contains identical terms for $\overline{A}B$ and $A\overline{B}$.

Example 1.12: Show that the function $f(ABCD) = \overline{A}\overline{C} + B\overline{C} + \overline{C}D + \overline{A}BD$ has equivalence symmetry about variables A and B, and non-equivalence symmetry about variables B and D.

Expanding the function and emphasizing terms in A and B gives

$$f(ABCD) = \overline{A}\overline{B}(\overline{C}) + \overline{A}B(\overline{C} + D) + A\overline{B}(\overline{C}D) + AB(\overline{C})$$

The entries for $\overline{A}\overline{B}$ and AB are both $\overline{C}$, indicating equivalence symmetry about A and B. Now rearranging the expansion to emphasize the terms in B and D gives

$$f(ABCD) = \overline{B}\,\overline{D}(\overline{A}\,\overline{C}) + \overline{B}D(\overline{C}) + B\overline{D}(\overline{C}) + BD(\overline{A} + \overline{C})$$

and we see that the terms in $\overline{B}D$ and $B\overline{D}$ both include $\overline{C}$, indicating non-equivalence symmetry about B and D.

When all the variables of a function act as variables of symmetry, the function is said to be *fully symmetric*. The simple AND and OR functions can be seen to be fully symmetric functions, since we are really only stating the commutative law in a different form, but there are many other functions which are also fully symmetric. One common form is the *n*-out-of-*m* function which gives an output of one only when *n* of the *m* input variables are at one. Thus, for example, a 2-out-of-3 function is given by

$$f(ABC) = AB\overline{C} + A\overline{B}C + \overline{A}BC$$

The symmetry is maintained if we extend the function to give, for example, a one output when two or more of the input variables are at one, as would be the case with a three-input majority logic function. Then

$$f(ABC) = AB + BC + CA$$

A shorthand notation has been developed to indicate the symmetric properties of a function and takes the form $S_n^m(ABC\ldots)$. Using this notation, the 2-out-of-3 function given above is expressed as $S_2^3(ABC)$. Similarly, the three-input majority logic function becomes $S_{2,3}^3(ABC)$.

1.6 Analysis and synthesis

We earlier defined the basic connectives, AND and OR, and the operator NOT. Practical systems carrying out logical operations use devices which exhibit the logical properties of AND, OR and NOT, termed *gates*. They are usually electronic circuits, though other types, notably fluidic, are used in certain applications such as automatic gearboxes. Here we shall introduce general symbols used to indicate a gate's logical properties regardless of how they are achieved, and leave a fuller discussion of logic circuits until chapter 3.

The symbols we shall use for the three basic gates are shown in figure 1.20 and these symbols may be used to represent any circuit or arrangement which is used to generate, or realize, the appropriate value for the function as the input variables change. At this stage we can simply include an appropriate gate for each logical operation in the function.

Example 1.13: Draw the logic circuit necessary to realize the function $f = (AB + C)(BC + \overline{D})$.

The circuit is shown in figure 1.21.

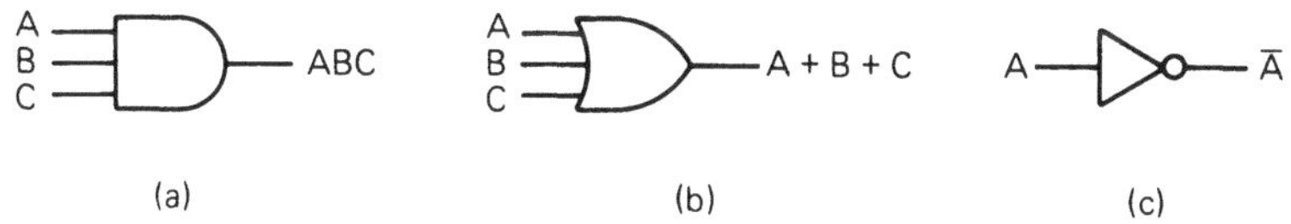

Figure 1.20 (a) Three-input AND gate; (b) three-input OR gate; (c) inverter. These symbols are based on the popular American Standard ANSI Y32–14–1973. Other standards are discussed in appendix A

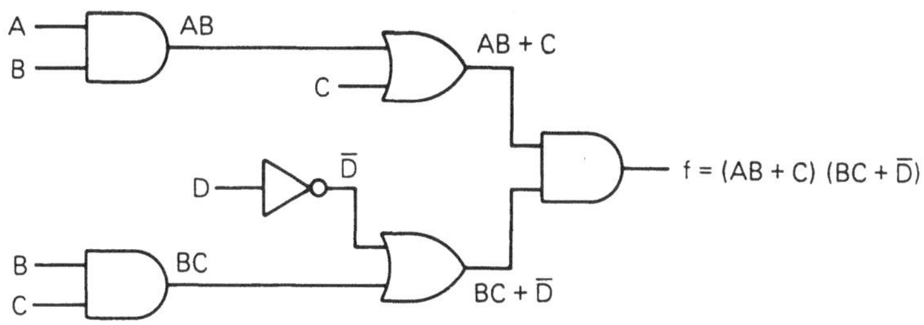

Figure 1.21 Example 1.13

There are two general areas in logic design: analysis of an existing circuit, and synthesis of a circuit to meet a given specification. Analysis of a circuit or system involves the derivation of the response to input stimuli and, for a logic circuit, is usually considered complete when we arrive at a K-map or truth table giving the output conditions for every combination of the input variables.

Example 1.14: Analyse the circuit shown in figure 1.22a.

The output function can be built up as shown, giving $f = \bar{B}(AB + D + AC)$. The aim now is to draw the map for this function, so we first simplify the expression algebraically. Thus, $f = \bar{B}D + A\bar{B}C$, which gives the map and truth table of figure 1.22b and c.

Example 1.15: Analyse the circuit of figure 1.23a.

From the circuit,
$$f = \overline{(A + C)}D + (C + D)B\overline{CD}$$
$$= \bar{A}\bar{C}D + (C + D)B(\bar{C} + \bar{D})$$
$$= \bar{A}\bar{C}D + (C\bar{D} + \bar{C}D)B$$

The map is shown in figure 1.23b.

Let us now consider synthesis. Synthesis is the building-up of a system starting with the specification and, owing to the natural ambiguity in everyday language, it is essential that the specification be defined in the form of a truth table or K-map so that the information is presented in a precise form that does not permit misinterpretation.

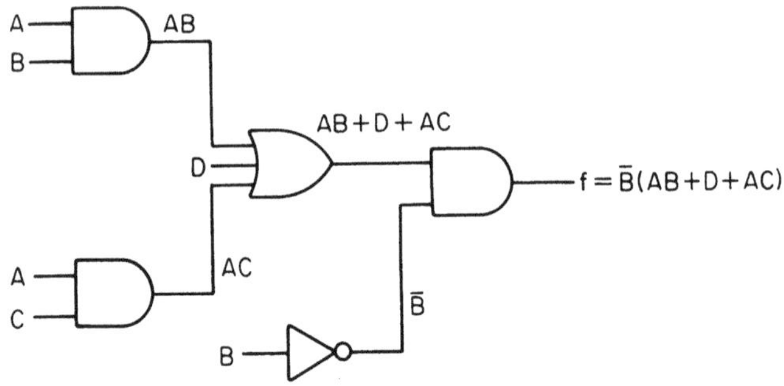

(a)

AB \ CD	00	01	11	10
00	0	1	1	0
01	0	0	0	0
11	0	0	0	0
10	0	1	1	1

(b)

A	B	C	D	f
0	0	0	0	0
0	0	0	1	1
0	0	1	0	0
0	0	1	1	1
0	1	0	0	0
0	1	0	1	0
0	1	1	0	0
0	1	1	1	0
1	0	0	0	0
1	0	0	1	1
1	0	1	0	1
1	0	1	1	1
1	1	0	0	0
1	1	0	1	0
1	1	1	0	0
1	1	1	1	0

(c)

Figure 1.22 Example 1.14

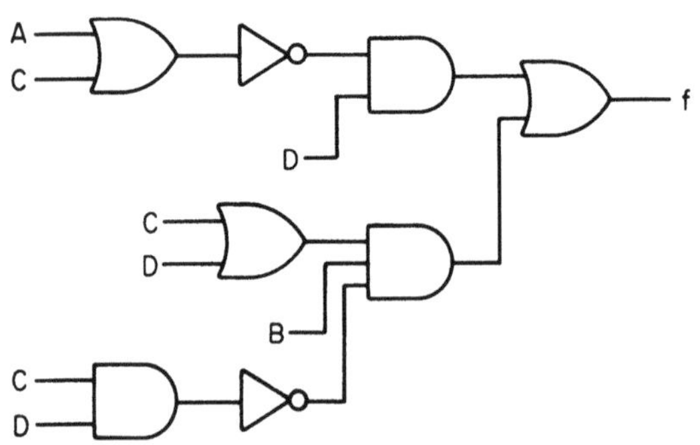

(a)

AB \ CD	00	01	11	10
00	0	1	0	0
01	0	1	0	1
11	0	1	0	1
10	0	0	0	0

(b)

Figure 1.23 Example 1.15

The truth table shown in figure 1.24a, for example, describes a system which gives an output every time two, and only two, of the input signals are present. The presence of a signal is indicated by a one and, to arrive at a circuit represen-

	A	B	C	f
0	0	0	0	0
1	0	0	1	0
2	0	1	0	0
3	0	1	1	1
4	1	0	0	0
5	1	0	1	1
6	1	1	0	1
7	1	1	1	0

(a)

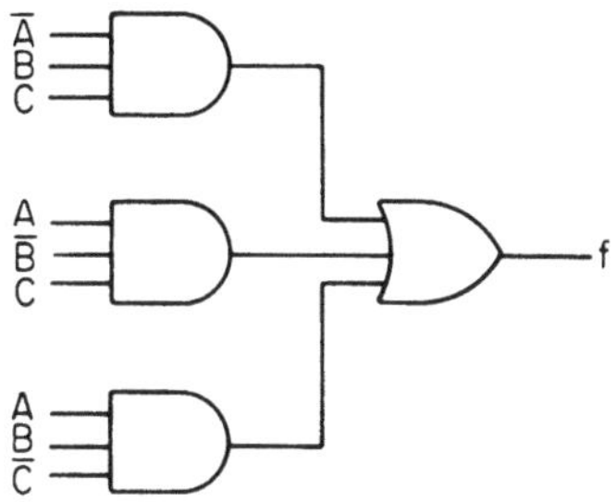

(b)

Figure 1.24

tation, we can write down minterms corresponding to each row of the truth table which gives a one output. Thus

$$f(ABC) = \Sigma m(3,\ 5,\ 6)$$

or $$f = \bar{A}BC + A\bar{B}C + AB\bar{C}$$

A circuit can be drawn directly from this equation, in terms of AND/OR gates, as shown in figure 1.24b. By inspection or algebraic manipulation, alternative circuits can be found, though not necessarily leading to a simpler circuit.

The system described by the table of figure 1.24a could also be realized from the maxterms. Referring to the zero output rows, we have $f(ABC) = \Pi M(0, 3, 5, 6, 7)$ since we subtract each from seven.

This gives

$$f = (A + B + C)(A + B + \bar{C})(A + \bar{B} + C)(\bar{A} + B + C)(\bar{A} + \bar{B} + \bar{C})$$

which leads directly to the circuit shown in figure 1.25a.

The function f can be rearranged using the absorption law. Thus

$$\begin{aligned} f &= (A + B)(B + C)[\bar{B} + (A + C)(\bar{A} + \bar{C})] \\ &= (AC + B)[\bar{B} + (A + C)(\bar{A} + \bar{C})] \end{aligned}$$

Using this form for the circuit, figure 1.25b, needs more gates but not the five-input AND gate of the previous circuit. In general, manipulation from either canonical form will lead to a circuit requiring more gates but fewer inputs, and may be useful for that very reason.

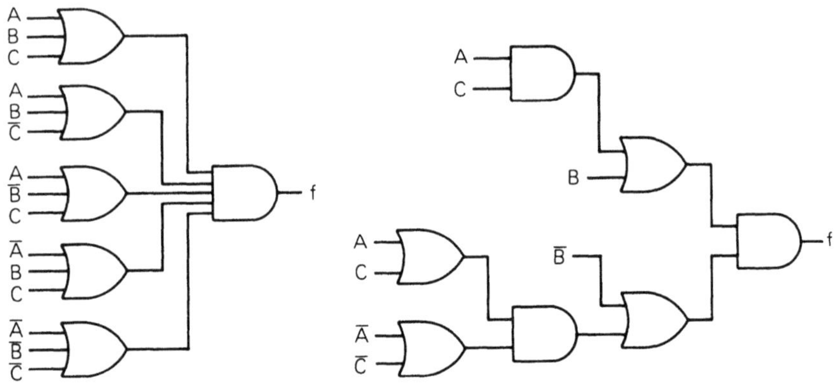

(a) (b)

Figure 1.25

Example 1.16: A logical circuit with inputs ABCD yields an output f(ABCD), given in maxterm form by

$$f(ABCD) = \Pi M(0,\ 2,\ 5,\ 7,\ 10,\ 12,\ 13,\ 14,\ 15)$$

Synthesize the circuit using AND/OR gates.

Using the maxterms, the truth table can be written down by entering zeros in the rows indicated by the maxterms, figure 1.26a.

		A	B	C	D	f
→	0	0	0	0	0	0
→	1	0	0	0	1	0
→	2	0	0	1	0	0
→	3	0	0	1	1	0
	4	0	1	0	0	1
→	5	0	1	0	1	0
	6	0	1	1	0	1
	7	0	1	1	1	1
→	8	1	0	0	0	0
	9	1	0	0	1	1
→	10	1	0	1	0	0
	11	1	0	1	1	1
	12	1	1	0	0	1
→	13	1	1	0	1	0
	14	1	1	1	0	1
→	15	1	1	1	1	0

(a)

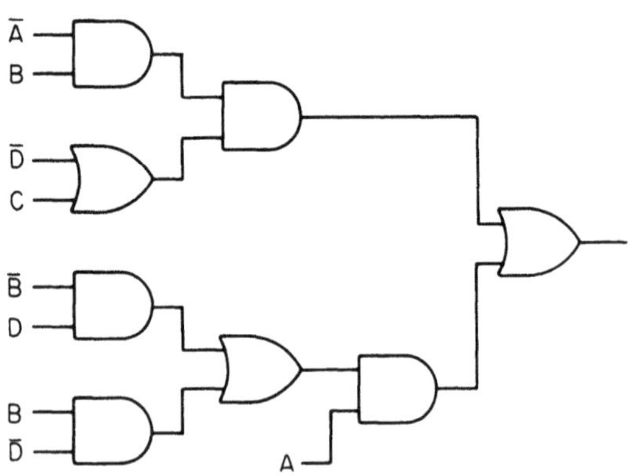

(b)

Figure 1.26 Example 1.16

We see that the function f is given in S of P form by

$$f = \bar{A}B\bar{C}\bar{D} + \bar{A}BC\bar{D} + \bar{A}BCD + A\bar{B}\bar{C}D + A\bar{B}CD + AB\bar{C}\bar{D} + ABC\bar{D}$$

This can be simplified to

$$f = \bar{A}B\bar{C}\bar{D} + \bar{A}BC + A\bar{B}\bar{C}D + A\bar{B}CD + AB\bar{D}$$

and hence

$$f \equiv \bar{A}B(\bar{D} + C) + A(\bar{B}D + B\bar{D})$$

The circuit follows directly, figure 1.26b.

As a further example, suppose we require a circuit to accept two inputs and to give an output if the two input values are not equivalent. The truth table is shown in figure 1.27a and, by inspection

$$f = \bar{A}B + A\bar{B}$$

or, alternatively,

$$f = (A + B)(\bar{A} + \bar{B})$$

The corresponding circuits are shown in figures 1.27b and c.

A	B	f
0	0	0
0	1	1
1	0	1
1	1	0

(a)

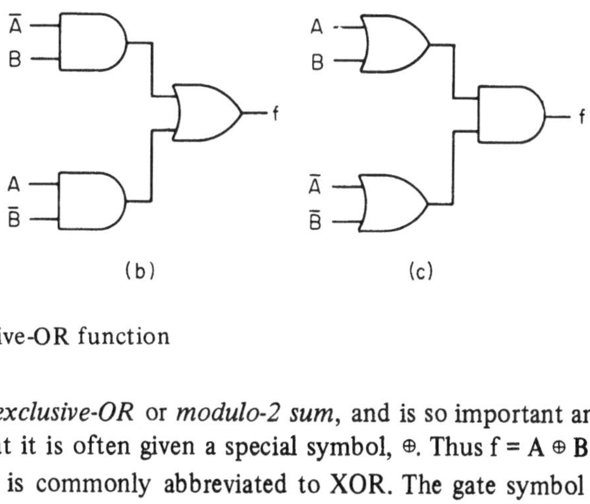

Figure 1.27 The exclusive-OR function

This function is the *exclusive-OR* or *modulo-2 sum*, and is so important and occurs so frequently that it is often given a special symbol, ⊕. Thus $f = A \oplus B = \bar{A}B + A\bar{B}$. Exclusive-OR is commonly abbreviated to XOR. The gate symbol is shown in appendix A.

Now let us include a second output, f′, to give a one only when both A and B are present. This additional requirement could be met by including a separate two-input AND gate, since $f' = AB$. However, by reorganizing the original expression we can include the AND function in the exclusive-OR circuit.

We have $\quad f = (A + B)(\bar{A} + \bar{B})$

by duality $\quad \bar{f} = \bar{A}\bar{B} + AB$

giving $\quad f = \overline{\bar{A}\bar{B} + AB}$

Thus the complete circuit could be as shown in figure 1.28. This circuit is referred to as a *half-adder*, where output f gives the binary sum of A and B, and output f′ gives the carry value.

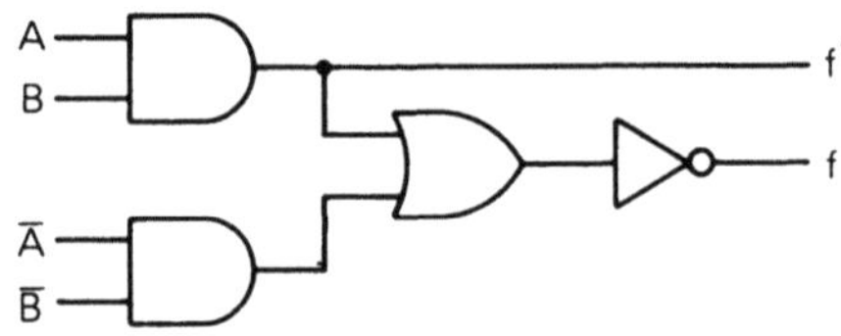

Figure 1.28 The half-adder circuit

1.7 Minimization

We have seen that the logical first step in synthesis is to derive either of the two canonical forms from the truth table or K-map. This may not lead directly to an optimum circuit since the canonical form is the most highly redundant form of the expression. There are minimization techniques which allow us to reach an optimum solution, but individual requirements vary according to circumstances. For example, we may wish to minimize the number of gates used, to limit the number of inputs to any gate, or to satisfy any of a number of different requirements. Experience shows that a good solution can often be obtained by using some composite expression for the function; that is, one that is neither wholly sum of products nor wholly product of sums. An ability to shape a solution can give a better balance between such factors as circuit complexity, redundancy, cost, number of gates, and so on.

From a logical point of view, during synthesis we must define the output value required for every possible input variable combination, even though in practice, owing to external restrictions, some of these combinations may never occur. Such combinations give rise to *don't care* conditions, in which we can assume the output value to be either zero or one, whichever is the more convenient in our design. Because they introduce the element of choice, don't care conditions are often helpful in providing simpler expressions.

Bearing these points in mind, a careful inspection of the truth table can lead to an optimal circuit.

Example 1.17: In a certain system with four input signals A, B, C, D, only ten combinations of the signals ever occur. These are shown in the table of figure 1.29. For half of these combinations the output W is to be zero, and for the other half the output is to be one, as shown. Derive a minimal expression for W.

	A	B	C	D	W	
0	0	0	0	0	0	
1	0	0	0	1	0	
2	0	0	1	0	0	
3	0	0	1	1	0	
4	0	1	0	0	0	Specified
5	0	1	0	1	1	conditions
6	0	1	1	0	1	
7	0	1	1	1	1	
8	1	0	0	0	1	
9	1	0	0	1	1	
10	1	0	1	0	Ø	
11	1	0	1	1	Ø	
12	1	1	0	0	Ø	Don't care
13	1	1	0	1	Ø	conditions
14	1	1	1	0	Ø	
15	1	1	1	1	Ø	

Figure 1.29 Example 1.17

By following the synthesis procedures outlined earlier we obtain the expression

$$W = \bar{A}B\bar{C}D + \bar{A}BC\bar{D} + \bar{A}BCD + A\bar{B}\bar{C}\bar{D} + A\bar{B}\bar{C}D$$

or

$$W = (A + B + C + D)(A + B + C + \bar{D})(A + B + \bar{C} + D)$$
$$(A + B + \bar{C} + \bar{D})(A + \bar{B} + C + D)$$

dependent on our selection of the ones or zeros on the table. The expressions could be reduced algebraically, but the alternative empirical approach is to examine the table and seek out the relationships between the input variables for the positions of the desired outputs.

In this particular case, if input A is present at all an output is required (rows 8 and 9). Thus, including A in the output expression covers two of the ones on the table. To cover the remaining three ones we include input B, but that also brings in row 4. However, if we stipulate that input C or D must be one, we can control B and exclude row 4, arriving at the simple expression

$$W = A + B(C + D)$$

Successful minimization depends to a large extent on experience and intuition, and the K-map with its graphical representation is an invaluable aid in presenting the alternatives concisely.

We recall that the cells on the K-map are defined by a form of binary reference or address built up from the input variable values, and the cell content indicates the value of the mapped function for those particular input variable values. If we move from any cell on the map to an adjacent cell, only one variable in the address changes. This is a very important feature and leads to the simplifying method known as *grouping* or *looping*. Also, since the addresses form a cyclic code, we note that the map can be considered as extending in all directions, so that a cell on one end of a row or column is adjacent to the cell at the other end of that row or column, as if the map were drawn over the whole surface of a sphere.

The function shown in the map of figure 1.30a is

$$f = \bar{A}B\bar{C}D + \bar{A}BCD + ABC\bar{D} + A\bar{B}C\bar{D}$$

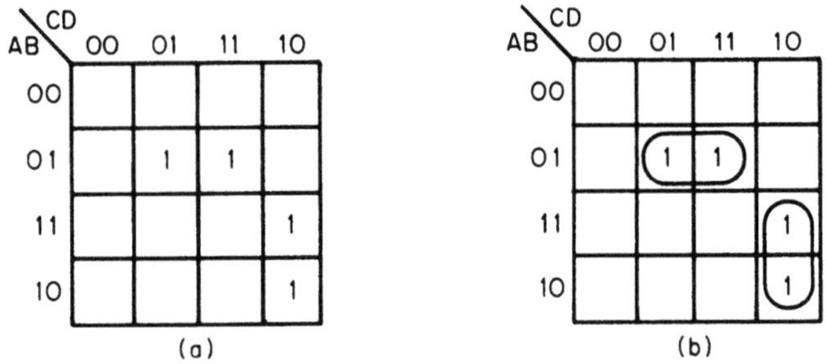

Figure 1.30 Grouping of adjacent entries

Rearranging the function slightly

$$\begin{aligned} f &= \bar{A}BD(\bar{C} + C) + AC\bar{D}(\bar{B} + B) \\ &= \bar{A}BD + AC\bar{D} \end{aligned}$$

Thus the C variable in the first two terms and the B variable in the last two terms can be removed. Use of the Karnaugh map allows us to see directly the terms that can be paired and, thus, the variables that change and can be removed, figure 1.30b.

The idea can be extended where groups of two ones are adjacent, giving a group of four ones and allowing two variables to be removed. Likewise, groups of four ones may be paired, and so on. The larger the group, the fewer the variables necessary to define it. When pairing any small groups to make a larger group, a change in only one variable is allowed, or the function defined by the larger group cannot be simplified to the single term. Typical groupings are shown in figure 1.31.

To sum up, the minimising procedure involves finding the largest possible groups; the set of these maximally coupled groups is called the set of *prime implicants* of the function. Having arrived at the set of prime implicants, the final step is to determine the fewest prime implicants necessary to cover the function completely. In some cases the prime implicants overlap and some redundancy is unavoidable. Which redundant terms are included is governed partly by other requirements of the circuit and, as always, experience helps. The following examples develop the technique:

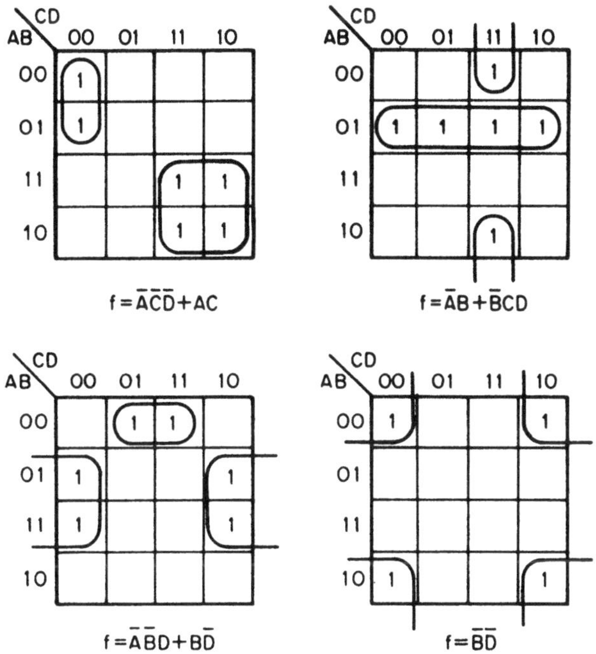

Figure 1.31 Typical groupings

Example 1.18: Minimize the function f(ABCD) = Σm(0, 1, 5, 6, 7, 8, 14, 15).

The map of the function, figure 1.32, shows

possible groups of four ones	(6, 7, 14, 15)
possible groups of two ones (not completely contained within larger groups)	(0, 1) (1, 5) (5, 7) (0, 9)

In algebraic terms the set of prime implicants is therefore

$$\{(BC)\ (\bar{A}\bar{B}\bar{C})\ (\bar{A}\bar{C}D)\ (\bar{A}BD)\ (\bar{B}\bar{C}\bar{D})\}$$

Figure 1.32 shows the selection of prime implicants sufficient to cover the function, and we see that the minimal function is given by

$$f = BC + \bar{A}\bar{C}D + \bar{B}\bar{C}\bar{D}$$

Example 1.19: Minimize the function f(ABCD) = Σm(0, 1, 5, 7, 9, 10, 11, 13, 15)

The map of the prime implicants is shown in figure 1.33a. The prime implicants are ABC, CD, BD, AD and $A\bar{B}C$, and there is considerable overlapping so the best final choice is not immediately clear. We can find a minimal set by means of a *prime implicant table* which lists the prime implicants as row

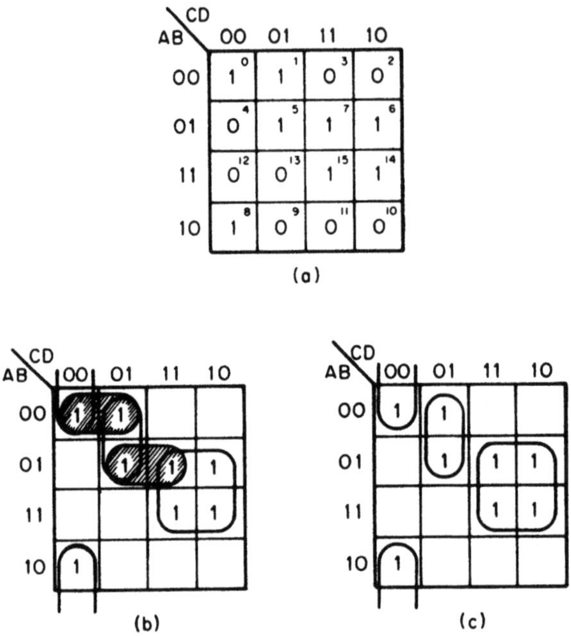

Figure 1.32 Selection of prime implicants of the function f(ABCD) = Σm(0, 1, 5, 6, 7, 8, 14, 15): (a) the map of the function; (b) the prime implicants; (c) minimum prime implicants necessary to cover the function

headings and the canonical terms as column headings. For each prime implicant a tick is entered in every column for which the canonical term contains the prime implicant, as in figure 1.33b. In choosing a set of prime implicants we must ensure that there is at least one tick in each column. As there is only one tick in columns 5, 8 and 9, no choice is possible, so $A\bar{B}C$, $\bar{A}\bar{B}\bar{C}$ and BD

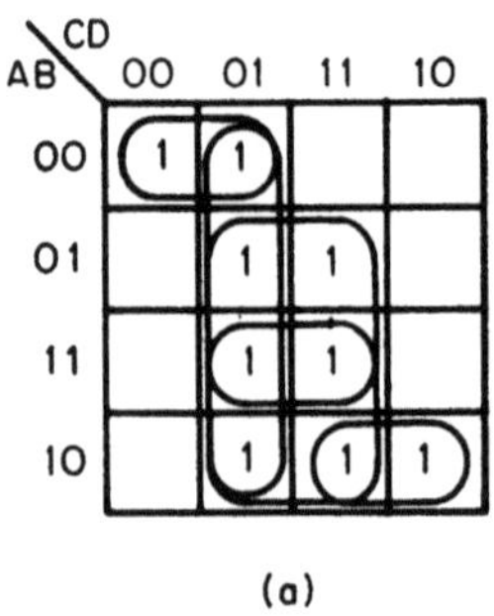

	1 $A\bar{B}\bar{C}D$	2 $A\bar{B}CD$	3 $AB\bar{C}D$	4 $ABCD$	5 $\bar{A}BCD$	6 $\bar{A}\bar{B}\bar{C}D$	7 $\bar{A}B\bar{C}D$	8 $\bar{A}\bar{B}\bar{C}\bar{D}$	9 $A\bar{B}C\bar{D}$
$A\bar{B}C$		✓							✓
$\bar{A}\bar{B}\bar{C}$						✓		✓	
AD	✓	✓	✓	✓					
$\bar{C}D$	✓		✓			✓	✓		
BD			✓	✓	✓		✓		

(b)

Figure 1.33 Example 1.19: (a) the prime implicants; (b) selection of essential prime implicants

must be included in the set. They are classed as *essential prime implicants.* In including the essential prime implicants we have also covered columns 2, 3, 4, 6 and 7, but the remaining column, column 1, allows a choice of AD or $\overline{C}D$. Thus, in this case, there is no unique minimal solution; two equal solutions are indicated

$$f = A\overline{B}C + \overline{A}\,\overline{B}\,\overline{C} + BD + AD$$

and $$f = A\overline{B}C + \overline{A}\,\overline{B}\,\overline{C} + BD + \overline{C}D$$

It is possible to arrive at the alternative P of S minimal form of a function using a similar approach. This time we consider the zero cells on the map, which define the inverse of the given function. Using the same function as figure 1.32 and grouping the zeros as shown in figure 1.34, the set of prime implicants can be seen to be a single group of four zeros and four groups of two zeros.

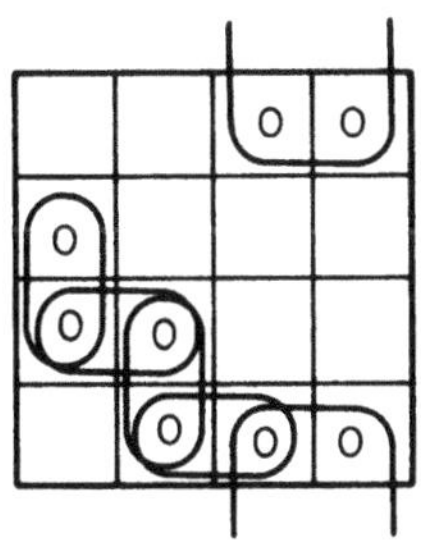

Figure 1.34 Grouping the zeros

The minimal set of these prime implicants, necessary and sufficient to cover this inverse function, leads to

$$\overline{f} = \overline{B}C + A\overline{C}D + B\overline{C}\,\overline{D}$$

and by duality

$$f = (B + \overline{C})(\overline{A} + C + \overline{D})(\overline{B} + C + D)$$

With practice, the final expression can be written directly from the groups of zeros, the necessary inversions being carried out mentally.

When we come to grouping terms on a variable-entered map we must ensure that we group only adjacent cells containing the same subfunction. This is not so restrictive as it appears, when we remember that $X + \overline{X} = 1$, and so a cell containing 1 can be grouped with an adjacent cell containing either X or $\overline{X}$, though, of course, the group covers only that part of the function.

Example 1.20: Draw a four-variable map of the function

$$f = ABC\overline{D}F + \overline{B}\,\overline{C}D\overline{F} + \overline{A}BCD\overline{E}\,\overline{F} + \overline{A}B\overline{D}\,\overline{E} + \overline{A}BC\overline{D}E + \overline{B}\,\overline{C}DF + A\overline{B}C\overline{D}E$$
$$+ \overline{A}B\overline{C}DF + \overline{A}BCDF + A\overline{B}\,\overline{C}\,\overline{D}E + \overline{A}CDE\overline{F} + AB\overline{C}\,\overline{D}\,\overline{E}\,\overline{F} +$$
$$\overline{A}B\overline{C}DF$$

using variable-entered techniques and hence derive a minimal form of the expression.

The least frequently occurring variables are E and F so we shall use them as the variables to be entered on the ABCD map. In canonical form

$$\begin{aligned} f = {} & \bar{A}\bar{B}\bar{C}D\bar{E}\bar{F} + \bar{A}\bar{B}\bar{C}D\bar{E}F + \bar{A}\bar{B}\bar{C}DE\bar{F} + \bar{A}\bar{B}\bar{C}DEF + \bar{A}\bar{B}CD\bar{E}F + \bar{A}\bar{B}CDE\bar{F} \\ & + \bar{A}\bar{B}CDEF + \bar{A}B\bar{C}\bar{D}\bar{E}\bar{F} + \bar{A}B\bar{C}\bar{D}\bar{E}F + \bar{A}B\bar{C}D\bar{E}F + \bar{A}B\bar{C}DEF + \bar{A}BC\bar{D}\bar{E}\bar{F} \\ & + \bar{A}BC\bar{D}\bar{E}F + \bar{A}BC\bar{D}E\bar{F} + \bar{A}BC\bar{D}EF + \bar{A}BCD\bar{E}\bar{F} + \bar{A}BCDE\bar{F} + A\bar{B}\bar{C}\bar{D}EF \\ & + A\bar{B}\bar{C}\bar{D}EF + A\bar{B}\bar{C}D\bar{E}\bar{F} + A\bar{B}\bar{C}D\bar{E}F + A\bar{B}\bar{C}DE\bar{F} + A\bar{B}\bar{C}DEF + A\bar{B}C\bar{D}E\bar{F} \\ & + A\bar{B}C\bar{D}EF + AB\bar{C}\bar{D}\bar{E}\bar{F} + ABC\bar{D}\bar{E}F + ABC\bar{D}EF \end{aligned}$$

Rearranging and grouping in ABCD terms, we get

$$\begin{aligned} f = {} & \bar{A}\bar{B}\bar{C}\bar{D}(0) + \bar{A}\bar{B}\bar{C}D(\bar{E}\bar{F} + E\bar{F} + EF) + \bar{A}\bar{B}C\bar{D}(0) \\ & + \bar{A}\bar{B}CD(\bar{E}F + E\bar{F} + EF) + \bar{A}B\bar{C}\bar{D}(\bar{E}\bar{F} + \bar{E}F) + \bar{A}B\bar{C}D(\bar{E}F + EF) \\ & + \bar{A}BC\bar{D}(\bar{E}\bar{F} + \bar{E}F + E\bar{F} + EF) + \bar{A}BCD(\bar{E}\bar{F} + E\bar{F}) + A\bar{B}\bar{C}\bar{D}(E\bar{F} + EF) \\ & + A\bar{B}\bar{C}D(\bar{E}\bar{F} + \bar{E}F + E\bar{F} + EF) + A\bar{B}C\bar{D}(E\bar{F} + EF) + A\bar{B}CD(0) + AB\bar{C}\bar{D}(\bar{E}\bar{F}) \\ & + AB\bar{C}D(0) + ABC\bar{D}(\bar{E}F + EF) + ABCD(0) \end{aligned}$$

So

$$\begin{aligned} f = {} & \bar{A}\bar{B}\bar{C}\bar{D}(0) + \bar{A}\bar{B}\bar{C}D(1) + \bar{A}\bar{B}C\bar{D}(0) + \bar{A}\bar{B}CD(E + F) + \bar{A}B\bar{C}\bar{D}(\bar{E}) \\ & + \bar{A}B\bar{C}D(F) + \bar{A}BC\bar{D}(1) + \bar{A}BCD(\bar{F}) + A\bar{B}\bar{C}\bar{D}(E) + A\bar{B}\bar{C}D(1) + A\bar{B}C\bar{D}(E) \\ & + A\bar{B}CD(0) + AB\bar{C}\bar{D}(\bar{E}\bar{F}) + AB\bar{C}D(0) + ABC\bar{D}(F) + ABCD(0) \end{aligned}$$

The resulting map is shown in figure 1.35a and the possible groupings are shown in figure 1.35b. Most are straightforward and involve an adjacent one. The $\bar{E}$ in cell 4 can be grouped with cell 6, but the $\bar{E}$ can be considered as $\bar{E} + \bar{E}\bar{F}$ so the $\bar{E}\bar{F}$ of cell 12 can also be grouped with cell 4. There are, therefore, eight essential prime implicants, leading to the minimal function

$$\begin{aligned} f = {} & B\bar{C}\bar{D}(\bar{E}\bar{F}) + \bar{A}B\bar{D}(\bar{E}) + \bar{A}\bar{B}D(E + F) + \bar{A}\bar{C}D(F) + \bar{A}BC(\bar{F}) + A\bar{B}\bar{D}(E) \\ & + BC\bar{D}(F) + \bar{B}\bar{C}D(1) \end{aligned}$$

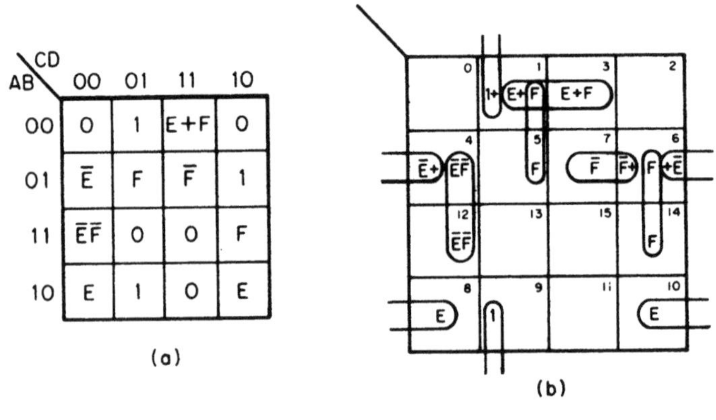

Figure 1.35 Groupings on the variable-entered map

It is often easier to find suitable groupings by using any possible don't care combinations of the imputs. This can be shown by considering, by way of example, a duodecimal counter where the output is required to indicate when the number of pulses received is a multiple of 2 or 3. The map of the required output function is shown in figure 1.36a and it is apparent that very little minimizing is possible.

AB \ CD	00	01	11	10
00	0	0	1	1
01	1	0	0	1
11	0	0	0	0
10	1	1	0	1

(a)

		1	1
1			1
∅=1	∅=1	∅=0	∅=1
1	1		1

(b)

Figure 1.36 (a) Output function of the duodecimal counter; (b) output function making use of don't care conditions

However, if our specification allows us to insist that in practice the four unused input combinations 11XX will not occur, we can take those positions to be don't care entries. Assigning values to the don't care positions, as in figure 1.36b, gives us larger groups and the minimized function

$$f = \bar{A}\bar{B}C + C\bar{D} + B\bar{D} + A\bar{C}$$

In practical logic system design, as we have seen, a more economical form of minimal expression is often achieved by the use of composite functions. Also, economies are often achieved when the realization of the complete system allows one set of gates to be used to generate signals common to several sections of the system. Techniques for arriving at such solutions involve *factorization* which is based on the two distributive rules

$$(A + B)(A + C) = A + BC$$

$$AB + AC = A(B + C)$$

For limited numbers of variables the Karnaugh map is once again the most convenient presentation and the distributive rules themselves can be used to demonstrate the idea. The map of the function f = A + BC can, for example, be considered as the logical product, that is the AND, of the functions shown on the two additional maps in figure 1.37.

The steps involved in decomposing to the two maps are shown in figure 1.38. What we have done is to identify groups of ones corresponding to some simple

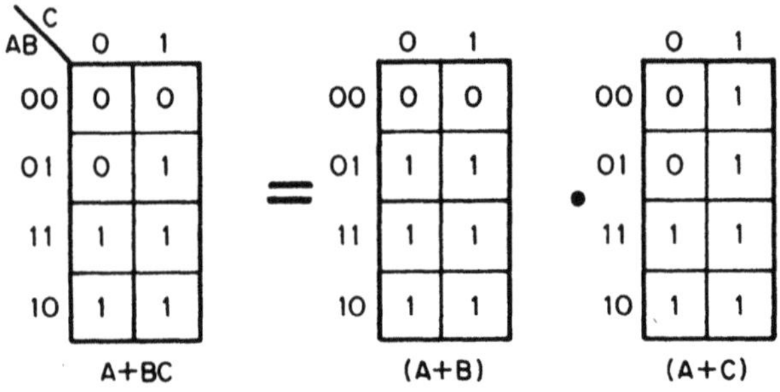

Figure 1.37 Logical product of two maps

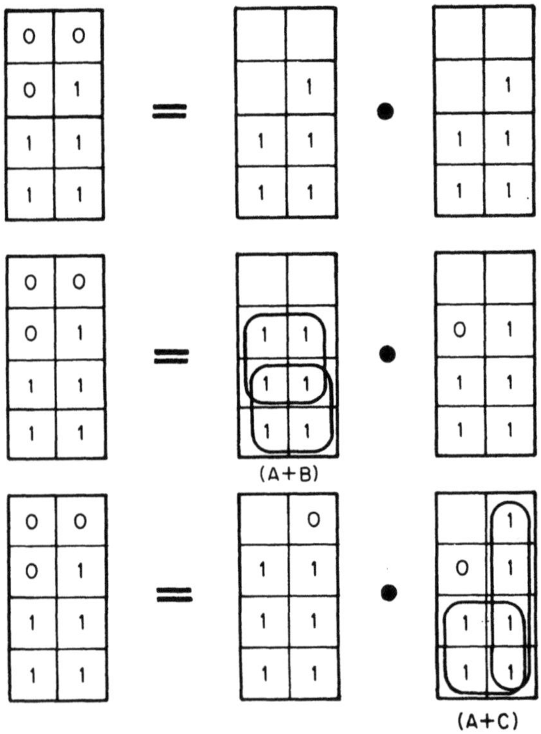

Figure 1.38 Map factorizing

or other desired function containing the given function. These are entered on a map which is then ANDed with another having zeros in those cells necessary to ensure that the composite function is the same as the original. The remaining cells on each map can now have zeros entered and the decomposition is complete. With practice some extremely elegant decompositions may be achieved.

Example 1.21: Show that the function $f = AD + \bar{A}\bar{B}C\bar{D}$ can be factorized into the form $f = (A + C\bar{D})(\bar{A}\bar{B} + D)$.

The mapping procedure is followed in figure 1.39, and we note that, unlike the original form, this expression needs only two-input gates, which may be desirable under certain circumstances.

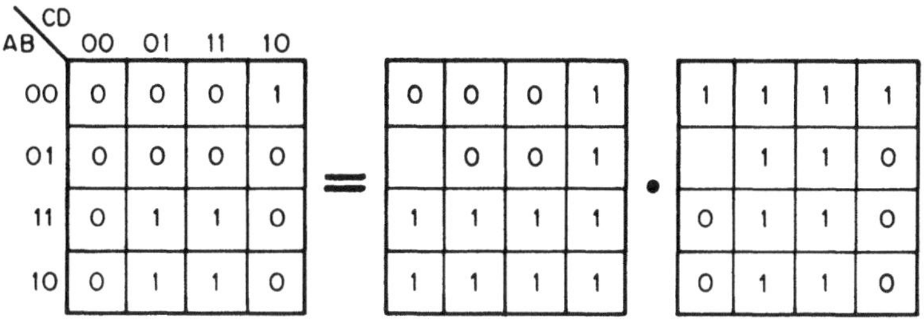

Figure 1.39 $AD + \bar{A}\bar{B}C\bar{D} \equiv (A + C\bar{D})(\bar{A}\bar{B} + D)$

Factorization can also be directed specifically at the extraction of particular functions which may already be available elsewhere, or which may form convenient subfunctions. Suppose we have the function

$$f = AB\bar{C}D + (A + \bar{B})C\bar{D}$$

We notice that the non-equivalence function $(\bar{C}D + C\bar{D})$ appears to be present, though not in a directly usable form. Factorization allows us to extract the required function intact giving the overall function $f = (\bar{C}D + C\bar{D})(\bar{B}\bar{D} + AB)$. The steps are shown in figure 1.40.

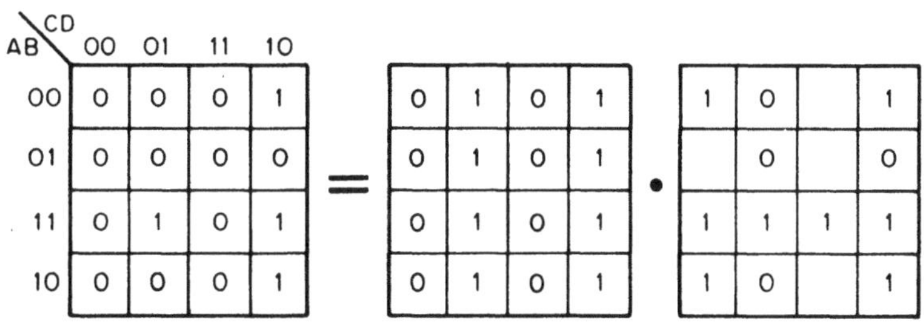

Figure 1.40 $AB\bar{C}D + (A + \bar{B})C\bar{D} = (\bar{C}D + C\bar{D})(\bar{B}\bar{D} + AB)$

Example 1.22: Show that the function $f = A\bar{B}\bar{C}D + \bar{A}BC\bar{D}$ has both $(A \oplus B)$ and $(A \oplus C)$ as factors.

We consider the function to be the result of ANDing the two given factors and some third factor which we are to find. By drawing $(A \oplus B)$ on the first map of the right-hand side of figure 1.41, and $(A \oplus C)$ on the second we see that we must allocate two zeros (cells 7 and 8) on the third map in order to prevent the AND function generating ones in those positions. We must also

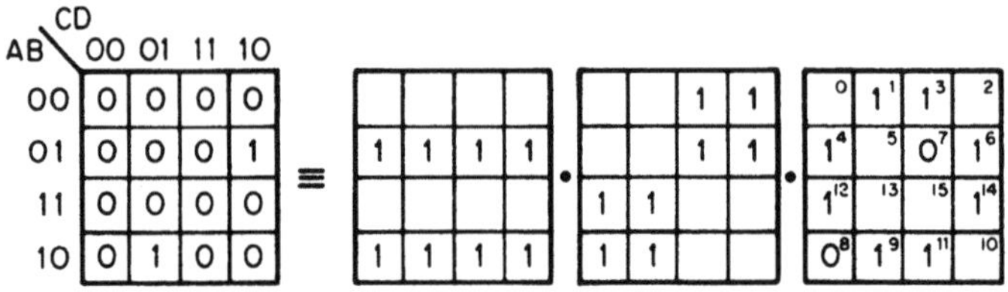

Figure 1.41 $A\bar{B}\bar{C}D + \bar{A}BC\bar{D} \equiv (A\bar{B} + \bar{A}B)(A\bar{C} + \bar{A}C)(B\bar{D} + \bar{B}D)$

ensure that cells 6 and 9 are set to one, but all other cells are don't care values. The simplest function, therefore, is another exclusive-OR, $(B \oplus D)$ as shown and

$$A\bar{B}\bar{C}D + \bar{A}BC\bar{D} \equiv (A\bar{B} + \bar{A}B)(A\bar{C} + \bar{A}C)(B\bar{D} + \bar{B}D)$$

The idea of combining maps can be extended to be of great use in dealing with problems of multiple output circuits by using multiple entries in each cell. Consider the two output functions

$$X = \bar{A}\bar{B}\bar{D} + BD + A\bar{B}\bar{C} + A\bar{B}\bar{D}$$
$$Y = C\bar{D} + A\bar{C}D + \bar{B}\bar{C}\bar{D} + A\bar{C}\bar{D}$$

The composite map of figure 1.42 shows the terms common to both outputs, and taking account of the remaining terms in the separate outputs we arrive at the expressions

$$X = \bar{B}\bar{D} + A\bar{C}D + BD$$
$$Y = \bar{B}\bar{D} + A\bar{C}D + A\bar{C}\bar{D} + C\bar{D}$$

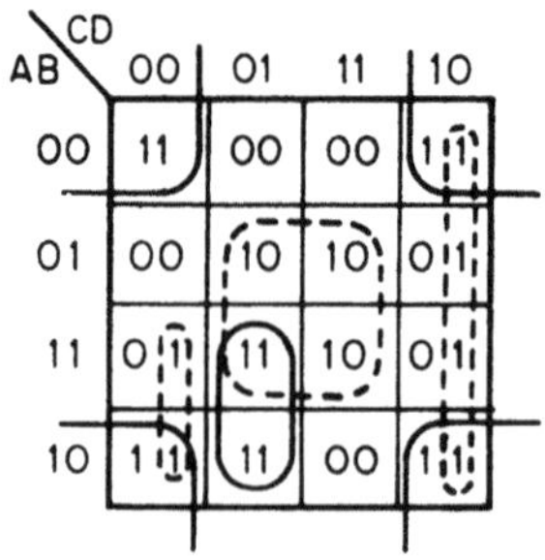

Figure 1.42 Composite map for multiple outputs

A more typical problem arises when it is not clear which selection of common terms gives the most economical solution.

Example 1.23: A circuit is to have three outputs X, Y and Z generating the functions

$$X = \bar{A}B\bar{D} + \bar{A}CD + A\bar{B}\bar{D} + ABCD + A\bar{B}D$$
$$Y = A\bar{B}\bar{C} + \bar{A}\bar{B}CD + \bar{A}B\bar{C} + A\bar{B}C + BCD$$
$$Z = \bar{A}\bar{C}\bar{D} + \bar{B}C\bar{D} + \bar{A}B\bar{C}D + A\bar{B}\bar{C}\bar{D}$$

Minimize the functions, making the best use of common terms.

The composite map is shown in figure 1.43 and, although three cells indicate terms common to all outputs, the positions of the other entries make a more limited selection of common terms a better choice in this case. Thus

$$X = \bar{A}B\bar{D} + A\bar{B} + CD$$
$$Y = A\bar{B} + CD + \bar{A}B\bar{C}$$
$$Z = \bar{A}B\bar{C} + \bar{B}\bar{D}$$

AB \ CD	00	01	11	10
00	001	000	110	001
01	111	011	110	100
11	000	000	110	000
10	111	110	110	111

Figure 1.43 Composite map for three outputs

Many different minimization methods have been developed. A tabular variation of the grouping method is due to Quine (1955). McCluskey (1956) extended Quine's method to cope with a larger number of variables by coding the minterms of the function. A method based on maxterms and using a form of inverse Karnaugh map has been suggested by Evans (1969). The advantage of tabular methods lies in their ability to handle a large number of variables when used as the basis for a computer program (Muroga, 1979).

1.8 Inverting logic and the NAND/NOR transform

The vast majority of logic gates are electronic circuits which represent the logic variable values by two well-defined voltage levels. Which level is chosen to indicate logical one and which logical zero is purely arbitrary, but must be known if we are to determine the gate's logical operation. The principle of duality ensures that if we change our representation, by converting all zeros to ones and ones to zeros, any OR function becomes an AND, and any AND function becomes an OR.

This can be shown quite simply by truth table: figure 1.44a indicates an AND function, but by interchanging all zeros and ones, we get the OR function of figure 1.44b. As we shall see later the most commonly used logic elements incorporate

A	B	Output
0	0	0
0	1	0
1	0	0
1	1	1

(a)

A	B	Output
1	1	1
1	0	1
0	1	1
0	0	0

(b)

Figure 1.44 (a) AND function. (b) OR function

an inverter with each logic gate so that the resulting logic element becomes the NOT-AND or the NOT-OR, abbreviated to NAND and NOR respectively. The resulting inverting logic has many features which make it a powerful form of logic, and circuit design carried out solely in terms of NAND/NOR operations is by no means restricted, as it might at first appear. In fact, any logic function synthesized in terms of the basic connectives AND and OR may readily be transposed into NAND/NOR terms and, conversely, any function generated by NAND/NOR circuits may be expressed in simple AND and OR terms. The transformation method follows directly from de Morgan's theorems

$$\overline{AB} = \overline{A} + \overline{B} \quad \text{Theorem for NAND}$$
$$\overline{A + B} = \overline{A}\,\overline{B} \quad \text{Theorem for NOR}$$

In generalized form, a NAND circuit appears in figure 1.45a where the inputs $f_1, f_2, \ldots$ may be either single variables or outputs from other NAND networks. Algebraically the function is given by

$$F = \overline{(\overline{(\ldots\overline{(\overline{(\overline{f_n\, f_{n-1}})\, f_{n-2}})\ldots f_3)}\, f_2)}\, f_1}$$

By repeated application of de Morgan's theorem for NAND we arrive at the expression

$$F = [(\ldots + \overline{f}_5)\, f_4 + \overline{f}_3]\; f_2 + \overline{f}_1$$

This form of the function can be generated by an AND–OR circuit, which is also shown in figure 1.45 and we note that the number of gates required and their interconnections bear a one-to-one relationship with the NAND gates required for the original form of the expression.

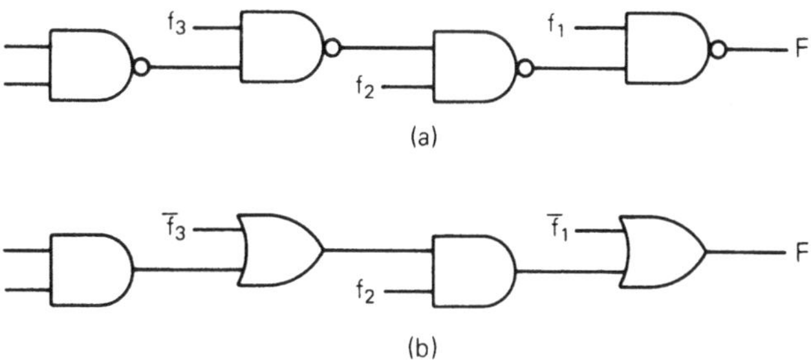

Figure 1.45 (a) NAND network. (b) OR-AND-OR network

However, variables entered at odd levels (with the output level taken as level one) are inverted when transformed to the AND-OR form. Using these features, the circuit described by any expression can be drawn directly in NAND form.

Example 1.24: Draw the NAND circuit necessary to generate the function $f = (C\overline{D} + \overline{C}D)E + A$.

The circuit is shown in figure 1.46, with the independent variable, A, inverted since it enters at an odd level. The intermediate AND–OR circuit is included in figure 1.46 but with practice that stage is soon discarded.

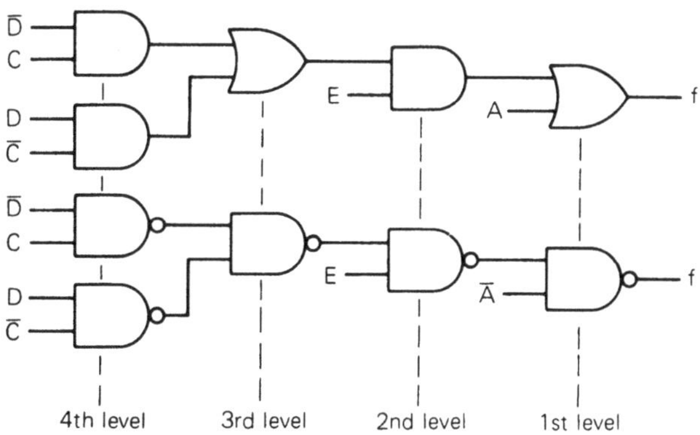

Figure 1.46 The NAND transform

By a similar exercise it can be shown that a function generated by NOR units can be expressed as an OR–AND sequence. Thus, if

$$F = \overline{(\ldots\overline{(\overline{(f_n + f_{n-1})} + f_{n-2})}\ldots + f_2)} + f_1$$

then, by de Morgan

$$F = [(\ldots + f_4)\,\overline{f}_3 + f_2]\,\overline{f}_1$$

The corresponding circuits are shown in figure 1.47.

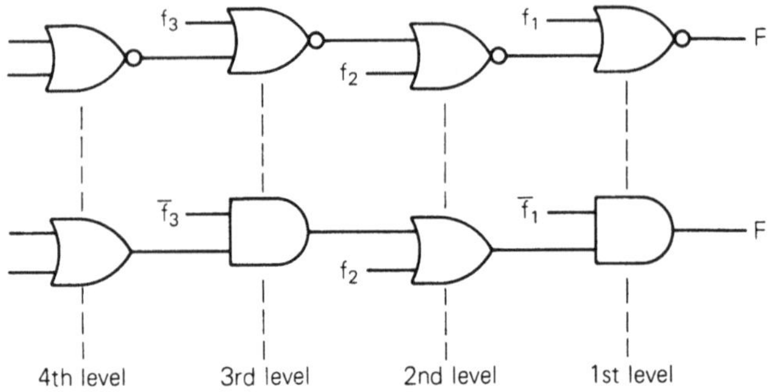

Figure 1.47 The NOR transform

From the generalized expression for NAND and NOR functions we can derive simple rules allowing us to write down directly, in AND-OR form, the function generated by any NAND or NOR circuit. These NAND/NOR transform rules are as follows:

Rule 1: The NAND/NOR gates act as alternate AND and OR gates. In a NAND circuit the first level gate (that is, the output gate) and any other ODD level gates act as OR; in a NOR circuit the first level gate and any other ODD level gates act as AND.

Rule 2: External variables entering the circuit at an ODD level appear in the output function INVERTED.

De Morgan's theorems act as a summary of the transform rules since they can be considered as the transforms for single two-input NAND and NOR gates respectively. Hence, for a single two-input NAND gate, $f = \overline{AB} = \bar{A} + \bar{B}$, indicating that the first, and in this case only level has transformed to OR, and the variables, A and B, at the odd level are inverted. Similarly, for a single NOR gate, $f = \overline{A + B} = \bar{A}\bar{B}$.

Two minor problems arise in applying the transform to a typical circuit, but they are readily handled by the transform if applied correctly. Firstly, how to cope with a NAND or NOR gate with only one input signal – such a gate is being used simply as an inverter and, as far as the transform is concerned, its only effect is to shift the gates feeding it on to a different level. The transform automatically takes care of such a gate but it is sometimes helpful to think of it as being an AND gate with all other inputs at one, or an OR gate with all other inputs at zero, as appropriate. Secondly, if a gate feeds two or more gates on different levels, how is its level to be determined? Again the transform copes with this case, since the gate can be considered as being on both an odd and an even level simultaneously, and it is the gate it is feeding that determines the level as far as that gate is concerned. The two circuits, (a) and (b), of figure 1.48, for example, are equivalent.

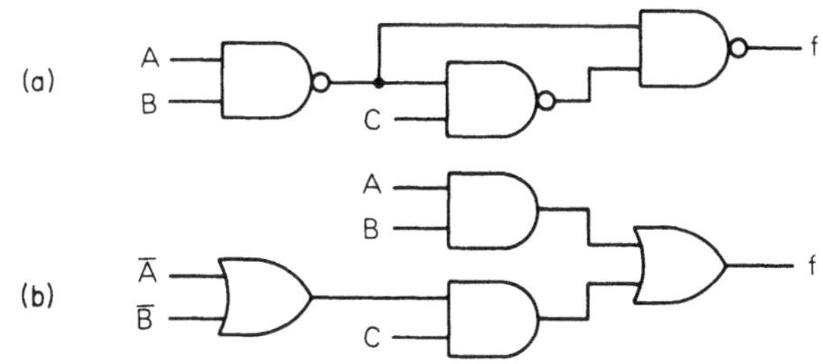

Figure 1.48 A gate driving two levels

Example 1.25: Show that the circuit of figure 1.49a can be simplified to need only five NAND gates for the generation of X and Y outputs from the same inputs.

The output Y is already of minimal form. By use of the transform the function X is seen to be

$$X = (A + (\overline{B} + \overline{C})CD)\,(D + (\overline{B} + \overline{C}))$$
$$\equiv AY + AD + CDY$$

Using the inverse transform a new circuit can be drawn directly, figure 1.49b, since the expression is in sum of products form.

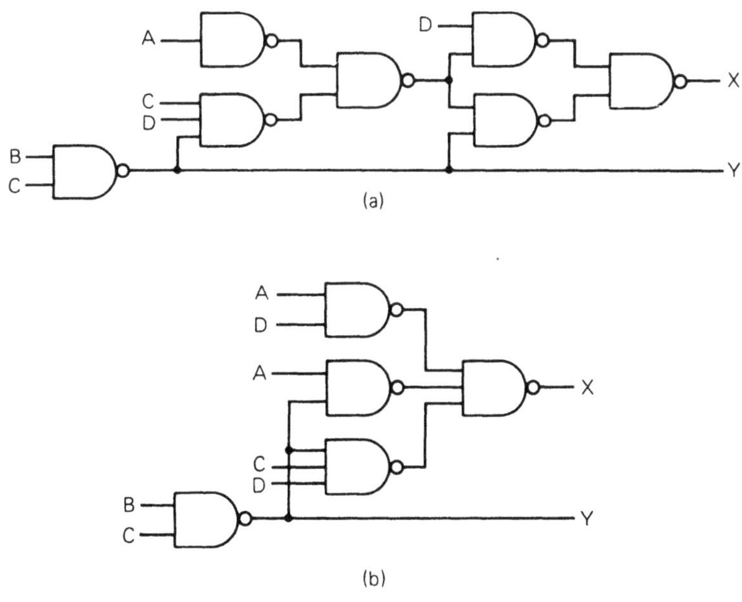

Figure 1.49 Use of the transform to simplify a circuit

Example 1.26 : (a) The exclusive-OR gate acts in the same way as a two-way switch, in that changing either input variable causes the output value to change to its opposite state. Assuming we have three input variables, A, B and C, use NAND gates to provide a circuit which acts as a three-way switch. That is, changing any one input variable causes the output to change.
(b) Logic functions can be generated by interconnecting switches or relay contacts. Using changeover switches, devise connections to give

(1) a two-way switch action

(2) a three-way switch action.

(a) If we assume that all inputs are initially at zero and the output is at zero, any combination showing a single change of input must give an output of one. Thus 001, 010 and 100 give a one output. A further change of a single input, to 101, 011 or 110, must give a zero output. Finally, a change to 111 must give a one output. The truth table is shown in figure 1.50a and is seen to give the function $f = A \oplus B \oplus C$.

Note that it is easy to check that this is true if we take the function to be $f = (A \oplus B) \oplus C$. Expansion of the function gives

Expansion of the function gives

$$f = [A(\overline{A} + \overline{B}) \ + B(\overline{A} + \overline{B})]\,\overline{C} + \overline{[A(\overline{A} + \overline{B}) + B(\overline{A} + \overline{B})]}\,C$$

which leads to the NAND circuit of figure 1.50b.

(b) Switches in series give an AND function, and switches in parallel give an OR function.

A	B	C	f
0	0	0	0
0	0	1	1
0	1	0	1
0	1	1	0
1	0	0	1
1	0	1	0
1	1	0	0
1	1	1	1

(a)

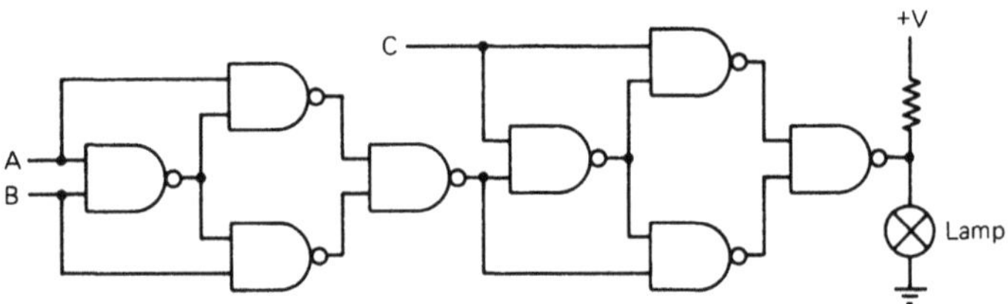

(b)

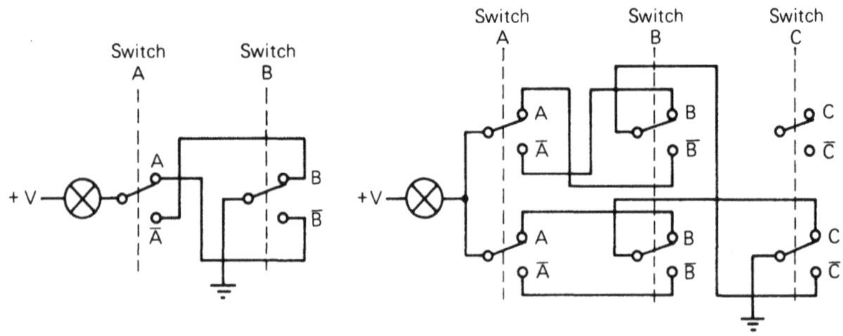

(c) (d)

Figure 1.50 Example 1.26

(1) The two-way switch action relies on the exclusive-OR function. In other words, the function defining when the lamp or indicator is to be on is given by $f = \bar{A}B + A\bar{B}$. The values of the variables A and B are indicated by the positions of the changeover switches. Using the logic expression, we connect contacts in series to form the required AND functions, and arrive at the circuit of figure 1.50c.

(2) The three-way function was defined in part (a) as

$$f = [A(\bar{A} + \bar{B}) + B(\bar{A} + \bar{B})]\,\bar{C} + [\overline{A(\bar{A} + \bar{B}) + B(\bar{A} + \bar{B})}]\,C$$

which can be rearranged to

$$f = (A\bar{B} + \bar{A}B)\bar{C} + (AB + \bar{A}\bar{B})C$$

and leads to the circuit of figure 1.50d.

Problems

1.1 Prove algebraically that $\bar{A}\bar{C}D + \bar{B}D + AB\bar{C} + BCD + A\bar{B}\bar{C}\bar{D} = A\bar{C} + D$.

1.2 Prove algebraically that if $f = (A + \bar{B})(A + \bar{C})$, then $\bar{f} = \bar{A}(B + C)$.

1.3 Given the expression $T = \bar{B}\bar{D} + \bar{A}\bar{B} + B\bar{C}D + AC\bar{D}$, find the canonical expansion in (a) S of P form, and (b) P of S form.

1.4 Each student on the Management course was required to spend a period in industry. The tutor had places available at three establishments but Undersea Mining would take no women, the Ministry of Secrets insisted on home students only, and anyone working at High-speed Deliveries was required to have a current driving licence. The tutor eventually managed to form three equal-sized groups satisfying all the requirements, but he was surprised to find that no two students were alike, and that none of the men students was married. Furthermore, both married women on the course held driving licences, although two of the students allocated to the Ministry of Secrets could not drive. The man in Mrs Smith's group was not her brother; he went to Undersea Mining. Which group did Mrs Smith herself join?

1.5 Express the function $f = AC(\bar{B}D + \bar{B}\bar{D}) + \bar{A}BD + A\bar{B}\bar{C}$ in a form using the B variable only in the exclusive-OR factor, $A \oplus B$.

1.6 A logic circuit with inputs ABCD yields an output function given in maxterm form by $f(ABCD) = \Pi M(0, 1, 2, 3, 5, 8, 10)$. Develop a simplified expression in P of S form (a) algebraically, and (b) using the K-map.

1.7 Convert the expression $f = \Sigma m(0, 1, 2, 3, 4, 5, 6, 7, 12, 13)$ into minimal P of S and S of P forms.

1.8 Find the set of prime implicants of the function $f = \Sigma m(2, 3, 4, 5, 6, 10, 11, 13)$.

1.9 Is the function $f(ABC) = AB\bar{C} + \bar{A}BC + A\bar{B}C$ fully symmetric?

1.10 The function $f(ABC) = \overline{B}C + \overline{A}\overline{B} + \overline{A}C$ has three symmetries. What are they?

1.11 Show by appropriate substitution that the function of example 1.12 does in fact contain partial symmetry about A and B, and C and D.

1.12 (a) Express the following functions in the minterm form:

(1) $S^4_{2,4}$

(2) $S^5_{1,3,5}$.

(b) Express the following functions in the symmetric notation form:

(1) $f(ABCDE) = \Sigma m(3, 5, 6, 9, 10, 12, 17, 18, 20, 24)$

(2) $f(ABC) = \Pi M(0, 7)$.

(c) Express the following functions in the simplest algebraic form:

(1) $S^3_3\ (ABC)$

(2) $S^3_{1,2,3}\ (ABC)$

(3) $S^3_{1,3}\ (ABC)$.

1.13 Devise a circuit to give a one output when the majority of its five inputs, ABCDE, is at one.

1.14 Show graphically that $W = \overline{A}B\overline{C}D + \overline{A}BC\overline{D} + \overline{A}BCD + A\overline{B}\overline{C}\overline{D} + A\overline{B}\overline{C}D$ can be reduced to $W = A + B(C + D)$ if don't care terms exist and are defined by $\emptyset = A(B + C)$.

1.15 A three-output logic circuit is defined by the following expressions:

$$X = \overline{A}B\overline{C} + AB\overline{C}D + A\overline{B}C + \overline{A}C\overline{D}$$
$$Y = B\overline{C}D + \overline{A}\overline{B}\overline{C}D + \overline{B}C + \overline{A}CD$$
$$Z = B\overline{C} + \overline{B}CD + BC\overline{D}.$$

Minimize the expressions and sketch a minimal circuit using not more than three inputs per gate.

1.16 Use the transform method to find the function generated by the circuit of figure 1.51 and show that the same function could be generated using only three NAND gates.

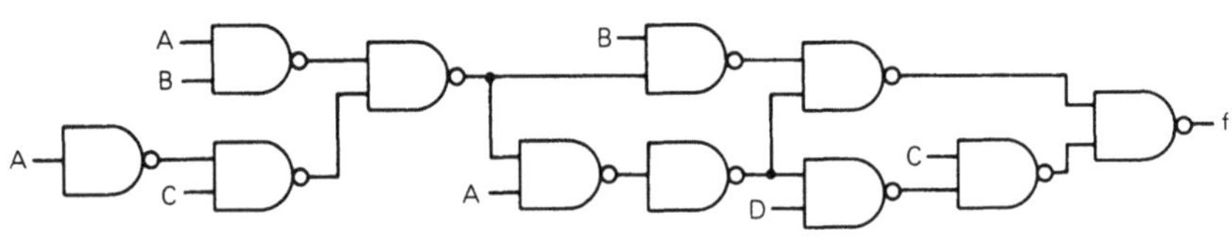

Figure 1.51 Problem 1.16

References

Boole, G. (1953). *An Investigation of the Laws of Thought*, originally published 1854, republished by Dover Publications, New York

Evans, F. C. (1969). Use of inverse Karnaugh maps in realization of logic functions, *Electronic Letters*, **5**, No. 21 (October), 670

Huntington, E. V. (1904). Postulates for the algebra of logic, *Trans. Am. Math. Soc.*, **5**, 288-309

Karnaugh, M. (1953). The map method for synthesis of combinational logic circuits, *Trans. Am. Inst. Elect. Engrs.*, **72 pt. 1**, 593-9

Lewin, D. (1985). *Design of Logic Systems*, Van Nostrand Reinhold, New York

McCluskey, E. J. (1956). Minimisation of boolean functions, *Bell Syst. Tech. J.*, **35**, No. 6 (November), 1417-44

Mead, C. and Conway, L. (1980). *Introduction to VLSI Systems*, Addison-Wesley, Reading, Massachusetts

Mukherjee, A. (1986). *Introduction to nMOS and CMOS VLSI Systems Design*, Prentice-Hall, Englewood Cliffs, New Jersey

Muroga, S. (1979). *Logic Design and Switching Theory*, John Wiley, New York

Quine, W. V. (1955). A way to simplify truth functions, *Am. Math. Mon.*, **62**, November, 627-31

Shannon, C. E. (1938). A symbolic analysis of relay and switching circuits, *Trans. Am. Inst. Elect. Engrs.*, **57**, 713-23

Veitch, E. W. (1952). A chart method for simplifying truth functions, *Proc. Ass. Comput. Mach.*, **May**, 127-33

2 Number Systems and Coding

Numeracy is a relative latecomer in man's intellectual development. The Greek and Roman civilisations had no need for anything other than a primitive numbering system and it was not until the twelfth century that the Arabic concepts of zero and positional notation were accepted. It is generally accepted that the decimal system became universally popular because man has ten readily available digits, but any other relatively small number would have served as well, if not better; twelve, for example, has more factors and would probably have been more useful.

The decimal counting system is an example of a *positional notation*. The name decimal implies ten unique symbols and the *radix* or *base* of the system is said to be ten. The value of the number represented is given by the sum of the digit values, each digit weighted by a power of ten determined by its position relative to the *decimal point*. For instance, the number 6278.35_{10} represents

$$6 \times 10^3 + 2 \times 10^2 + 7 \times 10^1 + 8 \times 10^0 + 3 \times 10^{-1} + 5 \times 10^{-2}$$

Note that the suffix 10 indicates the base value and is usually omitted where there is no possibility of confusion.

In more general algebraic terms the above expression is given by

$$N = d_n r^n + d_{n-1} r^{n-1} + \ldots + d_1 r^1 + d_0 r^0 + d_{-1} r^{-1} + \ldots \tag{2.1}$$

where $d_1, d_2, \ldots, d_n$ are the decimal symbols and r the base.

2.1 Numbering systems

The decimal notation is rarely used to represent numbers in digital systems, the need for devices capable of assuming any one of ten unique states being incompatible with reliable engineering. The binary system, with its base of two, requiring only two symbols or states, is much to be preferred. The symbols 0 and 1 are used to represent the two states. Using the general expression, equation 2.1, a number such as 10010.11_2 is equivalent in decimal terms to

$$1 \times 2^4 + 0 \times 2^3 + 0 \times 2^2 + 1 \times 2^1 + 0 \times 2^0 + 1 \times 2^{-1} + 1 \times 2^{-2}$$

taking the weightings relative to the position of the binary point.

When counting successively the digits are used in turn, increasing the next higher power by one when both have been used, as shown in figure 2.1. It is conventional in the binary system to keep the number of binary digits (abbreviated to *bits*) constant. This often involves the inclusion of redundant zeros in the higher-order bit positions. Note that for a given number of bits, K, the maximum number that can be represented is $2^K - 1$.

2^2	2^1	2^0	Decimal
0	0	0	0
0	0	1	1
0	1	0	2
0	1	1	3
1	0	0	4
1	0	1	5

etc.

Figure 2.1 Counting in binary

Example 2.1: A binary number contains 4 bits. How many different bit patterns can be portrayed?

The number of unique patterns is $2^4 = 16$. This is equal to the maximum count plus one since the all-zeros pattern must be included.

Although binary representation is pre-eminent in digital systems, numbering systems with bases of eight and sixteen find extensive use as a convenient shorthand notation. These systems are referred to as *octal* and *hexadecimal* (or just *hex*) respectively. Eight symbols are required for octal, and the first eight of the decimal series are taken. The hexadecimal notation requires sixteen symbols; the symbols 0 through 9 are therefore augmented by the letters ABCDEF.

The codes owe their popularity to the special relationship which they have with three- and four-bit binary numbers. In octal, the weightings are powers of eight relative to the octal point, as in figure 2.2a. Similarly, in hexadecimal, weightings are powers of sixteen relative to the hex point, as in figure 2.2b.

Example 2.2: Write down the octal and hex equivalents of 179_{10}.

$$179_{10} = 2 \times 8^2 + 6 \times 8^1 + 3 \times 8^0 = 263_8$$
$$179_{10} = \mathrm{B} \times 16^1 + 3 \times 16^0 = \mathrm{B}3_{16}$$

We see from figure 2.3 that the eight possible combinations of three binary digits can be represented by the eight symbols of the octal system. Similarly, the sixteen combinations of four bits can be represented by the sixteen symbols

8^2	8^1	8^0	Decimal
0	0	0	0
0	0	1	1
•	•	•	•
•	•	•	•
0	0	7	7
0	1	0	8
0	1	1	9
•	•	•	•
•	•	•	•
0	7	7	63
1	0	0	64
			etc.

(a)

16^2	16^1	16^0	Decimal
0	0	0	0
0	0	1	1
0	0	2	2
•	•	•	•
•	•	•	•
0	0	9	9
0	0	A	10
•	•	•	•
•	•	•	•
0	0	F	15
0	1	0	16
0	1	1	17
•	•	•	•
•	•	•	•
0	F	F	255
1	0	0	256
			etc.

(b)

Figure 2.2 (a) Counting in octol. (b) Counting in hexadecimal

8^1	8^0	8^1	8^0
0	0	000	000
0	1	000	001
0	2	000	010
0	3	000	011
0	4	000	100
0	5	000	101
0	6	000	110
0	7	000	111
1	0	001	000

(a)

16^1	16^0	16^1	16^0
0	0	0000	0000
0	1	0000	0001
0	2	0000	0010
0	3	0000	0011
•	•	•	•
•	•	•	•
•	•	•	•
•	•	•	•
0	9	0000	1001
0	A	0000	1010
0	B	0000	1011
0	C	0000	1100
0	D	0000	1101
0	E	0000	1110
0	F	0000	1111
1	0	0001	0000

(b)

Figure 2.3 (a) Octal–binary and (b) hexadecimal–binary equivalence

of the hexadecimal system. The advantage of this equivalence is apparent when dealing with binary numbers having a large number of bits. For example, the binary number $1\,0\,1\,0\,1\,0\,0\,1\,1\,0\,1\,1\,0\,0\,1_2$ becomes

$$\begin{matrix} 1\,0\,1 & 0\,1\,0 & 0\,1\,1 & 0\,1\,1 & 0\,0\,1 \\ 5 & 2 & 3 & 3 & 1 \end{matrix} = 52331_8$$

or

$$\begin{matrix} 0\,1\,0\,1 & 0\,1\,0\,0 & 1\,1\,0\,1 & 1\,0\,0\,1 \\ 5 & 4 & D & 9 \end{matrix} = 54D9_{16}$$

Note the inclusion of a redundant zero when converting this number to its hex form, in order to obtain groupings of four bits.

Hexadecimal notation is commonly used in microprocessor work where the binary numbers dealt with are usually multiples of 8 or 16 bits.

2.2 Base conversions

It becomes necessary on occasions to convert a number from one base to another. The simplest conversion is binary to either octal or hexadecimal. As we have seen, all that is necessary to convert a binary number to its octal or hexadecimal equivalent is to write down the number in groups of three or four bits appropriately, starting from the binary point, and replacing each group with its equivalent symbol. Conversion to binary involves replacing each octal or hexadecimal digit with the corresponding three or four binary digits.

Binary-to-decimal conversion follows from consideration of the general expression for a number N given by equation 2.1, thus

$$N_{10} = d_n 2^n + d_{n-1} 2^{n-1} + d_{n-2} 2^{n-2} + \ldots + d_0 2^0$$

where $d_n, d_{n-1}, \ldots d_0$ are of value 1 or 0.

Example 2.3: Convert $1\ 1\ 0\ 1\ 1\ 0_2$ to decimal.

$$\begin{aligned} N_{10} &= 1 \times 2^5 + 1 \times 2^4 + 0 \times 2^3 + 1 \times 2^2 + 1 \times 2^1 + 0 \times 2^0 \\ &= 32 + 16 + 0 + 4 + 2 + 0 \\ &= 54_{10} \end{aligned}$$

Rearrangement of equation 2.1 gives

$$\begin{aligned} N &= r(d_n r^{n-1} + d_{n-1} r^{n-2} + \ldots) + d_0 \\ &= r[r(d_n r^{n-2} + d_{n-1} r^{n-3} + \ldots) + d_1] + d_0 \\ &= r\{r[r(d_n r^{n-3} + d_{n-1} r^{n-4} \ldots) + d_2] + d_1\} + d_0 \\ &= r\{r[\ldots r(d_n r^1 + d_{n-1}) \ldots + d_2] + d_1\} + d_0 \end{aligned}$$

Interpreting this, we can see that the conversion could have been made by taking the most significant digit of the binary number d_n, multiplying by two, adding the next most significant digit, and so on until the least significant digit is added. Thus converting $1\ 1\ 0\ 1\ 1\ 0_2$ once more gives us

$1\quad 1\quad 0\quad 1\quad 1\quad 0_2$

$1 \times 2 + 1$

$3 \times 2 + 0$

$6 \times 2 + 1$

$13 \times 2 + 1$

$27 \times 2 + 0$

54

Octal-to-decimal conversion may be carried out in a similar manner. Thus

$$
\begin{array}{llll}
6 & 0 & 7 & 2_8 \\
6 \times 8 + 0 & & & \\
 & 48 \times 8 + 7 & & \\
 & & 391 \times 8 + 2 & \\
 & & & 3130_{10}
\end{array}
$$

By the conventional method, the conversion is

$$
\begin{aligned}
N_{10} &= 6 \times 8^3 + 0 \times 8^2 + 7 \times 8^1 + 2 \times 8^0 \\
&= 3072 + 0 + 56 + 2 \\
&= 3130_{10}
\end{aligned}
$$

In general, when converting from a higher to a lower base, it is easier to divide by the new base.

$$N = d_n r^n + d_{n-1} r^{n-1} + \ldots + d_1 r^1 + d_0 \tag{2.1}$$

therefore

$$N = r(d_n r^{n-1} + d_{n-1} r^{n-2} + \ldots + d_1) + d_0$$

whence

$$N = rQ + d_0$$

where Q is the integer quotient and d_0 the remainder.

By repeated divisions of the quotient by the base we obtain, as a series of remainder digits, d_n, d_{n-1}, d_{n-2}, etc., which are the digits of the required number.

Example 2.4: Convert 217_{10} to binary.

$$
\begin{array}{rll}
217 \div 2 = & 108 \text{ with remainder} & d_0 = 1 \\
108 \div 2 = & 54 & d_1 = 0 \\
54 \div 2 = & 27 & d_2 = 0 \\
27 \div 2 = & 13 & d_3 = 1 \\
13 \div 2 = & 6 & d_4 = 1 \\
6 \div 2 = & 3 & d_5 = 0 \\
3 \div 2 = & 1 & d_6 = 1 \\
1 \div 2 = & 0 & d_7 = 1
\end{array}
$$

Thus, $217_{10} \equiv 11011001_2$.

By successive divisions by 8, we could similarly convert 217_{10} to base 8.

So far only conversion of integers has been considered and we must now investigate fractional conversions. In all ordered number systems the magnitude of the increment between successive digits is unity and the symbols 0 and 1 have the same value in all systems. That means that integers in one base must convert to integers in another. Similarly fractions in one base will always convert to fractions in another since in any base a fraction, by definition, has a value between 0 and 1.

Basically the same procedures are adopted for all conversions. For integer conversions we have seen that we can divide by the new base and note the overflow, that is, the digit passing over the point.

When N is fractional, equation 2.1 reduces to

$$N = d_{-1}r^{-1} + d_{-2}r^{-2} + d_{-3}r^{-3} + \ldots \tag{2.2}$$

Note that the point comes before the first term of the expression.

Multiplying both sides of this expression by the base r yields

$$\begin{aligned} rN &= d_{-1} + d_{-2}r^{-1} + d_{-3}r^{-2} + d_{-4}r^{-3} + \ldots \\ &\equiv d_{-1} + N' \end{aligned}$$

where N' is a new fraction and d_{-1} is an integer. Further

$$\begin{aligned} rN' &= d_{-2} + d_{-3}r^{-1} + d_{-4}r^{-2} + \ldots \\ &\equiv d_{-2} + N'' \end{aligned}$$

This process is continued until sufficient accuracy has been achieved. The first digit (d_{-1}) obtained is the most significant of the new fraction.

Example 2.5: Convert 0.923_{10} to octal.

$$\begin{aligned} 0.923_{10} \times 8 &= 7.384 & d_{-1} &= 7 \\ 0.384_{10} \times 8 &= 3.072 & d_{-2} &= 3 \\ 0.072_{10} \times 8 &= 0.576 & d_{-3} &= 0 \\ 0.576_{10} \times 8 &= 4.608 & d_{-4} &= 4 \end{aligned}$$

Thus, $0.923_{10} = 0.7304\ldots_8$.

If we wish to convert the same decimal fraction to binary the procedure follows the same lines

$$\begin{aligned} 0.923_{10} \times 2 &= 1.846 & d_{-1} &= 1 \\ 0.846_{10} \times 2 &= 1.692 & d_{-2} &= 1 \\ 0.692_{10} \times 2 &= 1.384 & d_{-3} &= 1 \\ 0.384_{10} \times 2 &= 0.768 & d_{-4} &= 0 \\ 0.768_{10} \times 2 &= 1.536 & d_{-5} &= 1 \\ 0.536_{10} \times 2 &= 1.072 & d_{-6} &= 1 \\ 0.072_{10} \times 2 &= 0.144 & d_{-7} &= 0 \end{aligned}$$

Thus, $0.923_{10} = 0.1110110\ldots_2$.

Our method of conversion has been developed from equation 2.2 in terms of conversion from decimal. The expression is perfectly general but when converting from any base to another it is necessary to use the arithmetic of the initial base.

Example 2.6: Convert 0.7304_8 to binary.

$$
\begin{array}{ll}
0.7304_8 \times 2 = 1.6610 & d_{-1} = 1 \\
0.6610_8 \times 2 = 1.5420 & d_{-2} = 1 \\
0.5420_8 \times 2 = 1.3040 & d_{-3} = 1 \\
0.3040_8 \times 2 = 0.6100 & d_{-4} = 0 \\
0.6100_8 \times 2 = 1.4200 & d_{-5} = 1 \\
0.4200_8 \times 2 = 1.0400 & d_{-6} = 1 \\
0.0400_8 \times 2 = 0.1000 & d_{-7} = 0 \\
0.1000_8 \times 2 = 0.2000 & d_{-8} = 0 \\
0.2000_8 \times 2 = 0.4000 & d_{-9} = 0 \\
0.4000_8 \times 2 = 1.0000 & d_{-10} = 1
\end{array}
$$

Thus, $0.7304_8 = 0.1110110001_2$.

Recalling section 2.1 it is clear that in this particular case, octal-to-binary conversion, it is possible to write the answer by inspection. Thus

$$
\begin{array}{llllll}
0 & . & 7 & 3 & 0 & 4_8 \\
0 & . & 111 & 011 & 000 & 100_2
\end{array}
$$

Usually, the arithmetic involved in conversions is carried out in the initial base but it is possible, and in some cases more convenient (for example, in converting to decimal), to work in the final base.

We are now in a position to summarise this section in the form of the following two conversion algorithms.

(1) Conversion from base B_1 to base B_2 working in base B_1:
   (a) Whole numbers: Divide the number by the final base successively, noting the remainders (the first being the least significant digit) until the quotient is zero.
   (b) Fractions: Multiply by the final base successively, noting the overflow with the most significant digit first.

(2) Conversion from base B_1 to base B_2 working in base B_2:
   (a) Whole numbers: Multiply by initial base (most significant digit first) and add to the next most significant digit.
   (b) Fractions: Divide by initial base (least significant digit first) and add to the next digit. Continue until the point is reached.

2.3 Negative numbers

As we have seen, in digital systems engineering limitations restrict us to two states which are represented by the two binary symbols 1 and 0. We must therefore make use of these same symbols to indicate whether the number is positive or negative. Perhaps the simplest method is to add an additional bit to the number, normally at the most significant end, to represent the sign. A one usually indicates a minus sign. For example

$$-7_{10} = 1\,0\,1\,1\,1_2$$
$$+9_{10} = 0\,1\,0\,0\,1_2$$

This representation of numbers is called the *sign plus magnitude form* and finds its chief use in computer arithmetic circuits where sign information is inconvenient to handle, for instance in the multiplication process.

Two other methods of negative number representation are available to us, both of which involve a *complement* of the number. Every ordered number system has a *radix complement*. For instance, a decimal number has a *ten's complement*. The complement is found by subtracting the given number from the next higher power of ten. Thus

$$\text{Ten's complement of } 54_{10} = 100 - 54 = 46$$

Similarly, the *two's complement* of a binary number is found by subtracting that number from the next higher power of two.

$$\text{Two's complement of } 1010_2 = 10000 - 1010 = 0110_2$$

It is clear that the addition of a number and its radix complement results in all zeros and an overflow carry. In general terms, an n-digit number N_R in base R has a radix complement M_R given by

$$M_R = R^n - N_R \tag{2.3}$$

If the system in use limits all numbers to n digits any more significant digits are ignored. This form of limitation is called 'modulo' (or modulus) R^n arithmetic. For example, in the decimal system, if $n = 1$, giving modulo 10

$$6_{10} + 8_{10} = 4 \bmod 10$$

Similarly

$$497_{10} + 624_{10} = 121 \bmod 1000$$

Since the system ignores digits in excess of n, equation 2.3 leads us to

$$M_R = -N_R$$

because R^n always has n zeros following a one in the most significant (overflow) position. Thus we have succeeded in representing the negative of a number in the same number of digits. Using the complement the subtraction can be carried out by the addition of a complement number.

Example 2.7: Evaluate the expression

$$S = 47_{10} - 23_{10}$$

This is equivalent to

$$S = 47_{10} + (-23_{10})$$

Ten's complement of 23 = 100 – 23 = 77, therefore

$$\begin{array}{rr} S = & 47 \\ & +77 \\ \hline & 1\,24 \end{array}$$

1 24 = 24 ignoring the carry out.

A second complement form exists termed the *diminished radix complement* and is defined as

$$(R^n - 1) - N_R$$

This complement is always one less than the radix complement and takes the name of the highest digit value in the system. Thus in the decimal system, this is the *nine's complement*. In binary the term is the *one's complement* which is readily obtained by inverting each bit of the number. For example, the one's complement of 01101 is 10010.

Example 2.8: Use the radix complement method to subtract four from six (a) in base 7 and (b) in base 12.

(a) The radix complement of 4_7 is $7 - 4 = 3$. Thus, $6_7 - 4_7$ is carried out in base 7 arithmetic as $6 + 3 = 12$ which, ignoring the carry over, is 2.

(b) The radix complement of 4_{12} is $12 - 4 = 8$. Thus, $6_{12} - 4_{12}$ is equal to $6 + 8 = 12$ which is equivalent to 2, again ignoring the carry over and working in base 12 arithmetic.

2.4 Binary arithmetic

Binary addition is carried out in exactly the same way as addition in any other radix. The digits in columns corresponding to a given weighting are added, taking account of carries as they occur. Subtraction is also carried out by the conventional method. Figure 2.4 shows the sum and carry values and difference and borrow values resulting from all the possible values of the two binary variables X and Y.

X	Y	Sum	Carry	Difference	Borrow
0	0	0	0	0	0
0	1	1	0	1	1
1	0	1	0	1	0
1	1	0	1	0	0

Figure 2.4 Truth table for $(X + Y)$ and $(X - Y)$

Examples of addition in different bases are given below.

Addend	22_{10}	26_8	10110_2
Augend	15_{10}	17_8	01111_2
Sum	37_{10}	45_8	100101_2

Examples of subtraction in different bases give us:

Minuend	22_{10}	26_8	10110_2
Subtrahend	15_{10}	17_8	01111_2
Difference	07_{10}	07_8	00111_2

As we have noted previously, it is more convenient to carry out subtraction by an additive process. In binary this is achieved by the use of the two's complement. In section 2.3 we have seen the subtraction of two decimal numbers; the binary equivalent follows:

$$\begin{array}{rr} 47_{10} & 101111_2 \\ -23_{10} & -010111_2 \\ \hline 24_{10} & 011000_2 \end{array}$$

The two's complement of 010111 is 1000000 – 010111 = 101001. Thus the subtraction can be carried out as

$$\begin{array}{r} 101111 \\ +101001 \\ \hline 1\,011000 \end{array}$$

Note that a convenient way of arriving at the two's complement of any number is to copy out the number starting with the least significant digit until the first 1 has been copied. Thereafter invert each digit. Thus 0100100 may be written down directly as 1011100, the four most significant digits having been inverted. While this is a simple mental process, it can only be achieved in practical terms by finding the one's complement and adding 1.

Subtraction can be carried out by the use of the one's complement. However, to ensure that the result is consistent it is necessary to correct after each operation according to the rule: *If a carry overflow occurs add 1 to the result; if no carry overflow occurs do not add 1.* The rule leads to the name *end-around-carry* for the method.

We can show the validity of the end-around-carry process by considering the two numbers N_{R_1} and N_{R_2} where N_R is an n-digit number. The case where both numbers are positive is trivial since the complement is not involved. We define the problem as $N_{R_1} - N_{R_2}$ where N_{R_1} and N_{R_2} are positive numbers. Two possibilities arise:

(a) $N_{R_1} \geqslant N_{R_2}$
(b) $N_{R_1} < N_{R_2}$

Case (a) in complement form becomes

$$N_{R_1} + (2^n - 1 - N_{R_2}) \text{ or } 2^n - 1 + (N_{R_1} - N_{R_2})$$

The correct answer is $N_{R_1} - N_{R_2}$, a positive number which must have 0 in the most significant digit position. Since we are working in modulo 2^n, the uncorrected answer obtained is $N_{R_1} - N_{R_2} - 1$. To correct this we must add one. We note that since N_{R_2} is in complement form a 1 must have been present in its most significant digit position and, for the result to have a 0 in this position, a carry overflow must have occurred. Thus

```
 13      01101
 -6     +11001
 --     ------
 +7    ┌100110
       └─────►1   end-around-carry
        ------
         00111
```

Case (b) in complement form will also be

$$N_{R_1} + (2^n - 1 - N_{R_2}) \text{ or } 2^n - 1 + (N_{R_1} - N_{R_2})$$

However, since $N_{R_1} < N_{R_2}$ the answer must be negative and should be in one's complement form, which it clearly is. In this case it is not possible to produce an overflow carry since the magnitude of the resultant number must always be less than N_{R_2}. Thus

```
  3     00011
 -6    +11001
 --    ------
 -3    011100
```

If N_{R_1} is itself negative, the problem becomes

$$-N_{R_1} - N_{R_2}$$

and in one's complement form is

$$(2^n - 1 - N_{R_1}) + (2^n - 1 - N_{R_2})$$

which is

$$2^{n+1} - 2 - (N_{R_1} + N_{R_2})$$

The correct solution must be $2^n - 1 - (N_{R_1} + N_{R_2})$. Since we work in modulo 2^n, 2^{n+1} is equivalent to 2^n and the only correction needed is to add one. The numbers are both in one's complement form so that both must have a 1 in their most significant digit place, therefore there must always be a carry out when they are added. This is the end-around-carry; thus

```
 -3        11100
 -6       +11001
 --       ------
 -9      ┌1 10101
         └─────►1   end-around-carry
          ------
           10110
```

It is appropriate at this point to introduce the concept of the *number range* available to computing circuits – these being in most cases the arithmetic circuits of a digital computer.

In digital systems numbers are usually held in *registers* which consist of a series of bistable elements, each one indicating a bit value. The number of elements determines the *word length*. Within the limits imposed by the register we must be able to represent numbers that are positive or negative, fractional or integer. Normally it is arranged that numbers are interpreted throughout as either integers or fractions and any scaling necessary is left to the user. However, we must be able to represent positive or negative and the most significant digit is reserved for this purpose.

Consider a 4-bit register. Figure 2.5 shows the various ways in which the bit pattern can be interpreted. It is interesting to note that the allocation of a bit to represent the sign has not resulted in the loss of one-half of the number capacity, but has merely shifted the axis on the number scale. Figure 2.6 shows this in general terms for a register of n bits and number capacity $(2^n - 1)$.

An alternative method widely used in more complex digital systems is that of *floating point*, which allows representation of a much wider range of numbers without extending the register size. The descriptive term 'floating point' is used because the numbers are taken to be of the form

$$N_R = M \times R^c$$

with the values of M and C held in separate parts of the register. The part designated M is the *mantissa*, *argument*, or *fraction* and C the *characteristic* or *exponent*. The base, R, is implied, and, especially in the larger computers, is not necessarily equal to two, 16 being a common value.

Register bit pattern	Positive integers	One's complement	Two's complement	Sign and magnitude	Two's complement fractional
0 1 1 1	7	+7	+7	+7	+0.875
0 1 1 0	6	+6	+6	+6	+0.750
0 1 0 1	5	+5	+5	+5	+0.625
0 1 0 0	4	+4	+4	+4	+0.500
0 0 1 1	3	+3	+3	+3	+0.375
0 0 1 0	2	+2	+2	+2	+0.250
0 0 0 1	1	+1	+1	+1	+0.125
0 0 0 0	0	+0	+0	+0	+0.000
1 1 1 1	15	−0	−1	−7	−0.125
1 1 1 0	14	−1	−2	−6	−0.250
1 1 0 1	13	−2	−3	−5	−0.375
1 1 0 0	12	−3	−4	−4	−0.500
1 0 1 1	11	−4	−5	−3	−0.625
1 0 1 0	10	−5	−6	−2	−0.750
1 0 0 1	9	−6	−7	−1	−0.875
1 0 0 0	8	−7	−8	−0	−1.000

Figure 2.5 Number representation

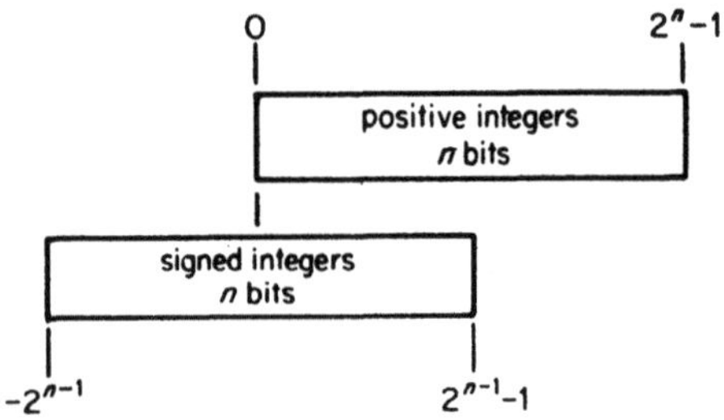

Figure 2.6 Effect of using most significant digit as sign bit

It is usual to represent the mantissa as a fractional quantity with a restricted range

$$\frac{1}{R} \leqslant M < 1 \quad \text{or} \quad M = 0$$

for positive fractions and

$$-\frac{1}{R} > M \geqslant -1$$

for negative fractions (one's complement).

Thus the mantissa of a floating point number, apart from the all-zeros special case, is of the form

0.1XXX . . . for positive values
1.0XXX . . . for negative values

The exponent is sometimes a simple signed number but by adding a *bias* to the exponent value, equal to 2^{n-1}, where n is the number of exponent bits, a number is obtained which can range from all zeros (representing the maximum negative value) to all ones (the maximum positive value). Biasing allows the zero state of both exponent and mantissa to be a recognizable logical state, that of a number smaller than the smallest value otherwise obtainable. Such a state commonly occurs when a digital machine is reset at switch-on or during an initialization phase.

Figure 2.7 shows the breakdown of a 12-bit floating point number having a four-bit exponent. The largest positive value that can be represented, assuming a base of two, is 0.1111111×2^7 or approximately 128.

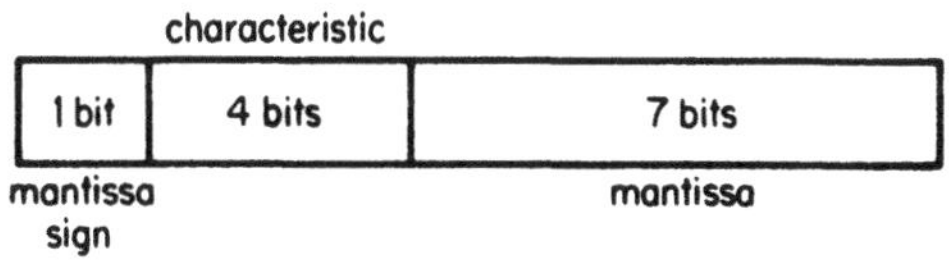

Figure 2.7 Floating point number construction

In some forms of floating point representation the most significant mantissa bit, excluding the sign bit, is omitted since its value can be inferred from the sign bit. It is taken into account, of course, during any arithmetic operation but its exclusion makes a further data bit available to extend the accuracy within a given mantissa width.

Example 2.9: Find the decimal equivalent of the biased floating point number having a mantissa value 011011 and an exponent 01101.

The exponent has five bits, therefore the bias is 2^{5-1} = 10000. The binary value of the exponent is 01101 − 10000 = 11101 which is equal to −3 (two's complement).

The mantissa, 0.11011, is equal to (1/2) + (1/4) + (0/8) + (1/16) + (1/32) = 0.84375. Thus the decimal value of the number is 0.84375×2^{-3} = 0.10546875.

Regardless of register size and number representation, it is inevitable that some computations involving numbers well within the number range of the registers will generate results outside that range. It is necessary to sense this occurrence, since the apparent result is incorrect. Thus

+12	01100
+ +7	00111
+19	0 10011

An *overflow* has occurred and the result in two's complement form would be interpreted as −13. Again

−11	10101
+ − 9	10111
−20	101100

An overflow (sometimes referred to in this case as an *underflow*) has occurred and this result appears as +12.

The problem occurs when the magnitude of the result exceeds the register capacity. This can only happen when two numbers of the same sign are added. Thus, overflow can be said to have occurred if the sign of the resultant differs from that of the original numbers. This argument leads to the simple overflow recognition circuit of figure 2.8.

An alternative approach, applicable to two's complement arithmetic, is to look for the non-equivalence of the carries from the two most significant bit positions. Thus, $V = C_n \oplus C_{n-1}$.

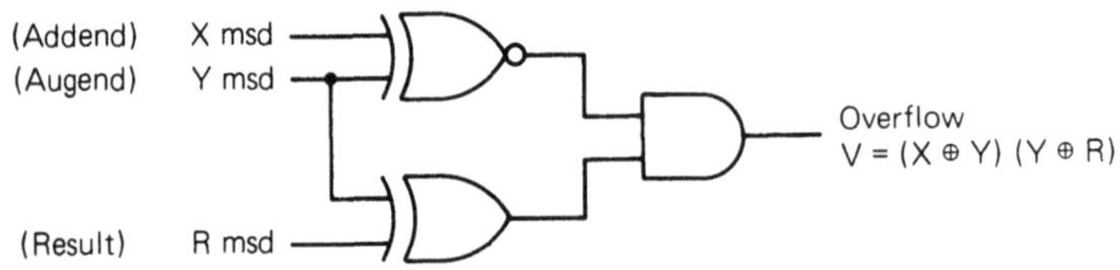

Figure 2.8 Overflow detector

2.5 Computer arithmetic

Binary addition of two digits can be accomplished using a *modulo-2 sum* or *exclusive-OR* circuit, and the truth table of figure 2.4 can be implemented by means of the circuit shown in figure 2.9. This *half-adder* provides the sum and carry signals required from the addition process, but such a circuit is not sufficient for the addition of multi-digit numbers as there is no means of including a *carry* from a previous addition. The circuit of figure 2.10 is termed a *full-*

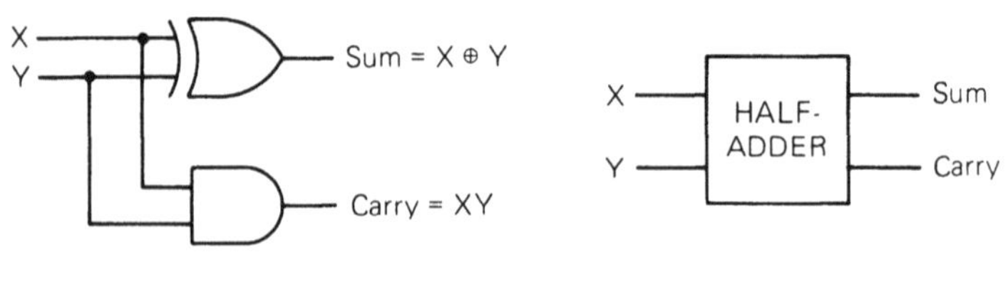

Figure 2.9 The half-adder: (a) circuit; (b) symbol

X	Y	C_{in}	Sum	C_{out}
0	0	0	0	0
0	0	1	1	0
0	1	0	1	0
0	1	1	0	1
1	0	0	1	0
1	0	1	0	1
1	1	0	0	1
1	1	1	1	1

(a)

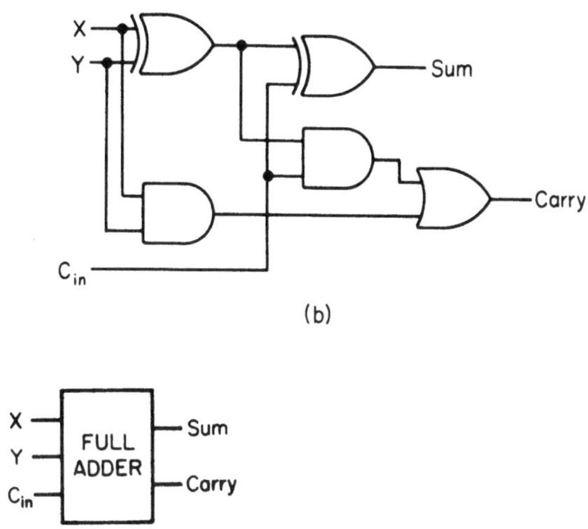

(c)

Figure 2.10 The full-adder: (a) truth table; (b) circuit; (c) symbol

adder and, as can be seen from the truth table, provides the sum and carry signals from the binary addition of three signals.

A unit capable of adding two binary numbers, each of, say, four bits in length, can be constructed from a series of such adders as the circuit in figure 2.11 shows. The registers X, Y, are needed to hold the two numbers to be added and a third register, R, accepts the result. It need not concern us yet how numbers are entered into these registers.

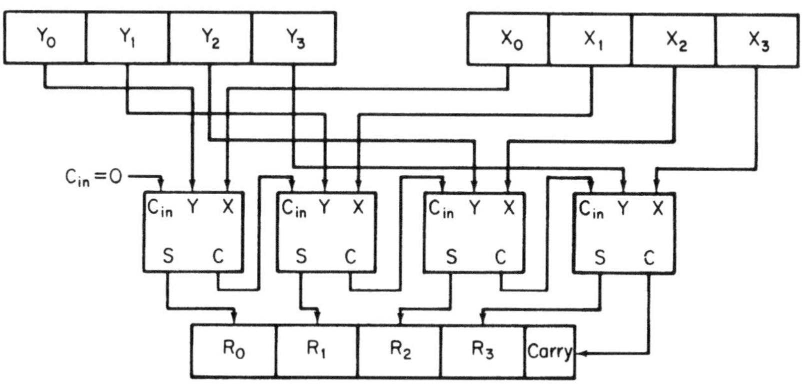

Figure 2.11 A four-bit adder

Although figure 2.11 depicts an adding circuit it is possible, with a little modification, to include subtraction by complement addition. If the Y register outputs are inverted before being presented to the adders and the carry in, C_{in}, is made equal to logical one, then subtraction by two's complement addition, X – Y, occurs. The inversion of register Y is most conveniently carried out using exclusive-OR gates, as in figure 2.12. The control signal, P, determines the function of the unit. Thus when P = 1 the Y register outputs are inverted and a one is presented to the carry in, at the least significant end of the adder.

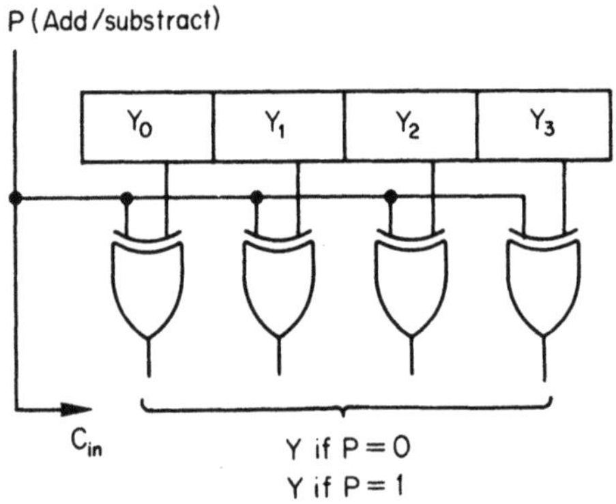

Y	P	Y ⊕ P	
0	0	0	
1	0	1	
0	1	1	Data inverted
1	1	0	

Figure 2.12 Generation of the inverse of Y

In all our circuits so far we have implicitly assumed that the output signal of each gate is generated instantaneously from the input signals, but when we come on to practical circuits in the next chapter we will see that the gate always introduces a *propagation delay*. Although these delays may be insignificant in many applications this is often not the case for arithmetic circuits. Examination of the full-adder circuit of figure 2.10 will show that the sum signal is generated by two exclusive-OR logical units, whereas the carry signal may require a signal contribution from three logical units, for example when X + Y = 1 and C_{in} = 1. The problem here is one of carry delays. The sum value from the second of the adders of figure 2.11 cannot be decided until the carry from the least significant stage has been taken into account. This carry also may affect the carry out from the second stage, which, in turn, may affect the third stage, and so on. The resulting carry *ripple-through* delay must be taken into account when such an adder is being used.

Example 2.10: Find the total add time for a 16-bit parallel binary adder constructed entirely from NAND units, each with a propagation delay of 10 ns.

A minimal full-adder constructed from NAND units alone is fairly complicated. Figure 2.13a shows the circuit. The maximum sum delay is four units and the maximum carry out delay three units. A parallel adder has the form shown in figure 2.13b, so the total carry delay path is 16 × 3 = 48 units. Thus, a time of 480 ns must elapse before we can be certain of the result from the adder.

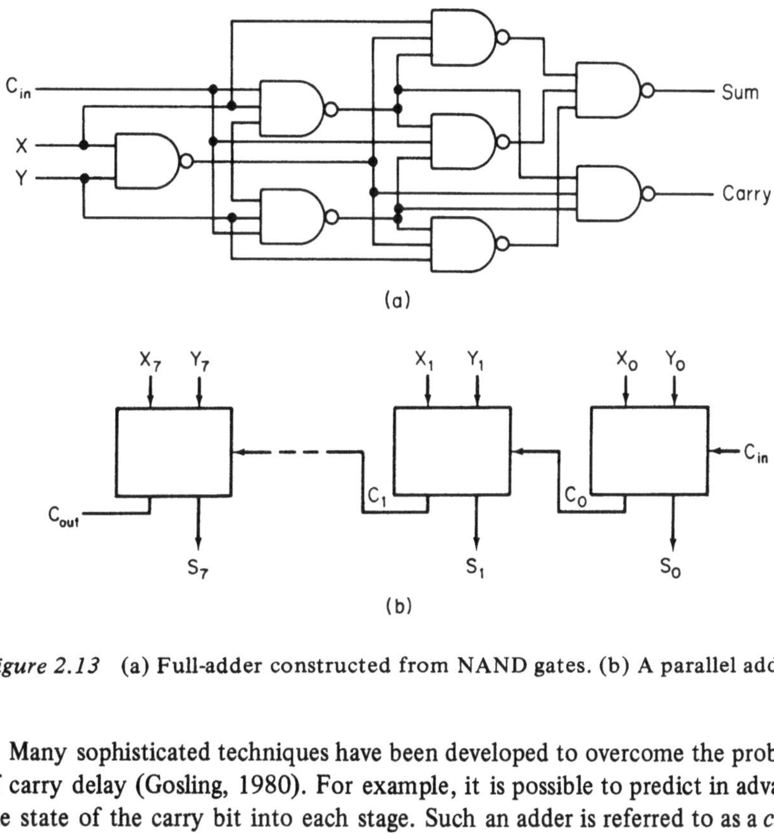

Figure 2.13 (a) Full-adder constructed from NAND gates. (b) A parallel adder

Many sophisticated techniques have been developed to overcome the problem of carry delay (Gosling, 1980). For example, it is possible to predict in advance the state of the carry bit into each stage. Such an adder is referred to as a *carry look-ahead adder* (CLA).

Three conditions govern the generation of a carry from any stage *n* of the adder.

(1) $X_n = Y_n = 1$	Carry is generated unconditionally.
(2) $X_n \neq Y_n$	Carry is propagated; that is, the carry is generated only if there is a carry into the stage.
(3) $X_n = Y_n = 0$	No carry is generated.

In algebraic terms, the first condition is $G_n = X_n Y_n$ and the second condition $P_n = X_n + Y_n$, so that carry out $C_n = G_n + P_n C_{n-1}$.

This expression can be expanded to include all the stages of the adder.

$$C_n = G_n + P_n[G_{n-1} + P_{n-1}C_{n-2}] = G_n + P_n[G_{n-1} + P_{n-1}\{G_{n-2} + P_{n-2}C_{n-3}\}] \text{ etc.}$$

and in the limit, for stage $n = 0$, G_{-1} becomes C_{in}, the carry in to the least significant stage of the adder. The series is

$$C_n = \sum_{j=0}^{n} \left(\prod_{i=j+1}^{n} P_i \right) G_j$$

Using the carry generate, G, and carry propagate, P, equations, an n-stage adder can be represented as in figure 2.14a. The carry circuits are shown in figure 2.14b and the complete circuit can be synthesized using two-level logic circuits. The complexity of the carry circuits increases rapidly but the unit delays remain constant, the maximum carry delay being three units and the sum four units. Unfortunately, a 16-bit adder needs a very wide parallel carry generation circuit involving a 17-input OR gate!

Four-stage carry-look-ahead adders such as the Texas Instruments SN74S181 are available as a single circuit. By arranging that the carry out of each block *ripples through* successive blocks, economical high-speed adders can be built. If higher speed is required, this ripple-through delay can be further reduced by generating carry-look-ahead over the successive blocks. Circuits providing group look-ahead over four 4-stage adders allow a 16-bit adder to be constructed in very compact form and having an add time of less than 20 ns.

Addition and subtraction are not the only operations required of a digital processor: multiplication and division are also essential functions and are often implemented in circuitry form. Most computers rely on successive additions to multiply and successive subtractions to divide, in a manner similar to 'long' multiplication and division by hand. It is in such cases that a high-speed adder is advantageous. A 16-bit multiply operation requires sixteen successive additions and, using the circuit discussed in example 2.10 for instance, a total time of 16×480 ns, 7.68 μs, is taken by the addition process. The carry-look-ahead technique can reduce this by a factor of twenty-four. Many techniques have been developed to speed up the process of multiplication and division but they are beyond the scope of this book and the reader is referred to specialist texts (Flores, 1963; Hwang, 1979).

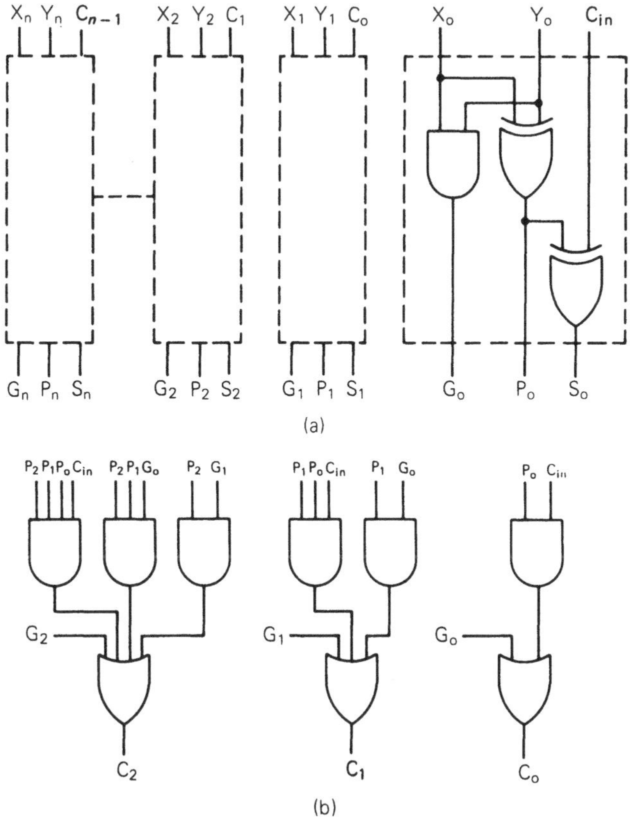

Figure 2.14 (a) An n-stage adder using carry, propagate and generate functions. (b) Generation of the carry signals

2.6 Binary coded decimal

While the binary system is the most convenient system for machines, people are more accustomed to the decimal notation and for many systems where the inputting and display of numerical information is a major feature, a common compromise is to retain the decimal digits but represent each in binary form. The resultant is termed *binary coded decimal* (BCD). There are numerous BCD forms, all requiring at least four bits per decimal digit. The 4-bit variants are usually referred to by the respective weightings of their bit positions. Thus we have the common code 8421, sometimes called 'straight' BCD since it uses the first ten of the natural binary numbers. Other codes that have found some application include the 2421, 5421 and 5311 codes.

Decimal	8421	Decimal	2421	Decimal	5311
0	0000	0	0000	0	0000
1	0001	1	0001	1	0001
2	0010	2	0010	unused	0010
3	0011	3	0011	2	0011
4	0100	4	0100	3	0100
5	0101	unused	0101	4	0101
6	0110	unused	0110	unused	0110
7	0111	unused	0111	unused	0111
8	1000	unused	1000	5	1000
9	1001	unused	1001	6	1001
unused	1010	unused	1010	unused	1010
unused	1011	5	1011	7	1011
unused	1100	6	1100	8	1100
unused	1101	7	1101	9	1101
unused	1110	8	1110	unused	1110
unused	1111	9	1111	unused	1111

On inspection it can be seen that the 2421 code has the useful property that the nine's complement of the decimal digit can be found by taking the one's complement of the binary code; that is, the code is *self-complementing*. For example, 7, 1101, has nine's complement 2, 0010. The latter codes indicate in their most significant bit position whether the decimal digit is 5 or greater, which can be of help in such arithmetic operations as 'rounding-off'.

It is common for computers to provide arithmetic working in BCD. Until carry considerations arise the arithmetic is straightforward, but then, owing to the six unused binary codes, it is necessary to introduce correction factors to adjust the result to the correct value in the appropriate code. Consider the following 8421 code examples.

Decimal	BCD	
47	0100	0111
+22	+0010	0010
69	0110	1001

The result is coded in BCD and no correction is necessary.

Decimal	BCD	
47	0100	0111
+25	+0010	0101
72	0110	1100

The number 0110 1100 is no longer correctly coded, since 1100 does not appear in the 8421 code. It is necessary to add binary 6 to this code in order to generate a carry and bring the result back to the correct value.

```
  0100        0111
+ 0010        0101
  ----        ----
  0110        1100
            + 0110
              ----
     1 <----- 0010
  ----
  0111
```

Note that the correction factor must still be added to the binary sum even if a carry signal is generated by the first addition, since the disallowed six states have been absorbed in generating the carry.

```
  47            0100          0111
+ 89            1000          1001
  --            ----          ----
 136            1100          0000
                   1 <-----'
                            + 0110
                ----          ----
                1101          0110
              + 0110
                ----
     0001 <---- 0011
```

Apart from the disadvantage of the correction factor itself, the correction process with the 8421 code is lengthy since the factor cannot necessarily be applied to each digit group or decade simultaneously. In some cases the carry resulting from the correction process in one decade causes the neighbouring decade to exceed 9 and therefore in turn to require correction. This time delay can become an appreciable handicap in high-speed computation.

The disadvantage of the 8421 form of BCD arises from the fact that the binary digits do not give a carry at the same time as the decimal digit. Codes have been devised which carry an excess of three, so that when two such numbers are added the six disallowed states are filled and a carry is simultaneously generated in both binary and decimal. The *excess-three* code is formed by adding 3 to each of the 8421 binary code groups.

Decimal	XS3
unused	0000 0001 0010
0	0011
1	0100
2	0101
3	0110
4	0111
5	1000
6	1001
7	1010
8	1011
9	1100
unused	1101 1110 1111

If the result of the addition of two decimal digits in excess-three representation is greater than 9 – that is a carry has been generated – the remaining code will be 8421 binary and a correction factor must be added. If the result of the addition is less than 9, no carry has occurred, the sum will be in 'excess-six' and a correction factor of binary three must be subtracted. An advantage over 8421 arithmetic is that under no circumstances does the application of a correction factor result in a carry, and corrections can be carried out on all decades simultaneously.

Decimal		XS3		
047	0011	0111	1010	
+ 089	+ 0011	1011	1100	
136	0110	0010	0110	
	1 ←	1 ←		carries
	0111	0011		
	− 0011	+ 0011	+ 0011	correction factors
	0100	0110	1001	

As with 8421 code, subtraction in excess-three can be achieved by nine's complement addition but with the added bonus that the code is self-complementing.

2.7 Unweighted codes

The excess-three code is an example of an *unweighted* code; that is a code in which the bit positions relative to the point do not carry a constant weighting. Other unweighted codes have been devised for specific applications. The best-known group is the Gray codes, which are often used when dealing with positional encoders. For instance, suppose we have a shaft position indicator whose output is in 421 binary. The transducer could take the form of a disc mounted on the shaft, having opaque and transparent regions as in figure 2.15a.

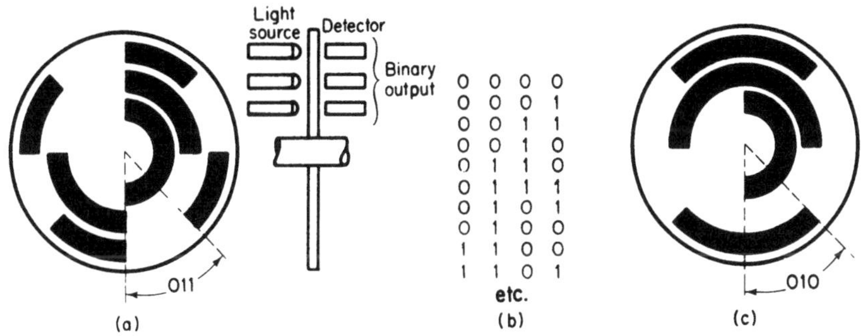

Figure 2.15 (a) Shaft encoding discs. (b) Four-bit Gray code. (c) Gray code disc

The binary output is taken from a set of photocells which sense whether an opaque or transparent region lies between them and the light source. The accuracy of measurement depends on the number of bits used in the code.

421	Range
000	0-45°
001	45-90°
010	90-135°
011	135-180°
100	180-225°
101	225-270°
110	270-315°
111	315-360°

At some point in time the transducer may indicate 011; that is the shaft lies between 135° and 180°. If the shaft now rotates slightly so that a segment boundary covers a photocell the output will be indeterminate, either 1 or 0. With the straight binary code several segment boundaries coincide, and in the example shown it is possible for the output to change to any value between 000 and 111 dependent on mechanical tolerances. This is clearly undesirable and can be overcome by the use of codes such as those developed by Gray (1953), and which bear his name, their important characteristic being that only one bit ever changes in going from any code pattern to the next.

The most commonly known Gray code takes the name 'Gray code' and the 4-bit version is given in figure 2.15b. This is *reflected* binary code, the name arising from the fact that if the highest-order bit is ignored, the remaining bits form a symmetrical pattern about the centre point of the code. We see now that in an optical shaft encoder there would be less likelihood of error if a Gray code were used since no segment boundaries coincide. The three-bit Gray code pattern is shown in figure 2.15c.

A great many Gray codes of various cycle lengths exist with a maximum cycle length of $2n$, where n is the number of bits. For example, two codes of cycle length six, referred to by their changing bit positions are (1, 2, 3, 1, 2, 3) and (1, 2, 1, 3, 2, 3).

(1, 2, 3, 1, 2, 3)	(1, 2, 1, 3, 2, 3)
000	000
001	001
011	011
111	010
110	110
100	100
000	000

The (1, 2, 3, 1, 2, 3) Gray code of cycle length six is an example of a *creeping code*, so called because a set of three ones creeps in at one end and out at the other as the code progresses. The cycle length c is determined by the number of bits, n, and $c = 2n$.

The most common requirement among codes is decimal representation, and a creeping code for decimal requires five bits:

00000	0
00001	1
00011	2
00111	3
01111	4
11111	5
11110	6
11100	7
11000	8
10000	9
00000	0

Although more bits are needed, the code has an advantage over 4-bit codes in that the decoding is straightforward and inexpensive. Since the cycle length of the creeping code is $2n$, using n ones allows only even-numbered cycle lengths, but if it is arranged that the creeping set of ones is $(n - 1)$ in length, the n-bit code has a cycle length of $(2n - 1)$ and thus codes of any cycle length can be generated.

Although Gray codes are relatively easy to convert to and from straight binary, arithmetic operations are cumbersome, and one would not, normally, attempt to use such a code for arithmetic purposes.

2.8 Error detection

In transmitting data from one system to another it is desirable to have complete confidence that the code received is in fact the code transmitted. To achieve this, some form of checking system is necessary and the efficiency of the system is determined by the expense one is willing to incur.

The simplest and therefore most popular method of error checking is to add a redundant bit to each word of the code, the bit value being determined by the number of ones in the code. The additional bit is called the *parity* bit. *Even parity* results if the code transmitted is arranged to contain an even number of ones; conversely for odd parity. At the receiving end the number of ones is counted to ensure parity is maintained.

8421	Odd parity	Even parity
0000	00001	00000
0001	00010	00011
0010	00100	00101
0011	00111	00110
0100	01000	01001
0101	01011	01010
0110	01101	01100
0111	01110	01111
1000	10000	10001
1001	10011	10010

The system described will not detect the less likely occurrence of two errors in the same word, nor will it indicate the position of the error.

Matrix parity – or *block* parity as it is sometimes known owing to its use in the block transfer of data, for instance with magnetic tape units – overcomes this limitation. Parity checks are made on all rows of a block of information and an additional parity row is added at the end of the block to check the columns of the data, as in figure 2.16a. Suppose the information received contains an error, as in figure 2.16b. By checking horizontal and vertical parity, the error is located and may be corrected. This method also allows multiple error detection but not correction.

7	0	1	1	1	0
8	1	0	0	0	0
3	0	0	1	1	1
9	1	0	0	1	1
Check word	1	0	1	0	1

(a)

0	1	1	1	0	✓
1	0	1	0	0	×
0	0	1	1	1	✓
1	0	0	1	1	✓
1	0	1	0	1	✓
✓	✓	×	✓	✓	

(b)

Figure 2.16 Matrix parity

A further development of the matrix parity idea is used in the *diagonal* parity checking scheme. The system was devised mainly for use with magnetic tape recording in which a speck of dust or other impurity can cause errors in adjacent columns extending over several rows. The parity check is made over a diagonal path as indicated in figure 2.17.

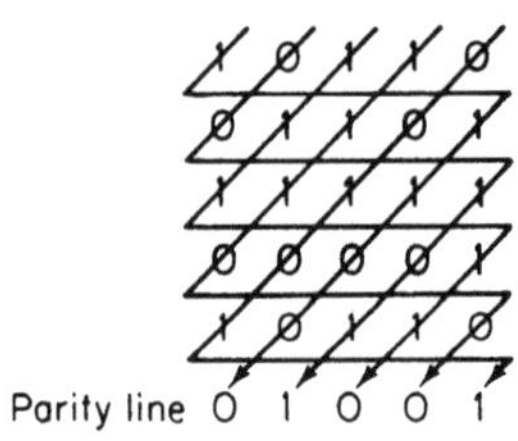

Figure 2.17 Diagonal parity

A class of codes making use of many of the features already introduced has been developed for use with bar codes in marking groceries, books and, in fact, any type of article. The original bar code was developed in the USA and is known as the Universal Product Code, UPC, but the international standard which grew from it is coordinated by the International Article Numbering Association, EAN (EAN, 1984).

The standard number for an article consists of thirteen decimal digits, and has the general structure shown in figure 2.18a. The UPC standard number has twelve digits but is made compatible with the EAN standard by considering it to have a leading zero.

The check digit is calculated by digit addition, modulo-10, in a specified manner: all digits in odd positions in the number (before the check digit is appended) are added together and the sum is multiplied by three. This is then added to the sum of all the digits in even positions, and the check digit is found by subtracting the grand total from the next higher multiple of ten.

Example 2.11: Calculate the check digit for the code given in figure 2.18a.

The sum of the odd position digits is	0 + 0 + 7 + 0 + 6 + 8 = 21
Multiplied by three	= 63
The sum of the even position digits is	5 + 1 + 6 + 2 + 1 + 3 = 18
Giving a grand total	63 + 18 = 81
Subtracted from next higher multiple of ten,	90 – 81 = 9

Thus the check digit is 9.

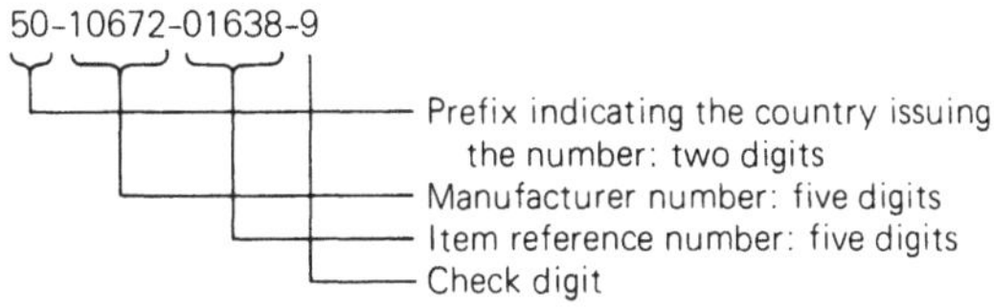

(a)

	Left characters		Right characters
Digit value	Set A	Set B	Set C
0	0 0 0 1 1 0 1	0 1 0 0 1 1 1	1 1 1 0 0 1 0
1	0 0 1 1 0 0 1	0 1 1 0 0 1 1	1 1 0 0 1 1 0
2	0 0 1 0 0 1 1	0 0 1 1 0 1 1	1 1 0 1 1 0 0
3	0 1 1 1 1 0 1	0 1 0 0 0 0 1	1 0 0 0 0 1 0
4	0 1 0 0 0 1 1	0 0 1 1 1 0 1	1 0 1 1 1 0 0
5	0 1 1 0 0 0 1	0 1 1 1 0 0 1	1 0 0 1 1 1 0
6	0 1 0 1 1 1 1	0 0 0 0 1 0 1	1 0 1 0 0 0 0
7	0 1 1 1 0 1 1	0 0 1 0 0 0 1	1 0 0 0 1 0 0
8	0 1 1 0 1 1 1	0 0 0 1 0 0 1	1 0 0 1 0 0 0
9	0 0 0 1 0 1 1	0 0 1 0 1 1 1	1 1 1 0 1 0 0

(b)

Figure 2.18 Bar code format and code sets: (a) standard number format; (b) code sets

In converting the standard numbers into bar codes a seven-bit binary code is allocated to each digit, and each character of the bar code is composed of seven modules, light or dark, with a dark module corresponding to logic one in the code. In addition, guard bars and centre bars or strips are included to indicate beginning, centre and ending of the whole number. Several constraints have been satisfied in building the seven-bit codes:

the ones are grouped to form dark bars with widths of 1,2,3 or 4 modules:
each character contains two dark bars and two light spaces;
the symbols are readable in either direction, so each contains sufficient information for the decoding circuitry to decipher its value regardless of the direction of reading.

Of the 128 possible combinations of seven bits, the three sets shown in figure 2.18b have been selected. All the codes contain only two groups of ones, giving two dark bars, and all codes contain only three changes, from zero to one and one to zero. Codes in sets A and B all start with zero and end with one; those in set C all start with one and end with zero. This ensures that the boundary between two characters can always be distinguished visually. Set A codes have odd parity and sets B and C have even parity. Reversing any code does not give another valid code, except for codes in sets B and C but then the two forms represent the same character. Code sets A and B are used for digits to the left of the centre indicator, and code set C is used for those to the right.

The complete bar code pattern is made up of a guard pattern, 101, six left characters, the centre pattern, 01010, six right characters and a final guard pattern, 101, as illustrated in figure 2.19. But this format, as a quick check confirms, allows for only twelve digits, not thirteen, and we find that the thirteenth, leftmost, digit value is implied by the choice of left characters from set A or set B. The value of the thirteenth digit is then printed in the left margin. If the thirteenth digit is zero, all six left digits are taken from set A and compatibility with the UPC standard is maintained. All other digit values for the thirteenth position have specific mixtures of codes, three from set A and three from set B, defined in the following way:

Figure 2.19 Bar code pattern

Value of 13th digit	Choice of code set for left digits 1	2	3	4	5	6
0	A	A	A	A	A	A
1	A	A	B	A	B	B
2	A	A	B	B	A	B
3	A	A	B	B	B	A
4	A	B	A	A	B	B
5	A	B	B	A	A	B
6	A	B	B	B	A	A
7	A	B	A	B	A	B
8	A	B	A	B	B	A
9	A	B	B	A	B	A

Thus, although the right-side codes always maintain even parity, the left-side codes have mixed parity, but, because the thirteenth digit forms part of the nationality indicator, the parity mix for a particular country is always the same.

2.9 Alphanumeric codes

Although numerical pieces of data form a large part of information processed digitally, an increasingly large proportion of the data consists of alphabetic and punctuation characters. Also, automatic processing equipment uses control signals which must have unique codes. Accordingly there exist various codes of which a few have been chosen as standards. Among the most popular are the ISO code (International Standards Organization), the ASCII code (American Standard Code for Information Interchange), and the Extended Binary-Coded-Decimal Interchange Code (EBCDIC). The first two are seven-bit codes and differ only in minor details. They often include an eighth, parity check, bit. EBCDIC is an eight-bit code.

The codes are shown in full in appendix B. Note that in common with most practical codes the character indicating 'error' or 'delete' is the 'all ones'. The reason is historical: when punched paper tape was the predominant medium for holding data any punching errors could be removed very simply, it being easier to punch out the remaining bit positions rather than to refill the offending holes!

Example 2.12: Messages in ASCII code are received as a succession of characters, each having seven bits in parallel. Assuming that a data valid, DV, control signal is generated when the received data signals have stabilized, design a logic circuit to indicate when numerals are received.

Each character consists of seven bits, b_6-b_0, and b_6-b_4 at 011 indicate a numeral if b_3-b_0 have any value from 0000 through 1001. Using the Karnaugh map of figure 2.20b, we find the minimal expression to detect a numeral, which, when the data valid control signal is included, becomes

$$N = \overline{b_6}\, b_5\, b_4\, (\overline{b_3} + \overline{b_2}\,\overline{b_1})\, DV$$

The corresponding circuit is shown in figure 2.20c.

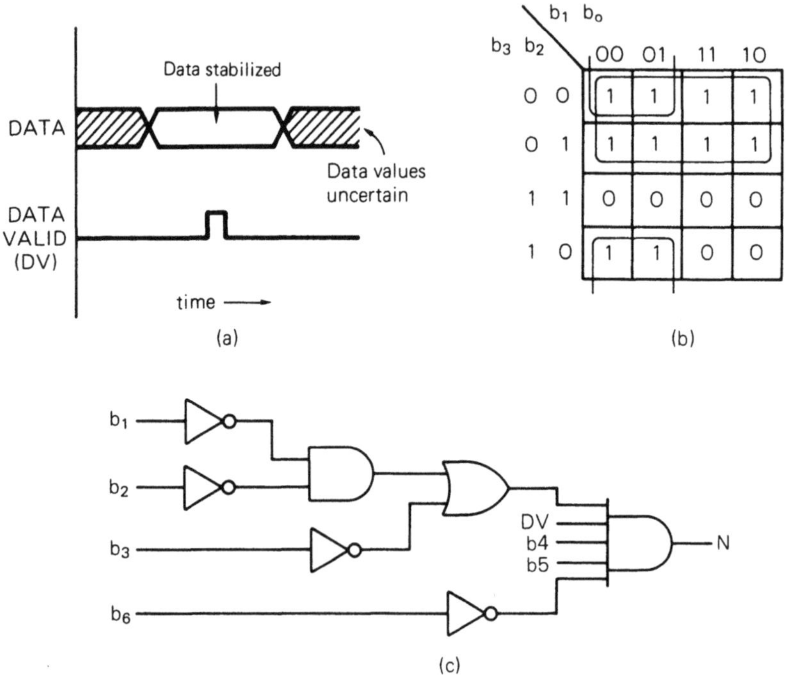

Figure 2.20 ASCII numeral detector circuit: (a) timing waveforms; (b) Karnaugh map; (c) circuit

2.10 Encoding and decoding

Encoding, and its converse decoding, are the names given to the translation of one code to another. We have seen previously how to synthesize a circuit from a consideration of the truth table defining the function. The same technique can be used in encoding.

Example 2.13: Devise a logical circuit to convert straight BCD to 2421 code.

It is necessary to define each code in the form of a truth table, as in figure 2.21. As each code has four bits, the circuit will have four inputs A, B, C, D and four outputs W, X, Y, Z.

8 4 2 1	2 4 2 1
A B C D	*W X Y Z*
0 0 0 0	0 0 0 0
0 0 0 1	0 0 0 1
0 0 1 0	0 0 1 0
0 0 1 1	0 0 1 1
0 1 0 0	0 1 0 0
0 1 0 1	1 0 1 1
0 1 1 0	1 1 0 0
0 1 1 1	1 1 0 1
1 0 0 0	1 1 1 0
1 0 0 1	1 1 1 1

Figure 2.21 Example 2.13

Considering the output W, we could write from the truth table the five canonical terms relating W to A, B, C, and D. Alternatively, we can take into account the fact that BCD codes have, by definition, only ten different bit patterns and the codes not used are *don't care* conditions. They can be allocated either one or zero and by drawing Karnaugh maps for the outputs W, X, Y, Z and including the don't care states, $\emptyset$, more effective minimal groupings may be achieved, shown in figure 2.22a.

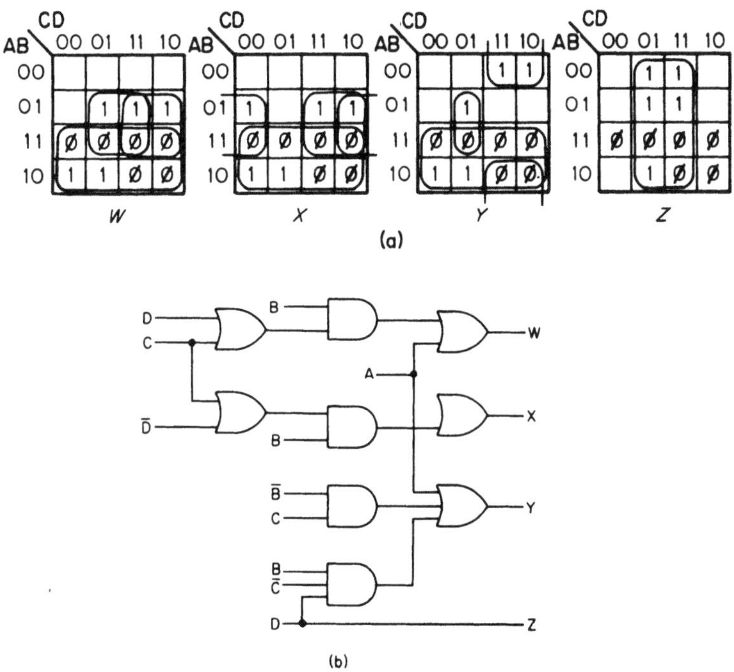

Figure 2.22 (a) Use of don't care conditions. (b) Circuit for example 2.13

In the case of output W, if all the don't care entries are made equal to logical one, minimal groupings give us

$$W = A + B(C + D)$$

and for the others

$$X = A + B(C + \bar{D})$$

$$Y = A + B\bar{C}D + \bar{B}C$$

$Z = D$ (here not all the don't care entries are used)

leading directly to the circuit of figure 2.22b.

The encoding of binary to Gray code is a good example of the use of exclusive-OR or *not-equivalence* circuits. The encoder is based on the algorithm: compare the binary digits in pairs; if the bits are equivalent write down a zero. If the bits are not equivalent, write down a one. The resulting code is the Gray code. For example

Binary 0 1 0 1 1

Gray 0 1 1 1 0

The circuit of figure 2.23 follows directly.

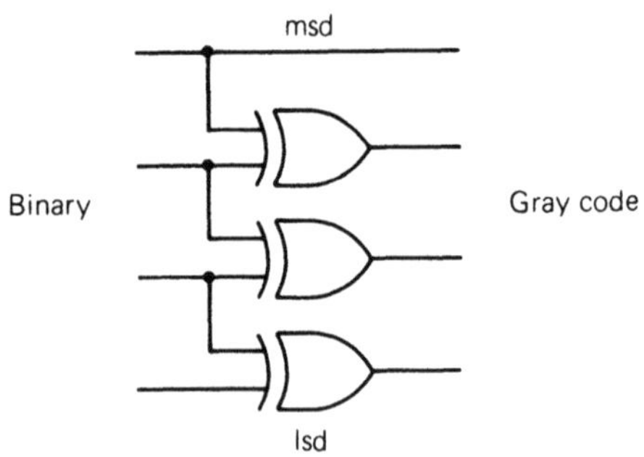

Figure 2.23 Binary-to-Gray encoder

Problems

2.1 Convert the following numbers to their decimal equivalents: (a) 1101.011_2, (b) 100110.0101_2, (c) 653.25_8.

2.2 Convert 110101011.111001101_2 into (a) octal, (b) decimal.

2.3 Evaluate the following by (a) one's complement addition, (b) two's complement addition: (i) $100110_2 - 011011_2$, (ii) $110101_2 - 111010_2$.

2.4 Evaluate the following using (a) BCD, (b) excess-three arithmetic: (i) $5243_{10} - 1165_{10}$, (ii) $423_{10} - 576_{10}$.

2.5 Design a circuit to give excess-three code for a four-bit BCD input.

2.6 Design a circuit to produce a one output only when the four-bit input contains an odd number of ones.

2.7 Devise a circuit to accept a three-bit number, n, and give a four-bit output code approximately equal to $n^2/4$.

References

EAN (1984). *Article Numbering and Symbol Marking; Operating Manual*, Article Numbering Association (UK) Ltd, London

Flores, I. (1963). *The Logic of Computer Arithmetic*, Prentice-Hall, Englewood Cliffs, New Jersey

Gosling, J. B. (1980). *Design of Arithmetic Units for Digital Computers*, Macmillan, London

Gray, F. (1953). US Patent No. 2632058, 17 March

Hwang, K. (1979). *Computer Arithmetic: Principles, Architecture and Design*, John Wiley, New York

3 Semiconductor Devices and Circuits

No matter how carefully a digital system is designed, satisfactory operation depends very heavily on the reliability of the circuits and components used – the *hardware*. Ideas originating in the latter part of the nineteenth century with Charles Babbage (see Babbage, 1961) are still used today, but his 'computing engines' were never fully operative because the technology of the time was unable to meet the challenge of the close tolerances needed in his machinery. In fact the modern digital computer, which made its appearance in the last few years of the 1940s, would have been an impossible venture only ten years earlier, owing its development to the impetus given to electronics by the Second World War (Metropolis, 1980). The invention of the transistor in 1948, and the rapid development of other semiconductor devices, with their inherent reliability and cheapness, led to the present-day emphasis on digital techniques, almost invariably using solid-state circuits, with their dependence on diode and transistor action.

3.1 Diode and transistor action

The ideal diode is a *unilateral* device: current is allowed to flow only in the forward direction, and only when the anode voltage is positive with respect to the cathode. The diode is then said to be forward-biased. It is not appropriate here to explore the fundamental properties of semiconductor devices, which are best covered in more specialized texts (Millman, 1979; Schilling and Belove, 1979), but it is important to appreciate the operation and limitations of the devices.

Figure 3.1 shows the forward and reverse characteristics of typical silicon and germanium diodes compared with those of an ideal diode. The characteristics can be described fairly accurately by the expression

$$I = I_s(e^{qV/kT} - 1)$$

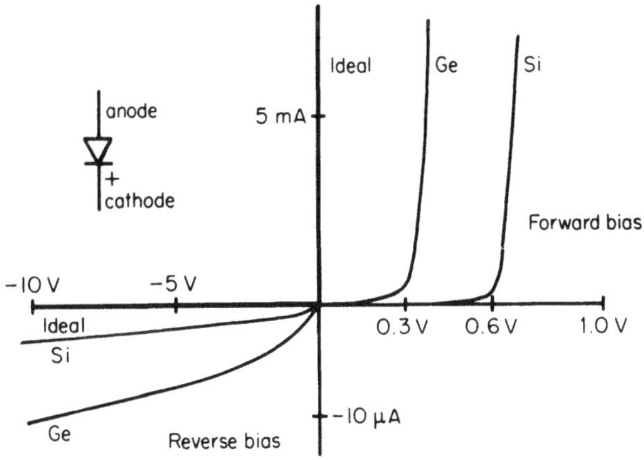

Figure 3.1 Diode characteristics

where I_s is the saturation or *leakage current*, q is the electronic charge, k is Boltzmann's constant, T is the absolute temperature and V is the applied voltage. This equation ignores certain factors relating to the structure of the diode material, but clearly shows that as V increases in the reverse direction, that is V is negative, the expression rapidly becomes $I = -I_s$. In the forward direction of applied voltage the diode exhibits an effective voltage step before the current increases to an appreciable amount. The magnitude of this step, or cut-in voltage, depends on the material used, being about 300 mV for germanium and about 600 mV for silicon.

The leakage current, I_s, is highly temperature dependent, and given approximately by the expression

$$I_s \propto T^{\frac{3}{2}} e^{-W/2kT}$$

where W is the work function. For germanium, I_s doubles for approximately every 10°C rise in temperature, whereas for silicon, I_s doubles for every 7°C rise. Though the leakage current in silicon shows a greater dependence on temperature, the absolute value of this current is very much less. Typically, at 25°C and with a reverse voltage of 10 V, I_s for germanium is 10 μA, whereas for a silicon diode under these conditions, I_s is 25 nA. Leakage current limits the range of temperatures over which the junction device can usefully be employed. For germanium devices the temperature range is usually restricted to below 75°C, whereas silicon devices can be operated at temperatures up to and in excess of 130°C.

Another feature of a junction device which limits its performance is the junction capacitance, the value of which is dependent on various factors, including the magnitude and polarity of the voltage across the junction. During current conduction, charge is stored in the capacitance, so that, when the junction polarity is reversed, the current does not immediately switch off, but continues until the stored charge has been used. Similarly in switching on, the need to charge up the junction capacitance delays the build-up of the current. It is this junction capacitance which defines the maximum switching speed of a logic gate. There is, however, a special type of diode which makes use of a metal-to-silicon bond, and which has negligible charge storage. This is the Schottky diode, which is otherwise like a junction diode, though with a lower cut-in voltage. Schottky diodes are extensively used in special *clamping* arrangements to improve the switching speed of logic gates employing junction devices (Horowitz and Hill, 1980).

The junction, or *bipolar*, transistor consists essentially of two diode junctions formed back-to-back in a single semiconductor crystal to create a current amplifying device. The characteristics of a typical *npn* silicon transistor in common-emitter configuration are shown in figure 3.2, and, when operated with a suitable load resistor, R_L, we see there are two well-defined operating states. These are

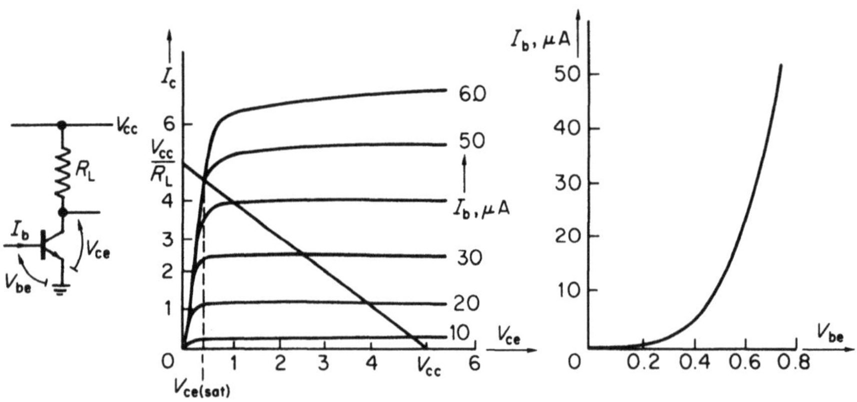

Figure 3.2 Common emitter characteristics for an *npn* transistor

termed *cut-off*, and *saturation* or *bottoming*. If the base–emitter voltage is sufficiently reduced to prevent the flow of base current, the collector current falls to zero and the collector voltage V_{ce} is equal to the supply voltage, V_{cc}. This is the *cut-off* state and, for silicon devices, a base–emitter voltage below about 0.2 V is sufficient. If the base–emitter voltage is increased, forward-biasing the base–emitter diode, the operating point of the transistor moves along the load line to a voltage $V_{ce(sat)}$ which, in the case of a modern silicon epitaxial transistor, can be less than 50 mV. In this condition the transistor is *saturated*, or *bottomed*. The design of switching circuits usually ensures that there is base current in excess of that required just to reach saturation, so that any production spread in the device parameters may be accommodated. Thus, although the transistor can be operated as a switch, there is some leakage current when cut off and a small voltage drop when saturated.

In the saturated state the transistor has an excess of charge stored in its base region, and the time taken to switch off depends on the efficiency with which this charge can be removed. Also a finite time, the *fall-time*, t_f, is required for the collector current to fall to approximately zero, shown in figure 3.3. For a typical *npn* switching transistor the turn-off time, t_{off}, is less than 20 ns and the charge-storage time, t_s, is in the order of 10 ns. The time taken to drive the transistor from the cut-off state into saturation, t_{on}, depends partly on the time required for the collector current to establish itself, the *rise-time*, t_r, and partly on the time taken for the base potential to reach the turn-on voltage of the base-emitter diode. This entails the charging of the capacitance associated with the base and is known as the *turn-on-delay*, t_d. It is in the same order as the charge-storage time, but t_{off} is always greater than t_{on} since the charge-storage time predominates.

An alternative indication of switching speed often quoted is the propagation delay from high to low, t_{PHL}, and from low to high, t_{PLH}, also shown on figure 3.3.

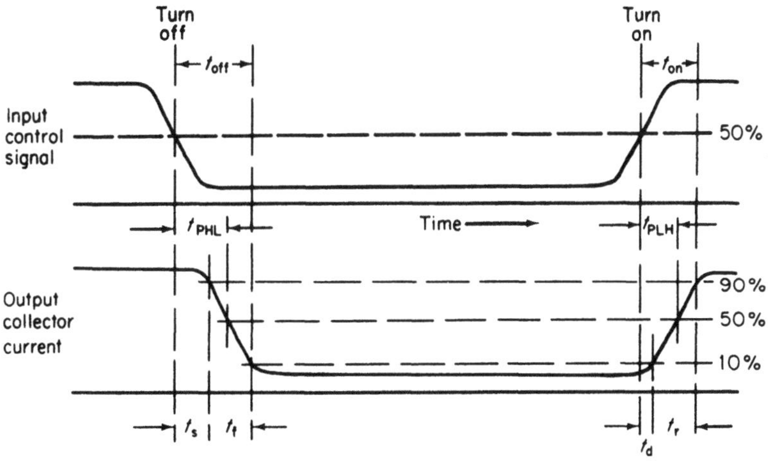

Figure 3.3 Switching times

Power dissipation in the transistor is low at cut-off and saturation since the product of V_{ce} and I_c is low, but care must be taken in switching large currents as the product reaches a maximum during the changeover and dissipation limits may be exceeded.

Metal-oxide-silicon, MOS, transistors are a more recent development. An MOS device is simple and very small, and so lends itself to large scale integration, LSI, methods, where high packing density is desirable. The MOS transistor is a *unipolar* device and basically consists of a semiconductor *channel* above which is an aluminium layer, the *gate*, insulated from the channel by a thin layer (100–1000 Å) of silicon oxide. Current flow between the two ends of the channel, *source* and *drain*, is determined by the electric field resulting from the potential maintained at the gate and consists entirely of majority carriers, hence the term 'unipolar'. Thus this device, unlike the bipolar transistor, is voltage controlled. If an increase in gate voltage causes an increase in drain current the device is known as an *enhancement* type; conversely, a reduction in drain current as the gate voltage increases indicates a *depletion* type.

Figure 3.4 illustrates an *n*-channel (nMOS) enhancement mode MOSFET (metal-oxide-silicon field effect transistor) and the complementary *p*-channel type (pMOS) is similar. nMOS has the lower on-resistance and can operate at

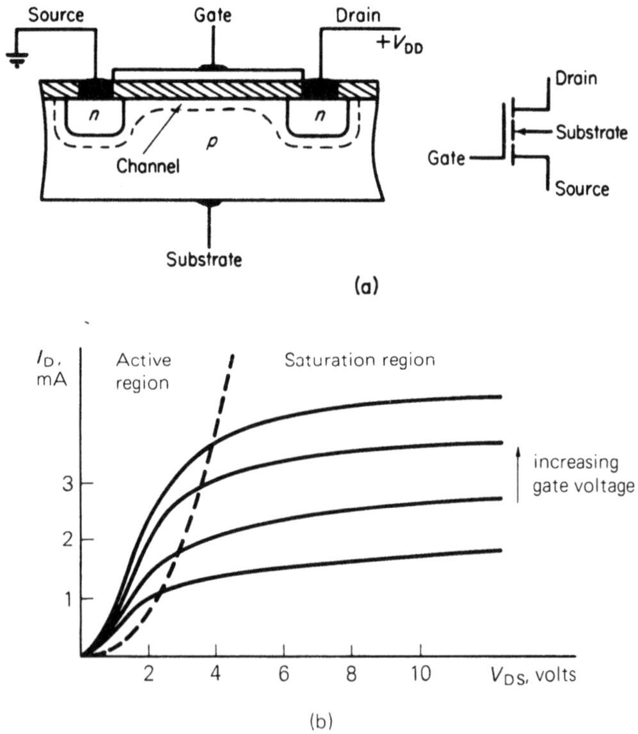

Figure 3.4 (a) MOSFET device and symbol. (b) Common-source characteristics

higher speeds since electron mobility is greater than that of holes. The bulk material is lightly doped *p*-type silicon into which two *n*-type regions are diffused. With the gate at the same potential as the source, the source and drain appear to be connected by two back-to-back diodes and no current can flow. If the gate is taken positive with respect to the source, the resultant field attracts minority carriers (electrons in *p*-type material) to form a conducting *inversion* layer, the channel, between source and drain.

For small values of V_{DD} (typically below four volts), an electron current will flow through this inversion layer, from source to drain. The current increases linearly with V_{DD} for a given gate voltage,

$$I_d \propto V_{DD}$$

but as the square of the gate voltage,

$$I_d \propto V_{gs}^2$$

This is the *active* region of the device (figure 3.4b).

For higher values of V_{DD}, a *cut-off* point is reached where field interaction between gate and drain results in a narrowing (*pinch-off*) of the inversion layer in the vicinity of the drain. This prevents I_d from increasing much further with increasing V_{DD}. When this occurs the transistor is said to be in *saturation*.

The shape of the I_d/V_{DD} characteristic is similar to that of the bipolar transistor (figure 3.2) and, with suitable drain loads, switching circuits can be designed to operate between similar output levels.

In the active region the FET functions as a voltage (gate voltage) controlled resistor and finds great use as an analog switch in sampling and multiplexing circuits. Note that the device, unlike the bipolar transistor, continues to conduct at very low source-drain voltages. Such switches compatible with TTL voltage levels can be obtained with on resistances lower than 100 ohms and off resistances greater than 10^8 Mohms (for example, Precision Monolithics' analog switch SW06).

By varying the doping levels and layout geometry, manufacturers can obtain extremely wide variations in parameters.

The MOSFET just described can be termed a *horizontal* transistor since the current flow is essentially along a horizontal path between source and drain. A *vertical* construction, VMOS, is also widely used and it can be seen in figure 3.5 that a V-groove has been created, by etching, in the body of the device. In operation, an *n*-channel is induced in both faces of the V notch, doubling the effective channel area. Also, the drain is formed on the substrate so no drain contact is

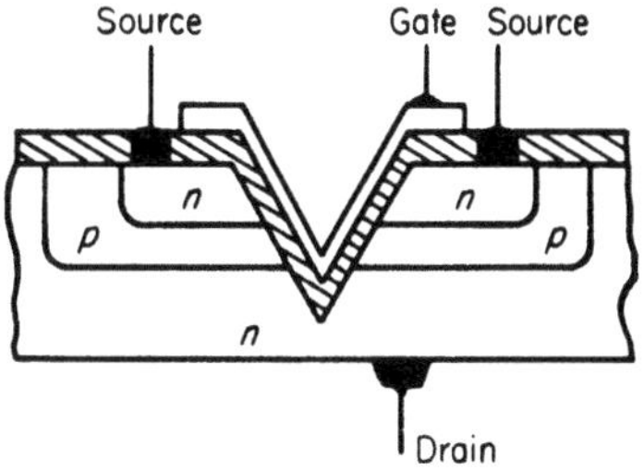

Figure 3.5 VMOS transistor

needed on the surface. This keeps the chip area small, saturation resistance low and provides good thermal conductivity. Because the length of the channel is determined by the diffusion depth, rather than a mask spacing, more controllable and therefore closer spacings can be obtained, increasing the possible current density. In the VMOS device the length of the channel is around 1.5 μm, compared with a typical 5 μm for the horizontal MOSFET. The result is a versatile device able to switch amperes and with a low gate capacitance measured in picofarads. It can be controlled directly from either MOS or bipolar transistor circuits and, unlike the bipolar transistor, the current capability can be doubled simply by paralleling two transistors, without the need of ballast resistors.

Thus we have a range of devices, each of which, for all practical purposes, can be considered as a switch controlled by a small base current in junction devices, or gate voltage in MOS devices. These form the basis of our logic circuits. Initially the circuits were built up from individual components but rapid developments in semiconductor technology during the 1950s led to the ability to construct a number of gates on a single silicon die. The levels achieved in these early *integrated circuits*, or *chips*, are now classed as small scale integration, SSI. As techniques have improved it has become possible to integrate more and more devices in each chip so that we have several levels of integration. As an indication of the complexity, small scale integration, SSI, normally indicates up to about ten devices per chip; medium scale integration, MSI, indicates up to 100; large scale integration, LSI, up to 1000. Very large scale integration, VLSI, and even super large scale integration, SLSI, carry the integration even further. At the time of writing, circuits with over a million devices are already available. No matter how many device equivalents exist in the integrated circuit, however, the logical action depends on the same fundamental operations.

3.2 Introduction to logic circuits

The basic AND and OR operations can be achieved with simple diode circuits. In figure 3.6, A and B are the inputs to the circuit, and voltages applied here must be at, or within a specific tolerance range of, either of two well-defined levels. It is usual for one of these levels to be ground potential, and we will assume that the other level is +V volts. Inspection of the circuit shows that if either (or both) input is at ground then that diode becomes forward-biased and,

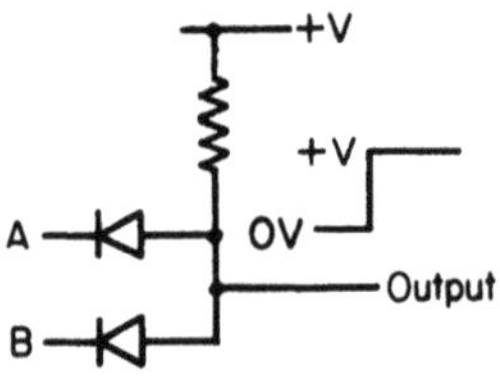

Figure 3.6 The diode gate

assuming perfect diode action for the moment, the output must also be at ground. The output can go to +V only when both inputs are themselves at +V.

A truth table can be drawn showing all possible combinations of input and output in terms of potentials, as in figure 3.7a, and if we define +V as the '1'

A	B	Output
0V	0V	0V
0V	+V	0V
+V	0V	0V
+V	+V	+V

(a)

A	B	Output
0	0	0
0	1	0
1	0	0
1	1	1

(b)

Figure 3.7 Truth table for the diode gate

logic level and ground potential as the '0' level the truth table becomes as in figure 3.7b. This is the truth table for an AND gate, but if we reverse the definitions of '0' and '1' the truth table indicates an OR function. The accepted convention is to define *positive logic* as logic in which the more positive level takes the value '1', and *negative logic* where the '1' value is assigned to the more negative potential. Thus the circuit of figure 3.6 acts as an AND gate in positive logic terms, and an OR to negative logic. By rearranging the circuit to its dual form we obtain the OR function with positive logic, and, therefore, AND with negative logic.

The usefulness of these diode circuits is limited in that, no matter how the circuit is rearranged, inversion cannot be achieved. Also, since the diodes are not perfect elements, power loss occurs and this shows itself as a shift in voltage at the output, leading to a convergence of the logic levels. Such tolerance problems severely limit the number of diodes that can be cascaded. Use of an inverting switching amplifier provides us with the inverter and also overcomes the signal loss by providing power gain. With the advent of the transistor it became feasible to provide each gate with its own inverting amplifier, ensuring well-defined output levels as it is switched between cut-off and saturation. A simple circuit of this type, termed a *diode-transistor-logic* (DTL) gate, is shown in figure 3.8. With

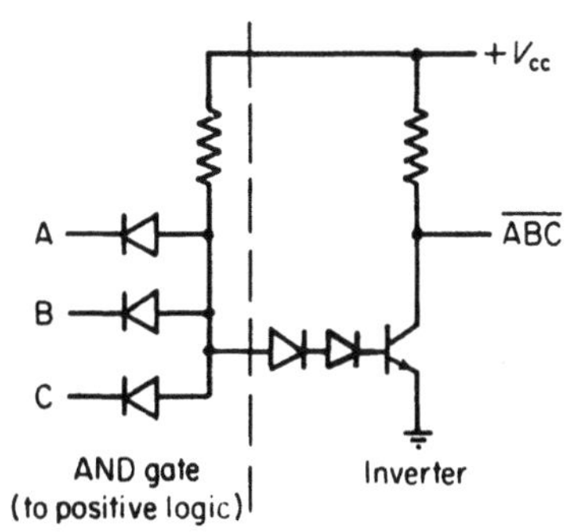

Figure 3.8 DTL gate

the inverter present the logic function is a NAND to positive logic or NOR to negative logic. Symbols for these gates are the simple AND and OR symbols with an additional small circle at the output denoting inversion (see appendix A).

3.3 Transistor–transistor logic (TTL)

DTL gates developed as discrete component circuits and the number and type of components had an important bearing on cost. Integrated circuit technology shifted this emphasis and, with increasingly sophisticated techniques, circuit complexity is not now a limiting factor. This led to the popularity of *transistor-transistor logic* (TTL) in which multi-emitter transistors replace the input diodes. These gates have the advantage of high-speed operation and are cheaper to produce. A basic TTL gate is shown in figure 3.9. With all the emitter inputs held positive, the base-collector diode of T1 becomes forward-biased, turning T2 on through R1 to V_{cc}. The collector potential of T2 then falls to $V_{ce(sat)}$. If any of the emitter inputs is taken low, transistor T1 turns on through normal transistor action and pulls the base voltage of T2 low. T2 turns off and its collector potential rises to V_{cc}. The circuit thus provides the NAND function to positive logic, and is the basis of the 74 series of TTL devices.

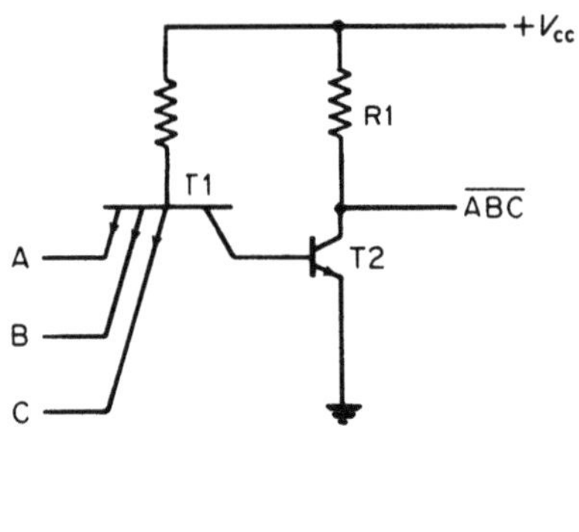

Figure 3.9 TTL gate

The simple TTL circuit suffers from a problem which arises when all the inputs are taken high. Under these conditions normal transistor operation ceases and current flows in at the input emitters with its magnitude governed by the effective input impedance of each emitter. When one input happens to have a lower impedance than others connected at the same point it tends to draw more than its share of the current available from the driving gate, and the higher input impedance gates may not reach full saturation at their output transistor. It is not possible to control the input impedances sufficiently closely and the only way to overcome this *current hogging* problem is to ensure that the driving gate has a very low output impedance in both logic states.

This is achieved by the *totem-pole* circuit, included in figure 3.10, which is in effect a modified push-pull power amplifier. Transistors T3 and T4 are driven in antiphase by the *phase-splitter*, T2, so that T3 is on when T4 is off, and vice versa. The voltage drop across diode D1 ensures that T3 is always off when T4 is on. TTL uses a standard 5 V supply, and the output therefore switches between two levels of approximately 0.2 V and 3.5 V.

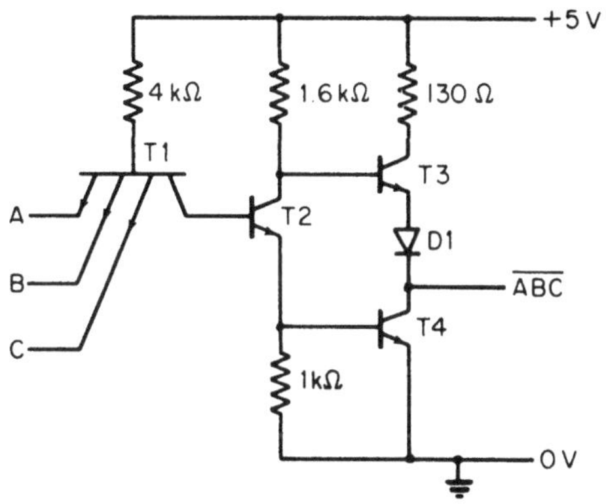

Figure 3.10 TTL gate with totem-pole circuit

Knowing the spread of values resulting from the production processes, it is possible for the manufacturer to specify the maximum number of gates any one gate can drive, and that is known as the *fan-out* of the gate. The value varies with the gate type but is around ten for TTL gates. A standard TTL gate, for example, can sink a maximum of 16 mA. That means that 16 mA can enter the gate output terminal when the output is in the low state. The standard TTL input is designed to switch 1.6 mA to the driving gate, so the fan-out is ten. Connecting more than ten gates may result in the driving gate being unable to switch correctly, and performance is then affected.

The 130 ohm resistor is included in the totem-pole circuit to protect T3 and T4 during the changeover when, for a brief period, both devices may be conducting simultaneously. Current surges occur during this changeover and voltage transients can be fed on to the supply line, which then interfere with the correct operation of other circuits. A well-stabilized power supply is therefore essential, with additional 0.1 μF decoupling capacitors from V_{cc} to ground every few packages on a printed circuit board.

The propagation delay through a TTL gate of the type shown in figure 3.10 is in the order of 10 ns. Most of this delay is caused by the saturation of transistors T1, T2 and T4 and it can be reduced by the inclusion of *antibottoming* diodes to prevent heavy saturation and limit the build-up of stored charge. A Schottky diode, which has negligible stored charge, is fabricated between the base and collector of each transistor. The Schottky diode has a lower voltage drop than the silicon base-collector junction and diverts excess base current into the collector. This technique is termed the 'Baker clamp' referred to earlier and described by Horowitz and Hill (1980). By this means propagation delays of less than 3 ns can be achieved, and TTL gates using the Schottky diode clamps are available as the 74S series.

Many TTL gates also have clamping diodes connected at their inputs to inhibit negative-going signal excursions which may damage the internal structure of the gate. They also prevent the ringing which may occur when interconnections appear as incorrectly terminated transmission lines to fast signal transitions.

TTL is characterized by a relatively low input impedance and a low-power version, with only a quarter of the current demand, is formed by increasing all the resistor values. Unfortunately an increased propagation delay then results, typically around 33 ns, and to overcome this a Schottky clamped version was developed, retaining the low-power feature and restoring the propagation delay to about 10 ns. This is the popular low-power Schottky, or 74LS, range.

More recently a third generation of TTL circuits has been introduced, known as *advanced TTL*, and making use of improved fabrication and isolation techniques within the integrated circuit. These improvements give very high switching speeds and very low parasitic capacitances, so that the speed of operation of the gates has been doubled, while the power dissipation is halved. Schottky clamping is used extensively, see figure 3.11, and diodes D1 and D2 are used at the inputs to minimize capacitance. Additional *speed-up* diodes, D3, D4 and D7, increase the operating speed still further by discharging internal capacitances very rapidly at the bases of T2 and T6. Diode D8 is included to divert some of the charge from the external load circuit to aid the output in switching to its low

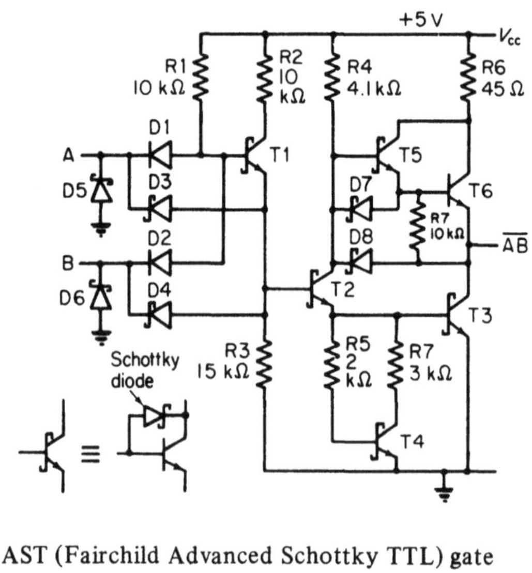

Figure 3.11 FAST (Fairchild Advanced Schottky TTL) gate

state; the charge passes through transistor T2 to increase the base drive to transistor T3. The output circuit is designed to sink at least 20 mA in this low state, and when in the high state the current in transistor T6 is limited by R6 to 1 mA. Diodes D5 and D6 protect the inputs from negative-going signals.

3.4 Integrated injection logic (I^2L)

Integrated injection logic, or *merged transistor logic*, MTL, as it is alternatively known, is another bipolar logic system with a high packing density and a very low power dissipation. It is therefore mainly used in large scale integrated circuits requiring low power consumption, such as digital watches and memory devices.

By shifting the logic connections to the output of the gate it is possible to merge the transistors in the construction of the circuit, as shown in figure 3.12.

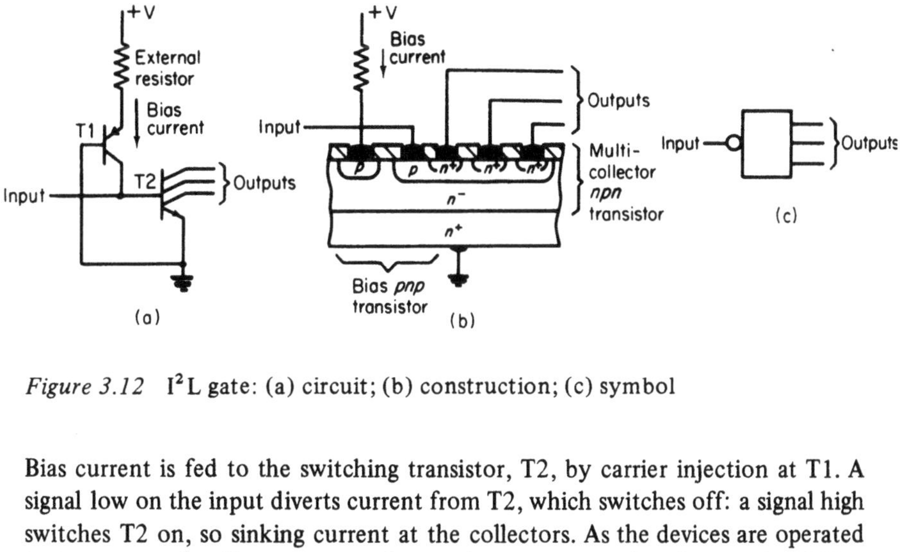

Figure 3.12 I^2L gate: (a) circuit; (b) construction; (c) symbol

Bias current is fed to the switching transistor, T2, by carrier injection at T1. A signal low on the input diverts current from T2, which switches off: a signal high switches T2 on, so sinking current at the collectors. As the devices are operated in inverse mode, $V_{ce(sat)}$ is very low and a power supply of around 1.5 V is adequate. A logic voltage swing of 0.6 V is then obtained, with a minimum static power dissipation of some 250 μW per gate, and propagation delays of around 10 ns are general. By introducing Schottky diodes at the outputs, and rearranging the circuit geometry slightly, a related family of devices called *integrated Schottky logic* (ISL) is achieved, with a propagation delay as low as Schottky clamped TTL and a power dissipation as low as I^2L.

The logic implemented in all cases is the wired-OR, which is discussed in the next chapter. It is necessary to use an external load resistor as shown in the OR gate of figure 3.13. The load resistor may be taken to a higher voltage to give suitable levels for external circuitry.

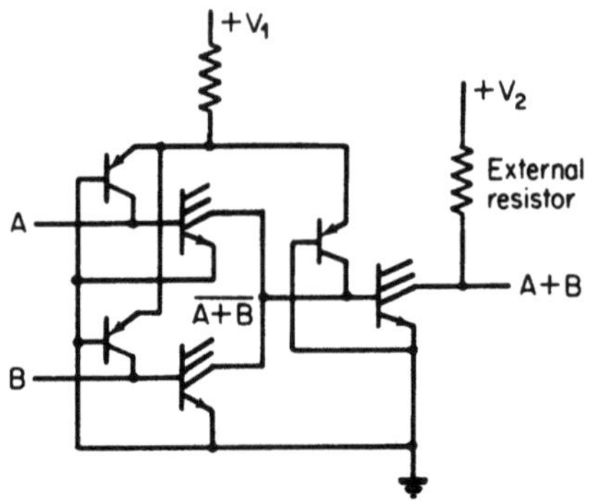

Figure 3.13 Illustrative I^2L circuit

3.5 Emitter-coupled logic (ECL)

Emitter-coupled logic has a faster response than any other type of gate available commercially, with switching times in the order of 1 ns. The basic circuit is that of a differential amplifier (Fitzgerald *et al.*, 1981), or long-tailed pair, in which the transistors do not saturate and the current through the emitter resistor R_E is switched from one transistor to the other, dependent on the input voltage being higher or lower than the reference voltage. The logic action is developed by paralleling several transistors on the input side. In addition to the high switching speed resulting from the non-saturation this type of logic gives antiphase outputs. In order to standardize input and output voltage levels emitter followers are provided at both outputs, and a typical circuit is shown in figure 3.14.

Any input A, B, C, taken to the high level causes that transistor to conduct, and the circuit is therefore a NOR/OR gate to positive logic and a NAND/AND

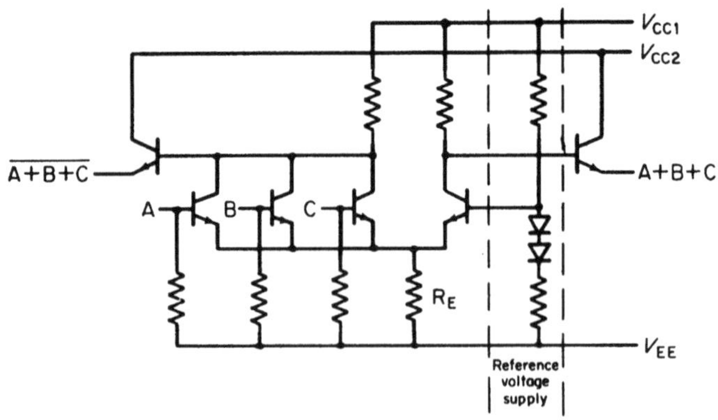

Figure 3.14 ECL gate

to negative logic. Absolute resistor values are not important since the action is dependent on resistor ratios which can be closely controlled in integrated circuits. Having emitter followers in effect at both input and output, the circuit has a high input impedance and a low output impedance, allowing a large number of inputs (high *fan-in*) and high fan-out. The reference voltage is normally generated internally and must lie approximately midway between the two logic voltage levels. The voltage swing needed to switch the emitter current is only about 0.6 V, giving a low immunity to unwanted, *noise*, signals. However, logic assemblies built to take full advantage of the high switching speed of these devices must use matched transmission lines as interconnections and these are of necessity low-impedance connections (50–120 ohms). The effect of noise is therefore minimized and signal levels in the order of a volt are adequate.

It is usual to provide a negative power rail, V_{EE}, of −5.2 V. V_{CC1} and V_{CC2} are then grounded at zero volts. The use of *ground planes* is recommended for this purpose (Matthews, 1983). Also separate pins are used for V_{CC1} and V_{CC2} to minimize internal crosstalk. The emitter-follower outputs are not normally provided with a load resistor, not just to facilitate wired-OR operation, but because the mandatory use of transmission lines with these very fast gates makes a load resistor in parallel unnecessary and wasteful. However, the use of emitter-follower outputs demands a certain amount of care if speed of operation is to be maximized, since the gates are adversely affected by load capacitance. When the gate output voltage goes high, the load capacitance is charged through the output impedance of approximately 5 ohms. However, when the input signals change so as to create a low voltage at the output, the capacitance maintains the output voltage at its previous level, the emitter cutting off. The capacitance must then discharge through the circuit impedance, usually 50 ohms for a correctly terminated connection. This problem is eased somewhat by the fact that the logic level change is small compared with the difference between the operating points and the supply voltage.

3.6 Metal-oxide-silicon logic (MOS)

One of the main advantages of MOS technology is the very small area required for each device, and in order to retain this size advantage transistors are used as dynamic loads instead of resistors. For a given value of load, the length of resistance channel when implemented in silicon is far greater than the size of an equivalent MOS transistor operating in its saturation region. The dynamic load is created from a depletion mode FET with its gate strapped to its source. By suitably doping the device, the fact that, externally, $V_{gs} = 0$ V, internally means that there is sufficient built-in field to maintain the device in its saturation region with an incremental resistance of several thousand ohms.

An nMOS NAND gate is shown in figure 3.15, in which transistors T2 and T3 act as simple switches in series and T1 is the load. Essentially, operation of the

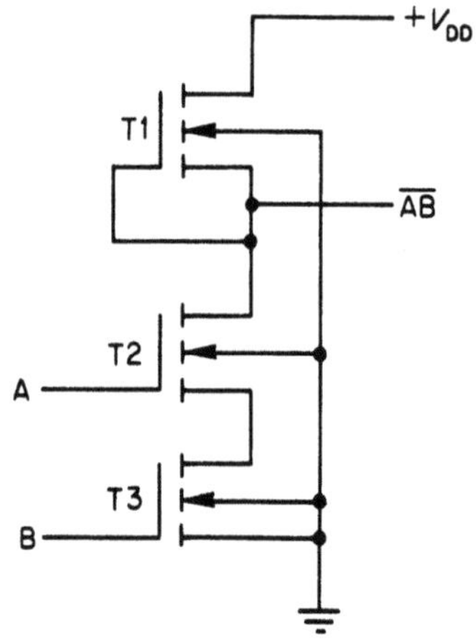

Figure 3.15 nMOS NAND gate

gate can be understood by considering the three devices as voltage-controlled resistors. With logical one voltage levels on T2 and T3, the on resistance of each transistor is arranged to be only a few tens of ohms. Consequently the voltage level at the output is pulled down almost to ground and current will flow through the load transistor. Logical zero on either gate input will switch that device off (high resistance) causing the output to be pulled up to $+V_{DD}$ by T1. NOR gates can be constructed by arranging for the switching gates to be in parallel, connected to a common load transistor.

Now, definition of the logic zero voltage level depends upon the ratio of the device resistances and logic gates constructed from these types of circuit are termed *ratioed*. The time taken for the gate to switch from logic zero to logic one will depend on the magnitude of the load resistance since external capacitances must be charged up to $+V_{DD}$. Too low a value for the load resistance will ensure a fast response but will give rise to wasteful power dissipation. Precise logic definition, fast response and low power consumption are therefore compromises.

It is possible to form both pMOS and nMOS on the same chip and logic families using the two forms are termed *complementary MOS logic*, CMOS. The basic circuit is the inverter, which consists simply of one *n*-channel and one *p*-channel enhancement-mode transistor as in figure 3.16. The source of transistor T1 is connected to the positive supply, $+V_{DD}$, and T2 source is taken to ground. When zero volts is applied at the input the gate of T1 is negative with respect to its source and, being *p*-channel, the transistor conducts. The gate of T2 is at the same potential as its source so T2 is off. There is therefore a low resistance path from the output to the positive supply via T1, with a very high resistance from output to ground, and the output voltage reaches almost $+V_{DD}$.

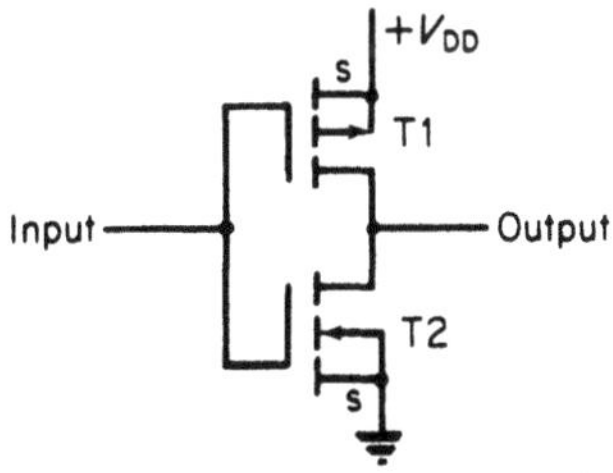

Figure 3.16 CMOS inverter

Conversely, if the input voltage is taken to $+V_{DD}$, T1 is off and T2 conducts since the gate of T2 is now positive with respect to its source, and T2 is an *n*-channel device. The output voltage levels are no longer dependent on the ratio of upper and lower channel resistances, and these circuits are termed *ratioless*. The output voltage is now governed by the load seen at the output. This is usually capacitive, in the form of other gate inputs, and thus the design of channel resistance is determined by the operating speed required. The circuit acts as an inverter and in either logic state one transistor is on while the other is off.

Because of the high impedance of the off transistor, the quiescent power consumption of a CMOS circuit is very low, being determined by the product of the leakage current in the off transistor and the supply voltage, $+V_{DD}$. It is in the order of a few nanowatts per gate. Note however that a much larger current flows during the switching of the transistors, and dynamic power consumption increases to the microwatt range.

The inverter circuit is extended to form logic gates by adding further transistors, as shown in figure 3.17. The MOS transistors approximate to ideal switches, the *n*-channel types being closed, that is on, when the gate voltage is $+V_{DD}$, and the *p*-channel types being closed when the gate voltage is 0 V. Thus in figure 3.17a the output is connected by a low impedance path to ground only

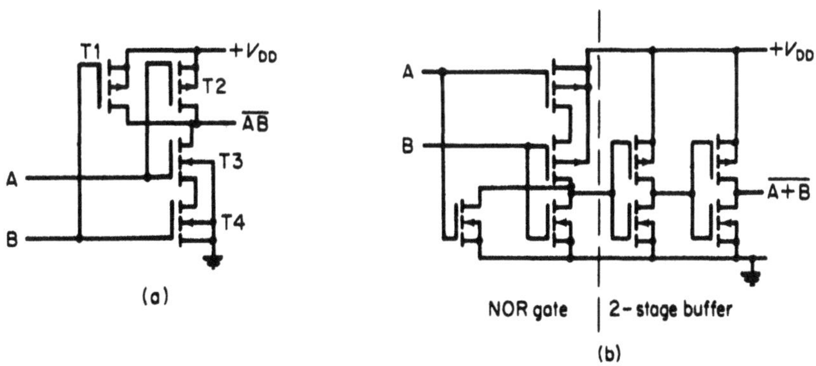

Figure 3.17 (a) CMOS NAND gate. (b) CMOS NOR gate with buffered output

when both inputs are at $+V_{DD}$. Then T1 and T2 are off, T3 and T4 on. In terms of positive logic this is the NAND function. With the alternative arrangement of figure 3.17b, either input at $+V_{DD}$ gives a low impedance path to ground and, of course, a high impedance to $+V_{DD}$. This is the NOR function.

The very high input impedance, of approximately 10^{12} ohms, necessitates an input current of only a few picoamperes. In other words the devices are voltage controlled and one gate can control at least fifty other CMOS gates. Each gate presents an almost purely capacitive load of approximately 5 pF, and current is required to charge and discharge the capacitance. The power dissipation in the driving circuit is directly proportional to the capacitive loading, and the frequency and rate of switching. The transition from one logic level to the other is progressively less abrupt as the supply voltage is increased. This is due to both the upper and lower output transistors being on for a longer period. Also a wide operating range of supply voltage, V_{DD}, can be tolerated, typically 3-18 V, with the logical transition occurring at about 45-50 per cent of V_{DD}, as in figure 3.23b. Since the transistor action does not rely on semiconductor junctions, the effect of temperature variations is very small over the entire range of $-55°C$ to $+125°C$.

It is not possible to achieve wired-OR logic with the simple CMOS gate because of the low output resistance in both logic states, so it is necessary to use the three-state gate for such applications. The three-state gate is designed to operate with the upper or lower half of the circuit off, as in normal logic operation, or both halves off by means of two additional controlling transistors in series with the output transistors.

The actual output resistance of a CMOS gate, and therefore the ability to drive external circuits, is dependent on the logic conditions at the input. To overcome this variation the more popular buffered CMOS (series B) devices standardize the output conditions by the inclusion of a buffer amplifier at each output. This gives a standard output impedance of 400 ohms, and also has the advantage of halving the propagation delay and improving the noise performance of the gate. The two-input NOR gate of figure 3.17b is shown buffered.

The very high impedance of a logic gate input means that high voltages can be built up on the gate of the MOS transistor even from very low energy sources, such as a static charge. Since the gate oxide is permanently damaged at about 100 V it is necessary for each input to have protective diodes built in to prevent the voltage building up. A typical protection circuit is shown in figure 3.18 and manufacturers further advise the user to be cautious in the handling of the

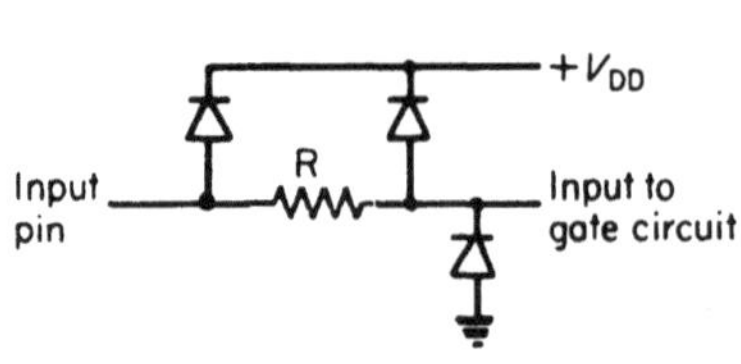

Figure 3.18 Input protection

device. To prevent smaller unwanted voltages developing on an unused input, causing it to be biased outside the usual logic regions, it must be connected to either $+V_{DD}$ or ground, dependent on the logic of the gate. Alternatively, the input can be connected to another used input. Although this has the effect of lowering the on resistance, and so increasing the speed slightly, it is offset by an increase in capacitance for the driving gate.

CMOS logic has always had the advantage of low power demands and smaller chip area, but for some years it was slower in operation and limited in drive capability when compared with the competing TTL circuits. Low-power Schottky TTL, in particular, provided an overall performance level which made it very attractive in a wide range of applications. Developments in manufacturing processes, however, have enabled the newer high-speed CMOS logic to surpass the performance of LS TTL. This is mainly due to the replacement of the metal gate of MOS transistors by a silicon gate.

The *silicon-gate process* is now one of the most widely used processes and enables smaller, high density circuits to be produced. A metal-gate device is formed by a repetitive depositing and etching process in which the buried structure is formed. Figure 3.19 shows the cross-section of a typical nMOS transistor. The gate oxide layer is 100–1000 Å thick. The important point to note is that great care must be taken in the alignment of the masks used to ensure that the gate extends fully to the source and drain regions. In practice the gate must overlap these areas. The result is a larger-than-necessary gate area, leading to increased size and capacitance and thus slower operation.

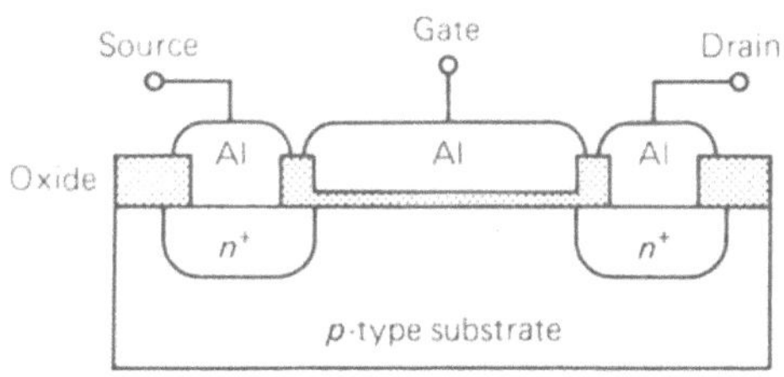

Figure 3.19 Cross-section of an nMOS transistor (metal-gate process)

The silicon-gate technique is a self-aligning process using as the conducting gate material *polycrystalline silicon* or 'polysilicon' which is deposited by chemical vaporization. Figure 3.20 shows the essential steps in the fabrication of an enhancement-mode transistor. Over a thin layer of oxide, which will eventually form the gate insulation, a layer of polysilicon is deposited. This is etched away to define a conducting gate region. The oxide is then removed in areas not covered by the polysilicon. The silicon wafer is then exposed to n-type impurities (arsenic or phosphorus), converting the exposed and underlying silicon to n type. The polysilicon and oxide layer prevent the impurities from reaching the *p*-type silicon region beneath the gate area, resulting in a perfectly aligned gate without the need for complex masking arrangements. Further layers of oxide and metal are then deposited to form the contacts. Depletion-mode devices are made by ion implantation of arsenic or phosphorus into the silicon to form a slightly conducting region. This can be carried out through the oxide, provided it is thin enough, though thick oxide, photoresist or metal areas can serve as masks to this process.

The smaller chip size of the transistor, made possible by this process, reduces capacitance, makes denser circuits possible and increases speed. In addition, the silicon-gate transistors have higher gains and the output buffers are able to provide 4 mA in either logic state – sufficient to drive ten LS TTL gates. High-speed CMOS is packaged to be pin-compatible with the 74-series TTL circuits and some earlier 4000 series CMOS. The designation 74HC is used to indicate high-speed CMOS designed to operate from a power supply between two and six volts, and some manufacturers provide a subset, designated HCT, intended for direct replacement in designs operating at TTL voltage levels and requiring a 5 volt ± 10 per cent power supply.

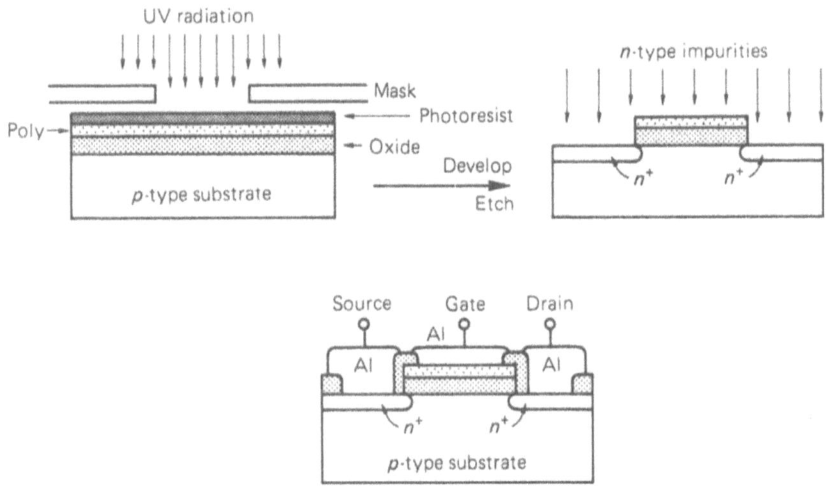

Figure 3.20 Cross-section of an nMOS transistor (silicon-gate process)

An ever-present problem with logic circuitry, and one which increases as the density of integration increases, is how to dissipate the heat generated internally. Many alternative approaches to manufacturing and mounting circuits have been suggested to overcome this problem. *Silicon-On-Sapphire* (SOS) technology provides a faster device with better heat dissipation properties. Sapphire is used as the substrate upon which a thin layer (1-2 μm) of epitaxial silicon is deposited and CMOS devices are then fabricated in this silicon. The low dielectric constant of sapphire, relative to silicon, results in a device lacking in substrate parasitic capacitance and with good dielectric isolation. Gates using this technology exhibit power-delay products (see section 3.8) of around 0.5 pJ, but the sapphire substrates are expensive compared with silicon. A compromise technology, which retains the advantages of both bulk CMOS and SOS, is *Silicon-On-Insulator* (SOI) in which silicon-oxide is used as the insulator (Pasa and Beresford, 1981).

3.7 Design of VLSI systems

One of the most exciting developments in LSI/VLSI technology in recent years has been the emergence of powerful computer-aided design, CAD, tools which enable the design of MOS circuits of considerable complexity to be carried out in a straightforward manner. Perhaps the most successful (and earliest) of these was the Mean-Conway design process (Mead and Conway, 1980) for nMOS. A set of rules was formulated, based on a *length unit* λ. The magnitude of λ depends on the state of the technology. In 1983, λ was 2 μm. Multiples of this length define a constraint on the width, separation and overlap of features on the silicon. For example, a transistor can be formed whenever a polysilicon conductor of width not less than 2λ passes over a diffusion region of width not less than 2λ, figure 3.21 (Mukherjee, 1986).

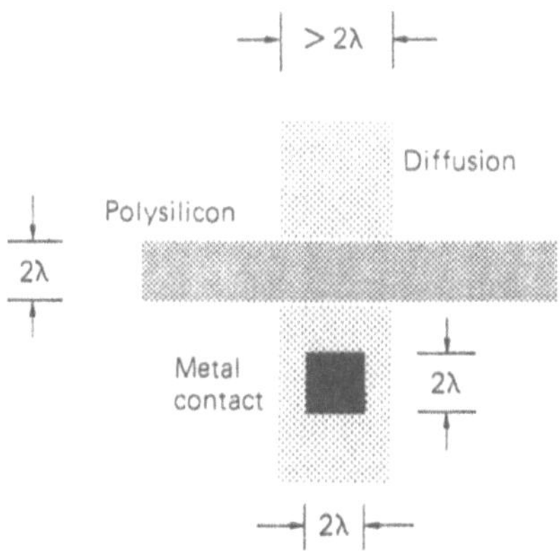

Figure 3.21 nMOS silicon-gate transistor (Mead–Conway rules)

Using standardized design rules it is possible to design circuits, which, though perhaps not the most economical in their use of the available silicon area, do stand a very good chance of working! One of the disadvantages of designing directly on to silicon is the impossibility of modification to the circuit once the chip has been produced. Because of this, there are available some very sophisticated simulator software packages which allow the designer to test a design as thoroughly as possible before committing the circuit to silicon. As circuits become more complex, the problem of *testability* becomes much more important and currently a great deal of effort is being devoted to this subject (Bennetts, 1984). The topic is discussed further in section 4.6.

3.8 Interconnection: practical considerations

Logic design is initially carried through assuming ideal gate inputs and outputs and ideal interconnections. When the design is implemented in practical terms the designer is obliged to consider inherent limitations and the consequential compromises. One of the main problems is the presence of noise signals, which in some cases can so disturb the voltage levels that they move outside the acceptable tolerance band and give erroneous operation of the system.

Electrical noise of some sort is inevitable in a practical system and, for convenience, it can be considered as originating from two distinct sources. Noise external to the system is governed by the environment and is usually outside the direct control of the system designer. For example, logic circuitry to control machine tools has to operate under severe electrical conditions owing to the switching of large inductive loads in the vicinity. For this type of work, where speed is not so important, CMOS is often preferred since its larger voltage swing demands more noise energy before logic level tolerances are exceeded.

Internal noise arises from a variety of causes. It can be introduced through the d.c. supply lines if they are not sufficiently decoupled, or via the ground connections if care is not taken to ensure proper ground returns. It can also be generated by reflections of energy on interconnections acting as unterminated transmission lines and giving rise to *ringing*, or by coupling between adjacent signal lines, referred to as *crosstalk*. Crosstalk arises when fast-changing voltage or current levels on one line cause transient signals to be injected into the other line. The internal noise problem is aggravated by the use of modern high-speed logic families such as TTL and ECL, and three parameters have been found to be useful in predicting the tolerance of a particular type of logic gate to noise. These are best defined with reference to the *transfer characteristic* of the gate, a generalized characteristic being shown in figure 3.22. Actual transfer characteristics differ according to gate type, as shown in figure 3.23.

If we specify the external operating levels as V_a and V_b in figure 3.22, then the line joining these two operating points cuts the characteristic in the *transition region* at the *threshold voltage point*, V_t. The transition region is considered to be bounded by the *unity gain points*.

Noise margin, NM, is defined as the maximum voltage disturbance a gate input will tolerate before its output registers a change of state. It is measured from the operating point to the nearer unity gain point and there are therefore two values, the gate not necessarily having the same noise margin in both states. Worst case noise margins are often quoted to take account of the variations in

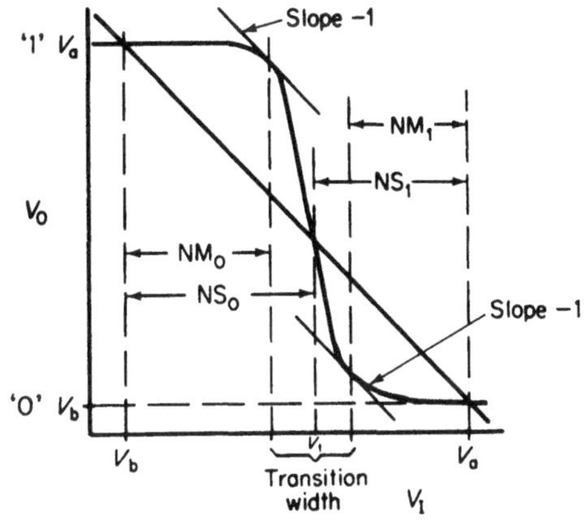

Figure 3.22 Logic gate transfer characteristic

logic levels within specific limits. These are found by subtracting the maximum value of V_{IH} of the driven gate from the minimum V_{OH} of the driver gate, and by subtracting the maximum V_{OL} of the driver gate from the minimum V_{IL} of the driven gate.

The *noise sensitivity*, NS, is similar to the noise margin but is defined in terms of the threshold voltage, V_t

$$NS_0 = V_t - V_b; \quad NS_1 = V_a - V_t$$

The third parameter provides a 'figure of merit', allowing comparisons between different logic families. This is the *noise immunity*, NI, which is useful when considering the effects of internally generated noise such as crosstalk, since it is dependent on the magnitude of the logic swing (see problem 3.2). Noise immunity can be defined as the ratio of noise sensitivity to logic voltage swing. Thus

$$NI = NS/(V_a - V_b)$$

whence

$$NI_0 = NS_0/(NS_0 + NS_1)$$

and

$$NI_1 = NS_1/(NS_0 + NS_1)$$

In many industrial digital systems it is necessary to take readings from sensors or to operate actuators which are situated in electrically noisy or dangerous areas. In order to prevent the injection of ground noise into the system solid-state relays or opto-isolators are used to provide isolation. The solid-state relay, SSR, performs all the functions of the conventional electromechanical relay but pro-

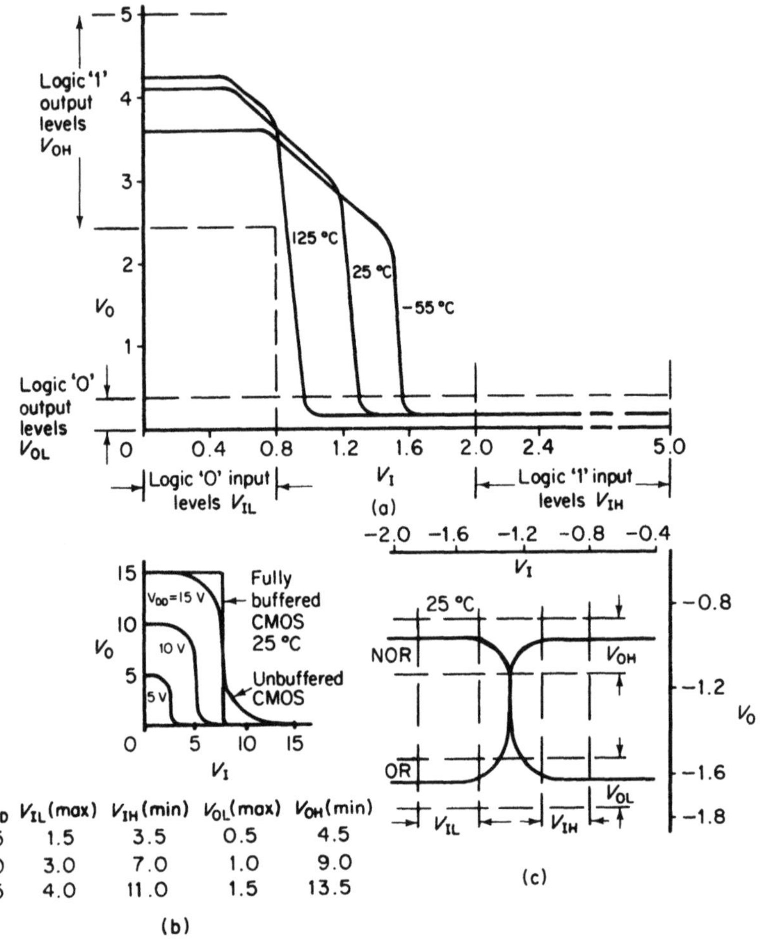

V_{DD}	V_{IL}(max)	V_{IH}(min)	V_{OL}(max)	V_{OH}(min)
5	1.5	3.5	0.5	4.5
10	3.0	7.0	1.0	9.0
15	4.0	11.0	1.5	13.5

Figure 3.23 Transfer characteristics: (a) TTL; (b) CMOS; (c) ECL

vides much greater reliability and, most importantly, a direct interface between the logic circuitry levels and the input or output line conditions, with complete electrical isolation at voltages up to, and in some cases in excess of, 1500 V. The basis of the solid-state relay is the opto-isolator in which the input photo-diode is coupled to the output photo-transistor only by means of the light passing from the diode to the transistor. Many versions of these devices are commercially available, including types with low input current requirements suitable for use with CMOS and other MOS circuits, and others with TTL-compatible inputs and outputs.

The slope of the transfer characteristic in the transition region indicates a very high gain and it is important, especially with TTL gates, to ensure that input signal transitions are rapid enough, otherwise instability may occur, giving oscillations at the output. A level change rate greater than about half a volt per nanosecond is recommended for standard TTL. Where slowly changing signals are inevitable, as from many input sensors, it is necessary to use gates incorporating Schmitt triggers on these inputs. The *Schmitt trigger* (Horowitz and Hill, 1980) uses an emitter-coupled bistable circuit which switches rapidly between states at defined input levels. These circuits exhibit a *hysteresis* effect in which the output changeover voltage as the input voltage rises is higher than the changeover voltage as the input voltage falls. The hysteresis voltage is typically 0.4 V for TTL gates, such as the 74LS14 hex inverter.

Several other parameters help to determine the overall performance of the different logic families. The main ones are switching speed or propagation delay, power dissipation and fan-out capability. These are all interrelated, and typical values for the popular logic families are tabulated in figure 3.24, though these will vary under different loading and operating conditions.

We have seen that any transistor has a propagation delay, and the performance of a logic circuit is determined not only by the speed of the transistors used, but also by the circuit conditions in which they operate. The circuit conditions also affect the power dissipation since increased switching speeds lead to higher dissipation. The product of power dissipation and propagation delay provides another convenient figure of merit which can be used in the comparison of different logic types. If the dissipation is measured in milliwatts and the delay in nanoseconds, the *power–delay product* is in picojoules. The dissipation of a gate differs between on and off states, with the on dissipation generally being the greater, and quoted dissipations are normally an average value assuming equal on and off periods in a cycle.

If the edge-speed of a waveform or logic signal is fast enough for the entire edge or a significant proportion of it to be present on the line at the same time, then the line must be considered as a transmission line and ground-plane techniques, or the use of *twisted pairs*, must be considered to avoid multiple signal reflections and ringing. A twisted pair consists of the signal wire and its return twisted tightly together. Signal lines such as these exhibit a characteristic impedance of approximately 120 ohms and provide efficient screening from electrical noise. Jarvis (1963) has formulated a useful expression for relating line length to signal rise time.

The use of transmission lines, with characteristic impedance in the order of 50–150 ohms, raises problems of matching. It is important to ensure that such interconnections are correctly terminated, otherwise reflections are produced which may lead to the spurious switching of logic gates. The input resistance of TTL and ECL gates is a few thousands of ohms, provided the input is not taken to a negative voltage with respect to the common line. This means that correctly terminated lines are not significantly unbalanced when gate inputs are attached. Unfortunately, the output resistance of the TTL gate is approximately 100 ohms when the gate is in the high state, thus to drive a transmission line without loss

Logic family	Supply voltage	Typical logic levels		Worst-case noise margin		Typical edge speeds		Typical propagation delay	Typical dissipation per gate	Power-delay product	Maximum clocking rate	Fanout
	Volts	Volts		Volts		ns		ns	mW	pJ	MHz	
		V_{OL}	V_{OH}	NM_0	NM_1	t_f	t_r	t_p				
CMOSB (3–18 V)	5	0	5	2	2	35	35	40	2*	80	8	50+
	10	0	10	4.5	4.5	20	20	30	7*	210	16	50+
	15	0	15	7	7	15	15	25	14*	350	20	50+
HCMOS (2–6V))	5	0.2	4.2	2	2	7	6	8	0.1	0.8	50	1000+ (10LSTTL)
TTL 74L	5	0.15	3.3	0.5	0.4	7	9	33	1	33	3	8
74	5	0.2	3.5	0.4	0.4	6	8	10	10	100	35	10
74H	5	0.2	3.5	0.4	0.4	5	7	6	22	132	50	10
74LS	5	0.25	3.4	0.3	0.7	6	7.5	9.5	2	19	45	10
74S	5	0.45	3.4	0.3	0.7	2	2.7	3	19	57	125	10
74ALS	5	0.35	3.4	0.4	0.7	5	5	4	1	4	50	10
74AS	5	0.43	3.2	0.4	0.6	1.8	2.3	1.5	22	33	100	10
ECL M10KH	-5.2	-1.63	-0.98	0.15	0.15	1	1	1	25	25	500	63
I^2L	1.5	0.9	1.3	0.15	0.2	1.5	1.5	15	0.25	3.75	100	3
ISL	1.5	0.9	1.5	0.2	0.2	1	1	3.5	0.25	0.9	100	4

*At 1 MHz frequency

CMOSB Buffered CMOS
74L Low-power TTL
74 Standard TTL
74H High-speed TTL
74LS Low-power Schottky TTL
74S Schottky TTL
74ALS Advanced Low-power Schottky TTL
74AS Advanced Schottky TTL

Figure 3.24 Comparison of logic families

of signal involves the use of a buffer stage. ECL gates, on the other hand, have an output resistance in the order of 25 ohms, which is sufficiently low to drive most transmission lines satisfactorily.

Standard transmission line theory (Taub and Schilling, 1977) shows that energy is reflected at the end of any line which is not correctly terminated in its characteristic impedance, Z_0. With any other impedance, Z_t, the reflection coefficient is given by $\rho = (Z_t - Z_0)/(Z_t + Z_0)$. Multiple reflections can occur, and the resulting voltages affect the signals we actually get on the line.

Example 3.1: A digital system is made up of seven printed circuit cards housed in a 19 inch (482 mm) rack. A certain bus line on the backplane connects to gates on every card but with negligible loading. The backplane line impedance is 120 ohms, and the driving gate on card 0 has an output impedance of 80 ohms. Sketch the waveform seen at card 1 when a positive-going voltage step, V, is generated by the driving gate.

In order to arrive at a reasonable expression we must assume that the first receiver card is very close to the driver, the backplane line acts as an open-circuit transmission line, and the loading effects of the receiver cards can be ignored. The problem then is essentially one of finding the voltage V_1, figure 3.25a, when an instantaneous voltage step, V, is applied and reflections occur from the open-circuit at the end of the line. Signals at the sending end of the line are attenuated by potential divider action, and these signals are reflected backwards and forwards along the line by an amount governed by the appropriate reflection coefficient and with a propagation time of τ seconds.

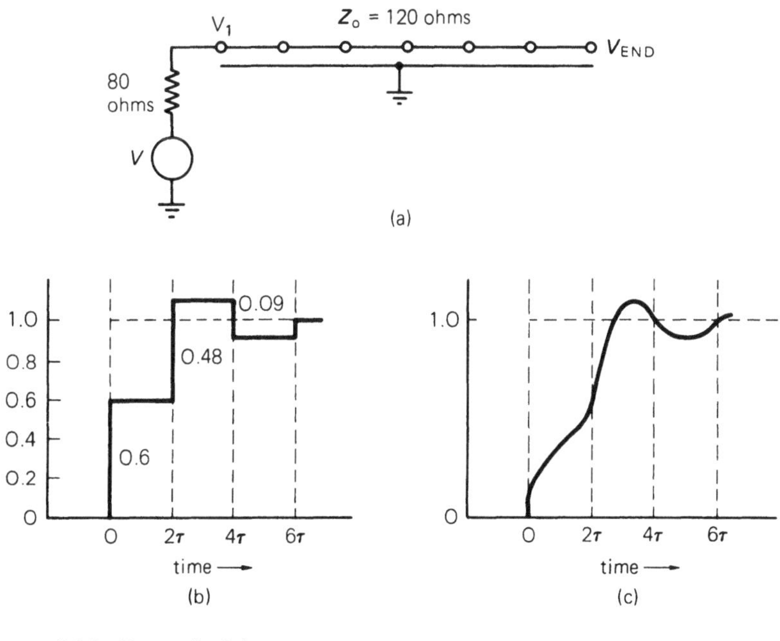

Figure 3.25 Example 3.1

The proportion of the initial drive voltage, V, reaching the line is $V' = V \times 120/(80 + 120) = 0.6V$.

The reflection coefficient at the end of the line is

$$\rho_{END} = (Z_0/c - 120)/(Z_0/c + 120)$$

$$\rightarrow +1 \qquad \text{(since } Z_0/c \rightarrow \infty\text{)}$$

The reflection coefficient when the signal arrives back at the driving end is

$$\rho_1 = (80 - 120)/(80 + 120) = -0.2$$

The received signal is then

$$V'' = (1 + \rho_1)V' = 0.8V'$$

and the retransmitted signal is V''.

Thus, at the first receiving card, the signal is the combination of applied and reflected voltages, giving a final level of

$$\begin{aligned} V_1 &= V' + V'' + (-0.2)V'' + (-0.2)V'' + (-0.2)V'' + \ldots \\ &= 0.6V + 0.48V(1 - 0.2 + 0.04 - 0.008 + \ldots) \end{aligned}$$

The time relationships and the waveform which would be seen in an 'ideal' arrangement are shown in figure 3.25b, but these assume a loss-free line and an infinitely fast voltage risetime which, in practical circuits, are not possible. Edge speeds are finite, and losses in the line lead to further degradation in edges to give the sort of waveform shown in figure 3.25c.

This method of analysis gives a good insight into the problems of interconnecting modern high-speed logic circuitry. It is important to note that, in the practical case, the voltage seen by card 1 after time 2, figure 3.25c, can remain for some time in the transition region of a gate input, dependent on the length of the line. Under these conditions the gate will be abnormally sensitive to external noise. Small perturbations will be greatly amplified and, if fed into counters or latches on the card, can cause erroneous operation. This type of dynamic fault is extremely difficult to detect and may often be the cause of seemingly random failure of equipment. These problems are discussed further in specialist texts such as Catt *et al.* (1979).

Because of the delays inherent in practical circuits all sections of a logic system do not react simultaneously to a change of input signal. The change must propagate through the circuitry. Where two signals control a single gate and they both change value as the result of a single variable changing, a *race* can occur dependent on different path lengths, and may lead to a transient change in the output value when no change should occur. The outcome of a race may be of no consequence in the overall action of the circuit, in which case the race is *non-critical*. A *critical* race, however, can lead to a malfunctioning of the circuit and the operation is then said to be *hazardous*. These hazards can be classified as *static* and *dynamic*; a third category, the *essential* hazard, occurs only in sequential circuits and will be dealt with later.

Figure 3.26a illustrates the type of circuit prone to static hazard. Gate 4 is controlled by the two signals from gates 2 and 3, but each of these gates is dependent on the value of the variable b. Suppose variables a and b are at 1 and c is at 0, then f is at 1 by virtue of $b\bar{c}$ being at 1. Now b changes to 0. Theoretically the function is still at 1, since $a\bar{b}$ is now at 1, but delays are present in a practical circuit and, until the change in the value of b is registered at the output of gate 1, gate 2 still registers ab while gate 3 registers $\bar{b}\bar{c}$. Thus the circuit gives a transient 0 output. This hazard is called a static 1 hazard. A static 0 hazard

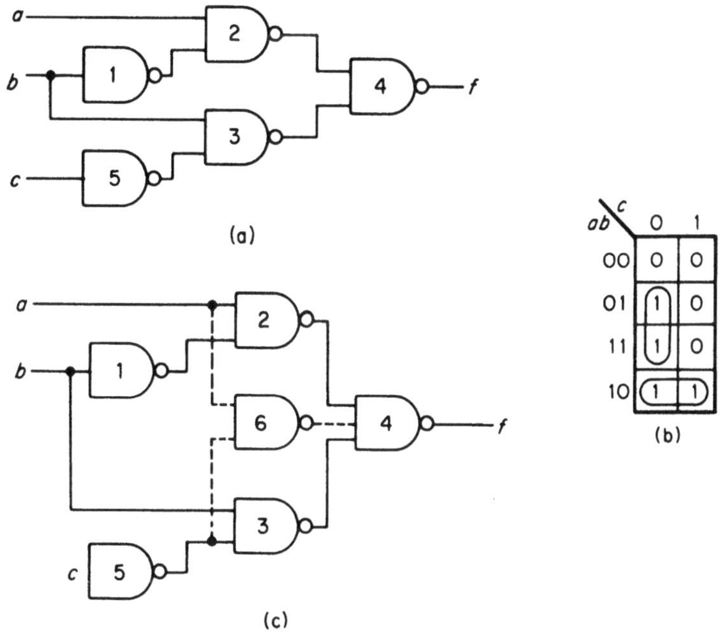

Figure 3.26 Example of static hazard: (a) $f = a\bar{b} + b\bar{c}$; (b) Karnaugh map; (c) hazard-free circuit

exists when a transient 1 output signal can occur in the transition between two adjacent 0 output conditions.

The presence of static hazards may be detected by inspection of the Karnaugh map of the output function, which for this circuit is given in figure 3.26b.

The prime implicants of the function, $a\bar{b}$ and $b\bar{c}$, are shown encircled and we see that a third prime implicant, $a\bar{c}$, is possible. The missing prime implicant contains the terms $a\bar{c}(b + \bar{b})$, which lead to the static 1 hazard in the circuit. Evidently the delays inherent in the practical circuit give a transient condition during which $(b + \bar{b})$ does not equal 1. To remove the hazard we must include all the prime implicants of the function, and the $a\bar{c}$ gate then generates a 'blanking' or 'holding' signal while the b variable change is transmitted through the various gates involved, as shown in figure 3.26c.

The result of a static hazard is a transient output spike, or *glitch*, giving a double change in level when no change is anticipated. A dynamic hazard is similar but gives three changes in output level when the output signal is meant to change level only once. The conditions necessary for the dynamic hazard to develop reduce to the presence in the circuit of three differing delay paths for a single variable, with an inversion in at least one path but not all three. Although additional delays can be introduced to remove the hazard, it is preferable to rearrange the circuit if possible.

All circuits within a particular logic family are designed to interface directly with any other circuit of the same family, and most nMOS circuits are buffered to drive one standard TTL load. However, a mixture of logic types is often justified in order to improve speed performance in a critical section of a system, or to lower the cost, or to reduce the power dissipation, and so on, and care must be exercised in connecting between gate types and to external circuitry. TTL will drive CMOS directly, but it is necessary to include a pull-up resistor to ensure an adequate high level, V_{OH}. When driving ECL a low impedance resistive network is needed to attenuate the TTL voltage swing and to shift the levels appropriately. CMOS can drive low-power TTL directly but requires a CMOS driver circuit to cope with the higher current levels of other TTL gates. CMOS can also drive ECL directly if the power supplies are arranged correctly, since ECL uses −5.2 V. It is possible to operate the CMOS with V_{DD} at 0 V and V_{SS} at −5.2 V, and a clamping diode should be used at the ECL input to prevent undershoot. ECL requires level shifting circuits in order to drive CMOS or TTL; when driving CMOS an active level shifting circuit may be used with the single −5.2 V supply or, alternatively, level shifting circuits are available allowing the CMOS to operate with a +5 V supply. When driving TTL, a voltage amplifier is necessary to increase the logic swing from 0.8 V to at least 2.5 V.

Problems

3.1 A logic gate has the following characteristic:

V_{in}	V_o	V_{in}	V_o	V_{in}	V_o
0 V	5 V	2.5 V	4.4 V	4.0 V	0.8 V
0.5 V	5 V	2.75 V	3.9 V	4.25 V	0.6 V
1.0 V	5 V	3.0 V	3.0 V	4.5 V	0.55 V
1.5 V	5 V	3.25 V	2.15 V	5.0 V	0.5 V
2.0 V	4.95 V	3.5 V	1.5 V	5.5 V	0.5 V
2.25 V	4.75 V	3.75 V	1.05 V	6.0 V	0.5 V

Calculate: (i) The worst-case noise margin, (ii) the noise immunity figures.

3.2 The ability of a logic system to reject internally generated noise, such as crosstalk arising from logic signal changes, can be quantified in terms of the noise immunity. Using the values shown in figures 3.23 and 3.24, compare and contrast the abilities of CMOSB, 74 series TTL and ECL families to reject both internal and external noise.

3.3 Show how to implement (a) OR, (b) AND, using (i) only NAND gates, (ii) only NOR gates.

3.4 Show that rearranging the circuit of figure 3.6 to its dual form gives an OR gate to positive logic.

3.5 Explain how open-collector TTL gates can be connected to generate the function $A \oplus B$ using distributed logic.

3.6 Demonstrate that a four-NAND gate circuit generating the function $f = AC + B\overline{C}$ includes a static 1 hazard and show how to remove that hazard.

References

Babbage, H. P. (1961). *Charles Babbage and his Calculating Engines*, originally published 1889, republished by Dover Publications, New York

Bennetts, R. G. (1984). *Design of Testable Logic Circuits*, Addison-Wesley, Reading, Massachusetts

Catt, I., Walton, D. and Davidson, M. (1979). *Digital Hardware Design*, Macmillan, London

Fitzgerald, A. E., Higginbotham, D. E. and Grabel, A. (1981). *Basic Electrical Engineering*, McGraw-Hill, New York

Horowitz, P. and Hill, W. (1980). *The Art of Electronics*, Cambridge University Press

Jarvis, D. B. (1963). The effects of interconnections on high-speed logic circuits, *IEEE Trans. Electronic Computers*, **EC12**, No. 5, 476–87

Matthews, P. L. (1983). *Choosing and Using ECL*, Granada Publishing, London

Mead, C. and Conway, L. (1980). *Introduction to VLSI Systems*, Addison-Wesley, Reading, Massachusetts

Metropolis, N. (Ed.) (1980). *A History of Computing in the Twentieth Century*, Academic Press, New York

Millman, J. (1979). *Microelectronics: Digital and Analog Circuits and Systems*, McGraw-Hill, New York

Mukherjee, A. (1986). *Introduction to nMOS and CMOS VLSI Systems Design*, Prentice-Hall, Englewood Cliffs, New Jersey

Pasa, G. and Beresford, R. (1981). *Electronics: Technology Update*, **54**, No. 21 (October), 114–241

Schilling, D. L. and Belove, C. (1979). *Electronic Circuits, Discrete and Integrated*, 2nd edn, McGraw-Hill, New York

Taub, H. and Schilling, D. L. (1977). *Digital Integrated Electronics*, McGraw-Hill, New York

4 Logic Design Techniques

We have seen previously how logical functions, developed from the basic equations or problem definitions, can be implemented using NAND/NOR elements and techniques for reducing the number of gates demanded by a function were also described. The introduction of devices such as the multiplexer and the read-only memory over the last decade has produced a change of emphasis in logic circuit synthesis. These complex LSI circuits make possible the realization of logic functions directly from their primitive canonical form, whilst still keeping the package count down to a minimum. Techniques for using such devices will be introduced in this chapter. However, the designer may be faced with the need to realize a logic function which he knows can be synthesized using only a few NAND/NOR gates, perhaps those contained in only one or two packages. In such cases LSI may not be the best approach and other techniques are more appropriate.

4.1 Signal assertion and the wired-OR

In the previous chapter the concept of positive and negative logic was introduced. We now show that the use of negative logic can lead to a particularly important facility, known as the *wired-OR* network.

Consider the circuit of figure 4.1. Taken separately, each gate operates as a

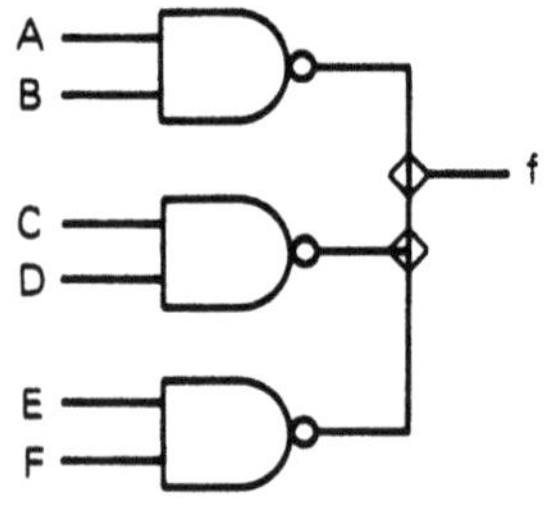

Figure 4.1 The wired-OR circuit

NAND to positive logic. When the units are connected as shown, the outputs cannot act independently. A useful catch-phrase to remember is 'ground takes control': any one gate output switching to ground controls the other gates, holding all outputs at ground. Thus a one output is only possible when all gate outputs are switched to the positive level which only occurs when A or B (or both) and C or D (or both) and E or F (or both) are at ground. In algebraic terms, we have

$$f = \overline{AB}\,\overline{CD}\,\overline{EF}$$

therefore

$$\overline{f} = AB + CD + EF \text{ by duality}$$

whence

$$f = \overline{AB + CD + EF}$$

This *distributed logic* is referred to as wired-OR or phantom-OR. Because of the usefulness of wired-OR, techniques have been developed which allow NAND circuits to be realized in a minimal form using this facility. Evans (1969) has shown that by mapping a logical function on to an *inverse map*, it is possible to derive an expression suitable for synthesis into the wired-OR configuration.

The function

$$f = AC\overline{D} + BD + AB$$

can be expressed in maxterm form as

$$f(ABCD) = \Pi M(4, 6, 7, 9, 11, 12, 13, 14, 15)$$

If this expression is mapped the inverse map of figure 4.2 results. The function so represented can now be minimized in the usual manner by grouping. It can be seen that a minimum of three loops is required to cover all the 1 cells.

Since this map represents the function in P of S form, the groups yield the minimal equation

$$f = (A + D)(B + C)(B + \overline{D})$$

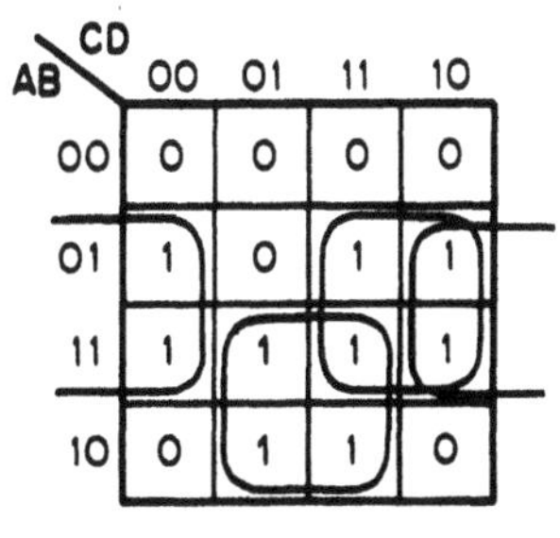

Figure 4.2 Inverse map of $f = AC\overline{D} + BD + AB$

By duality

$$\bar{f} = \bar{A}\bar{D} + \bar{B}\bar{C} + \bar{B}D$$

thus

$$f = \overline{\bar{A}\bar{D} + \bar{B}\bar{C} + \bar{B}D}$$

This form of the expression is obtained from a wired-OR circuit of the type shown in figure 4.1. Since logic is also performed at the outputs the wired-OR circuit, when minimized by a technique such as this, represents an absolute minimal network.

Example 4.1: Find a minimal NAND circuit for the function $f = \bar{C}\bar{D} +$ ABD + BCD using the wired-OR facility.

The minterm expression for the function is

$$f(ABCD) = \Sigma m(0, 4, 7, 8, 12, 13, 15)$$

giving the maxterm form

$$f(ABCD) = \Pi M(1, 4, 5, 6, 9, 10, 12, 13, 14)$$

The maximal groupings from the inverse map of the function, figure 4.3a, give

$$f = (\bar{C} + D)(B + \bar{D})(A + C + \bar{D})$$

whence, by duality

$$\bar{f} = C\bar{D} + \bar{B}D + \bar{A}\bar{C}D$$

and

$$f = \overline{C\bar{D} + \bar{B}D + \bar{A}\bar{C}D}$$

The circuit of figure 4.3b follows directly.

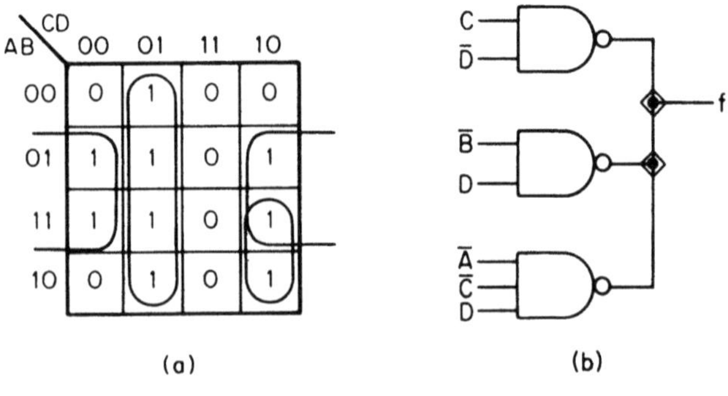

Figure 4.3 (a) Inverse map of $f = \bar{C}\bar{D}$ + ABD + BCD. (b) Minimal NAND circuit

Wired-OR logic is used in integrated injection logic, I^2L, where each gate has a single input but multiple outputs. Any required function is built up by connecting the appropriate gate outputs together.

Example 4.2: Design a full-subtractor circuit using integrated injection logic.

The full subtractor has two inputs, X and Y, and a borrow, B. The outputs must be the difference, D = X – (Y + B) and the new borrow, if any. The truth table is shown in figure 4.4a and from it we obtain the expressions

$$\text{Difference, } D = \overline{X}\,\overline{Y}B + \overline{X}Y\overline{B} + X\overline{Y}\,\overline{B} + XYB$$

and

$$\text{Borrow, } B' = \overline{X}\,\overline{Y}B + \overline{X}Y\overline{B} + \overline{X}YB + XYB$$

Rearranging the equations to a form suitable for wired-OR implementation gives

$$\overline{D} = \overline{(\overline{X}\,\overline{Y}B)}\;\overline{(\overline{X}Y\overline{B})}\;\overline{(X\overline{Y}\,\overline{B})}\;\overline{(XYB)}$$

and

$$\overline{B}' = \overline{(\overline{X}\,\overline{Y}B)}\;\overline{(\overline{X}Y\overline{B})}\;\overline{(\overline{X}YB)}\;\overline{(XYB)}$$

leading to the circuit shown in figure 4.4b.

An important limitation of TTL and CMOS gates is that the wired-OR facility cannot be obtained with the standard circuits because of their low output

X	Y	B	Difference	Borrow
0	0	0	0	0
0	0	1	1	1
0	1	0	1	1
0	1	1	0	1
1	0	0	1	0
1	0	1	0	0
1	1	0	0	0
1	1	1	1	1

(a)

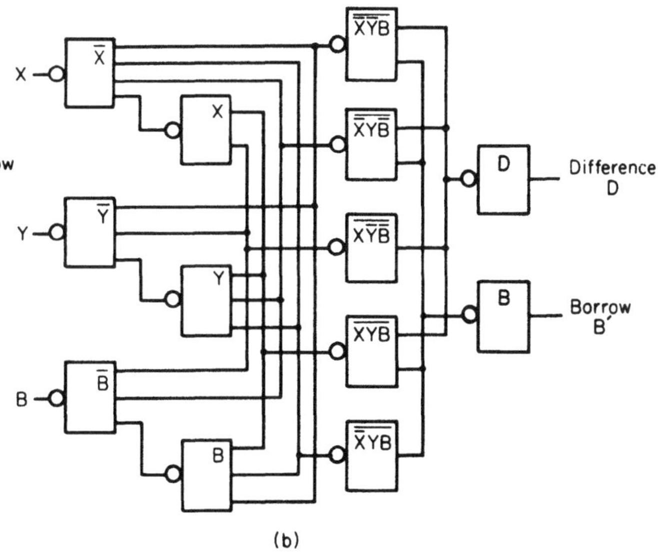

(b)

Figure 4.4 The full subtractor: (a) truth table; (b) circuit using I^2L gates

impedance in the high state. To overcome this problem TTL gates are available with the upper transistor of the totem-pole circuit omitted. The output stage is a single open-collector transistor, allowing several gate outputs to be connected in parallel. An external pull-up load resistor must be included with a consequent reduction in edge-speed. TTL gates are also available with a third transistor associated with the totem-pole circuit. This transistor acts as a disabling device by holding both of the totem-pole transistors off, so presenting a high impedance at the output. In this, the *third state*, the output is said to be *floating* and is effectively disconnected from the external circuitry. A similar modification of CMOS gates is available and these *three-state* or *Tri-state** gates are used extensively for the transmission of data from any of several sources on to a common data highway or bus.

The symbol of a three-state gate includes the controlling input and positive or negative assertion is indicated by the control signal name as well as the symbol. The symbols of figure 4.5, for example, all describe a three-state inverter which is enabled when the control signal is taken low. The controlling input is shown at the top edge of the gate symbol to emphasize that it is not part of the main logic of the gate.

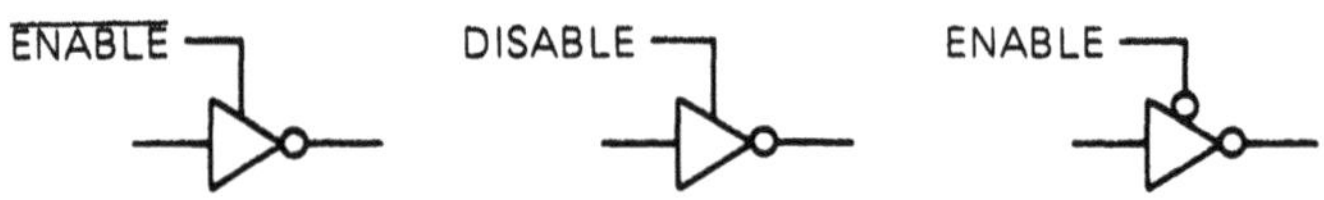

Figure 4.5 The three-state inverter

Unlike open-collector gates, the three-state gate is not restricted to negative assertion of a signal on the bus; either polarity can be asserted by the driving gate. Figure 4.6 shows three wired-OR gates connected to a common line or data bus. Any one of the gates can communicate with the receiver provided that the other two are held in the disabled state.

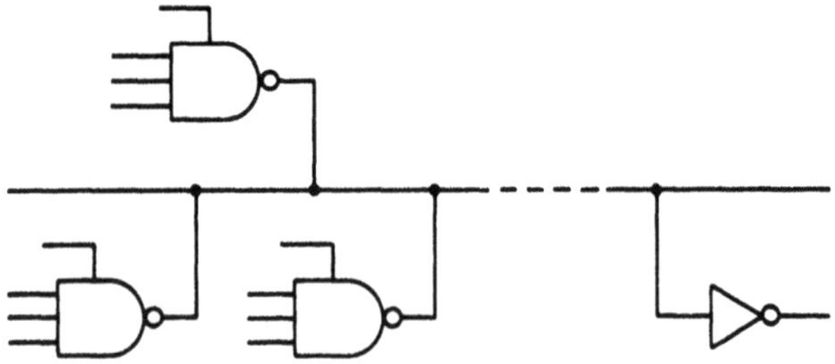

Figure 4.6 The use of three-state gates as data bus drivers

*'Tri-state' is a registered trademark of National Semiconductor Corporation.

4.2 Larger scale integration

The diode matrix is a variation of the simple AND/OR gate and can be used in encoding and decoding operations. For example, suppose we wished to generate the decimal output signals corresponding to the 2421 BCD. The code is shown in figure 4.7 along with the diode decoding matrix. It is necessary to have both phases of each input signal available, and inverters are shown in this particular example. The circuit considered is an AND gate decoding matrix since each decimal output is obtained from a four-input AND gate. As one might expect, in encoding, an OR gate matrix is required. Figure 4.8 illustrates by way of an example the encoding of decimal to excess-three code.

	2	4	2	1
Decimal	A	B	C	D
0	0	0	0	0
1	0	0	0	1
2	0	0	1	0
3	0	0	1	1
4	0	1	0	0
5	1	0	1	1
6	1	1	0	0
7	1	1	0	1
8	1	1	1	0
9	1	1	1	1

(a)

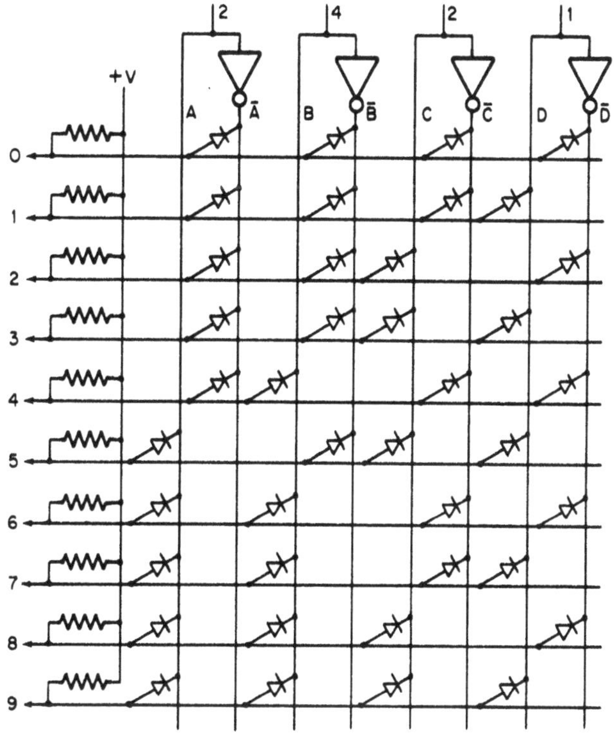

(b)

Figure 4.7 A 2421 BCD-to-decimal converter: (a) 2421 code; (b) diode matrix

The diode array described is an example of a *Read-Only Memory* (ROM). Using modern integrated circuit technology very large diode or transistor arrays can be produced in which the interconnection, or disconnection, of the nodal points can be made either by the circuit designer (or programmer) or, if many such identical matrices are required, by the manufacturer during the fabrication

Decimal	A	B	C	D
0	0	0	1	1
1	0	1	0	0
2	0	1	0	1
3	0	1	1	0
4	0	1	1	1
5	1	0	0	0
6	1	0	0	1
7	1	0	1	0
8	1	0	1	1
9	1	1	0	0

(a)

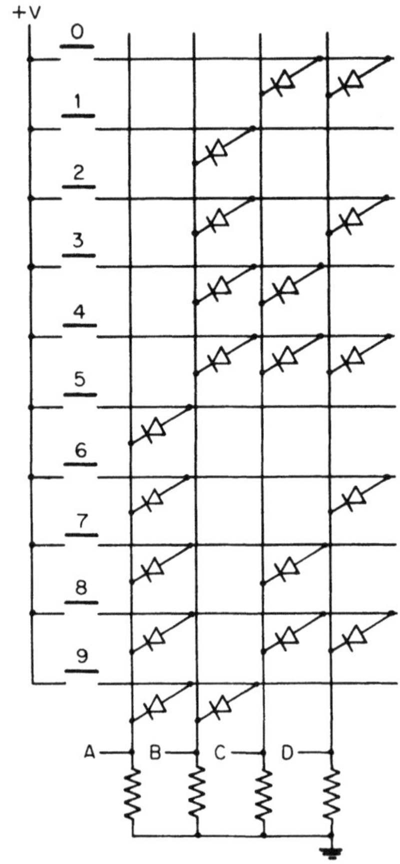

(b)

Figure 4.8 A decimal to excess-three code converter: (a) excess-three code; (b) diode matrix

process. These devices are discussed in detail in chapter 7 and suffice it to say at this stage that the read-only memory is an array which has a number of address-variables and a number of output-data lines.

Figure 4.9a illustrates such a device, in which a binary number placed on the address lines is decoded internally to select one unique row in the matrix. This row is connected via diodes or transistors to the external data lines, and the presence or absence of a diode connection in the selected row to the data lines determines the pattern of ones and zeros presented at the output. Read-only memories are available in various configurations as TTL-compatible, dual-in-line packaged devices. Typical of these, suitable for logic circuit synthesis, are the devices listed in the table of figure 4.9b. All are user-programmable, meaning

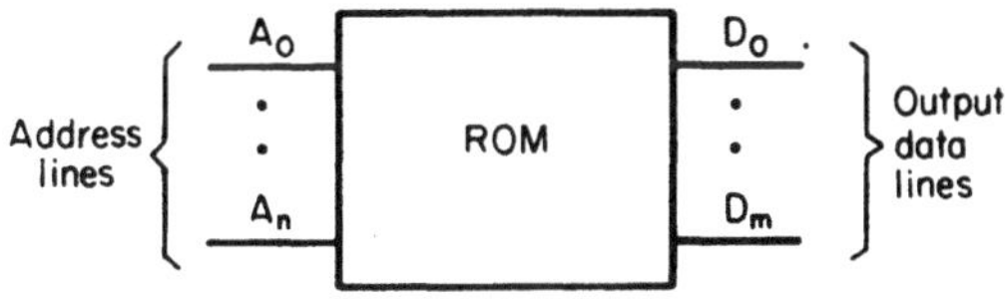

(a)

Type	Memory size	Format	Manufacturer
DM74S183	256 bits	32 words of 8 bits; open-collector	NSC
TBP24S10	1024 bits	256 words of 4 bits; 3-state outputs	TI
N825123	256 bits	32 words of 8 bits; 3-state outputs	Signetics

(b)

Figure 4.9 (a) The read-only memory. (b) Common ROM types

that the data pattern presented at the output in response to a particular address code is determined during a programming operation carried out by the user.

The use of read-only memories renders many logic design problems trivial. Suppose, for example, we wish to synthesize the functions

$$f_1 = A\overline{B}\overline{C} + \overline{A}B\overline{C} + \overline{A}\overline{B}C + ABC$$

and

$$f_2 = \overline{A}BC + A\overline{B}C + AB\overline{C} + ABC$$

which are, of course, the equations for a full-adder. The truth table is shown in figure 4.10a. If we consider the variables A, B, C as forming an address word, then for any such word there exists a required data output word, f_1f_2. This can readily be implemented in a ROM, see figure 4.10b. Admittedly the use of a standard 32-word, 8-bit device does seem to be excessive as only eight of the 32 words available are required and only two of the eight data output lines are used. However, the overall cost of a system depends not on the complexity of

Address			Data	
A	B	C	f_1	f_2
0	0	0	0	0
0	0	1	1	0
0	1	0	1	0
0	1	1	0	1
1	0	0	1	0
1	0	1	0	1
1	1	0	0	1
1	1	1	1	1

(a)

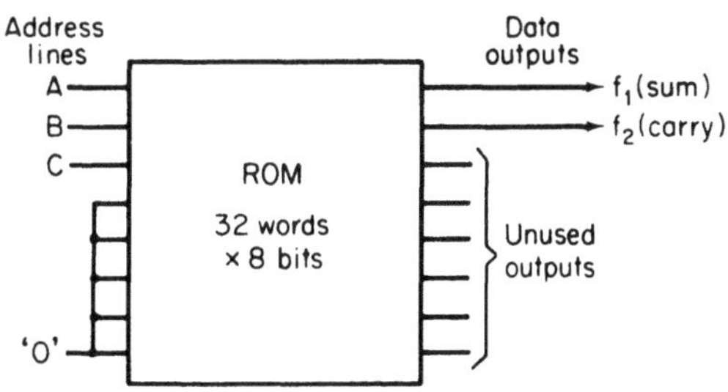

(b)

Figure 4.10 The full-adder: (a) truth table; (b) circuit using a ROM

the devices used, but more on the number of packages or, in other words, the pin count. Only one dual-in-line package is required with this solution.

The use of ROMs is particularly advantageous when dealing with multiple-output problems where several independent outputs may be required. As an example, the circuit of figure 4.10 could also provide a 'borrow' output, thus turning the ROM into a full-adder/full-subtractor circuit. (Remember that with the subtractor the 'difference' is identical to the adder 'sum'.) When implementing large functions, of five or more variables, it may be possible to utilize two smaller ROMs more efficiently than a single larger one. This is carried out by a judicious grouping of the input variables. At present there is no formal procedure, an heuristic approach being required. For example, consider the function

$$f(ABCDE) = \Sigma m(1, 8, 9, 10, 14, 15, 16, 17, 27, 31)$$

A straightforward implementation can be achieved by means of a 32-word ROM. However, listing the minterms in their binary form, it can be seen that the variables BCD appear only in four combinations, 000, 100, 101, 111, which we can code as combinations of two variables, X and Y. These variables can conveniently be generated in a small 8-word ROM and used, together with A and E, to provide the 4-bit address for a 16-word ROM which generates the required function, f. Figure 4.11 shows the process.

However carefully the coding is done, the use of a ROM to implement a logic function represents a considerable 'overkill'. In terms of cost-effectiveness this may not matter, but a 32-word ROM, for example, can encode any five-variable

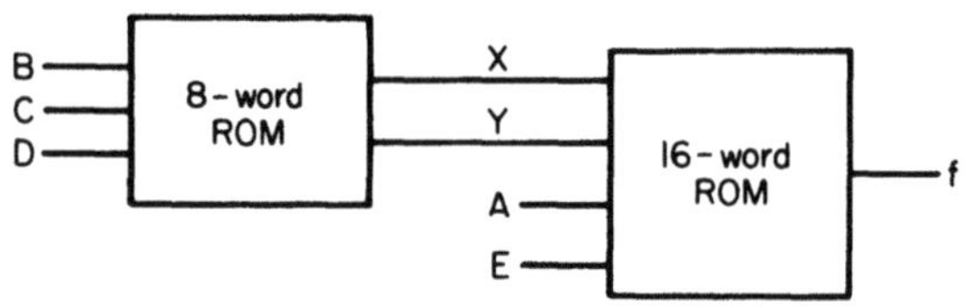

Decimal	A B C D E	A E	B C D
1	0 0 0 0 1	0 1	0 0 0
8	0 1 0 0 0	0 0	1 0 0
9	0 1 0 0 1	0 1	1 0 0
10	0 1 0 1 0	0 0	1 0 1
14	0 1 1 1 0	0 0	1 1 1
15	0 1 1 1 1	0 1	1 1 1
16	1 0 0 0 0	1 0	0 0 0
17	1 0 0 0 1	1 1	0 0 0
27	1 1 0 1 1	1 1	1 0 1
31	1 1 1 1 1	1 1	1 1 1

Rearranging the terms →

Decimal	A E	B C D	XY
1	0 1	0 0 0	00
16	1 0	0 0 0	00
17	1 1	0 0 0	00
8	0 0	1 0 0	01
9	0 1	1 0 0	01
10	0 0	1 0 1	10
27	1 1	1 0 1	10
14	0 0	1 1 1	11
15	0 1	1 1 1	11
31	1 1	1 1 1	11

Figure 4.11 The use of two small ROMs to decode a function in five variables

function and, with a standard 8-bit word ROM, provide up to eight different outputs. This capability is very rarely needed and certainly all the 32 input combinations, or addresses, would never be required to be logically true. If they were, the function would be better generated by a piece of wire! Considering the example given in figure 4.11 the 8-word ROM output is logically true for only four of its eight addresses. The 16-word ROM is more densely populated with ones, the output being logically true for ten of its sixteen addresses. Had we used a single 32-word ROM, only ten of its possible 32 input addresses would have set the output logically true. In the next section we will consider a circuit element, a natural development of the ROM, which, by embodying the ability not only to encode the required output function but also to specify the word address, offers a high bit density and hence in many cases a more attractive solution.

4.3 The programmable logic array

The *Programmable Logic Array*, PLA, is a flexible logic array which can be used to synthesize directly a sum of products logical expression. The device consists of a logic AND-OR array with user-programmable or user-definable interconnections. If the former, it is referred to as a field programmable logic array, FPLA. If the latter, then the user supplies a listing of the interconnections from which the manufacturer prepares the mask for the final metallization stages during manufacture. In general form, shown in figure 4.12, the PLA generates the logical function

$$f = P_0 + P_1 + P_2 \ldots P_n \quad \text{where } P_0 \text{ to } P_n \text{ are canonical terms}$$

The interconnections in both the product matrix and the sum matrix are programmable. As an example, consider a bipolar FPLA such as the Signetics

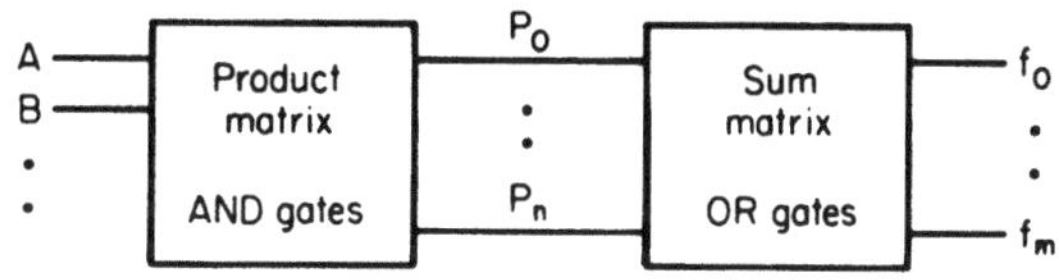

Figure 4.12 The programmable logic array

or Motorola 82100 (figure 4.13). This is a 16-input, 8-output device, providing up to 48 product terms. It is generally referred to as a 16 × 48 × 8 FPLA. The AND matrix is constructed from Schottky diodes, the OR matrix from emitter-follower transistors and the outputs are three-state circuits. The PLA is programmed by selectively blowing the nichrome fuse interconnections from the diodes or emitters. Carefully controlled voltages are applied to the outputs,

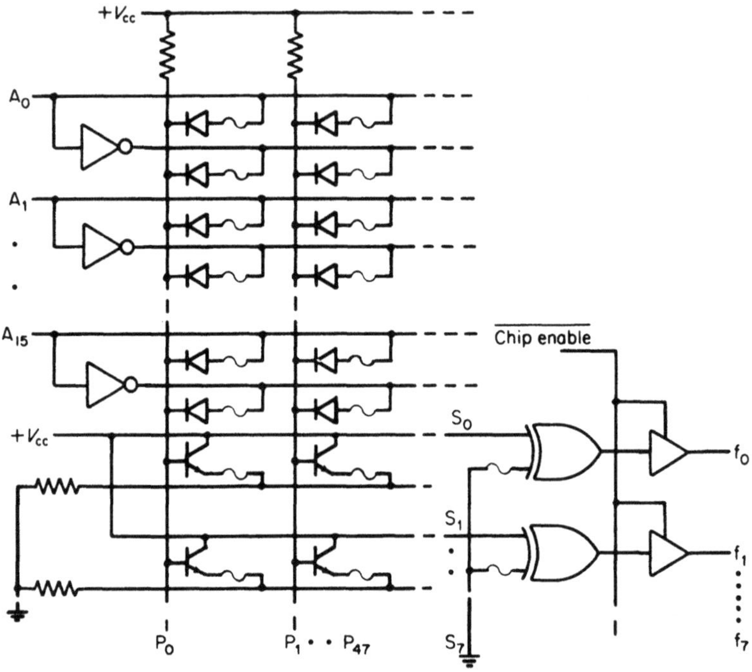

Figure 4.13 Logic diagram for the Signetics 82S100 FPLA

forcing accurately defined currents into the device during the programming stages. Note that inversion of the output is a choice available to the user and is achieved by blowing the fuse ground connection on the output exclusive-OR gate, effectively substituting logical one for logical zero at this input.

Let us consider the use of the PLA in generating the four-bit product of two 2-bit numbers. The truth table for this is shown in figure 4.14. Using standard minimizing techniques we arrive at the following expressions

$$W = ABCD$$
$$X = A\bar{B}C + AC\bar{D}$$
$$Y = A\bar{C}D + A\bar{B}C + \bar{A}BC + BC\bar{D}$$
$$Z = BD$$

For this problem a PLA as large as the one shown in figure 4.13 is not necessary. A 4-input, 4-output device with eight product terms would be more suitable. Such a device is shown diagrammatically in figure 4.15. The two matrices are assumed to be programmed, so that only where a dot is shown are the diodes and transistors wired into the matrix.

As in the case of synthesis with the ROM, minimization is not very important. Unless the number of product terms can be reduced sufficiently to enable a

A B	C D	W X Y Z
0 0	0 0	0 0 0 0
0 0	0 1	0 0 0 0
0 0	1 0	0 0 0 0
0 0	1 1	0 0 0 0
0 1	0 0	0 0 0 0
0 1	0 1	0 0 0 1
0 1	1 0	0 0 1 0
0 1	1 1	0 0 1 1
1 0	0 0	0 0 0 0
1 0	0 1	0 0 1 0
1 0	1 0	0 1 0 0
1 0	1 1	0 1 1 0
1 1	0 0	0 0 0 0
1 1	0 1	0 0 1 1
1 1	1 0	0 1 1 0
1 1	1 1	1 0 0 1

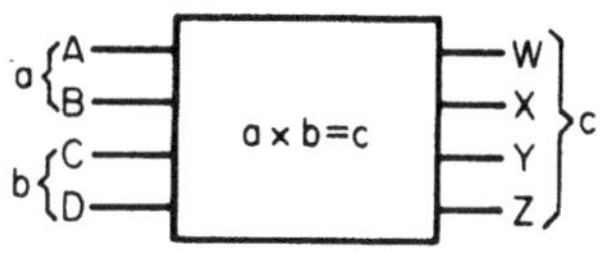

Figure 4.14 Truth table for the product WXYZ = AB × CD

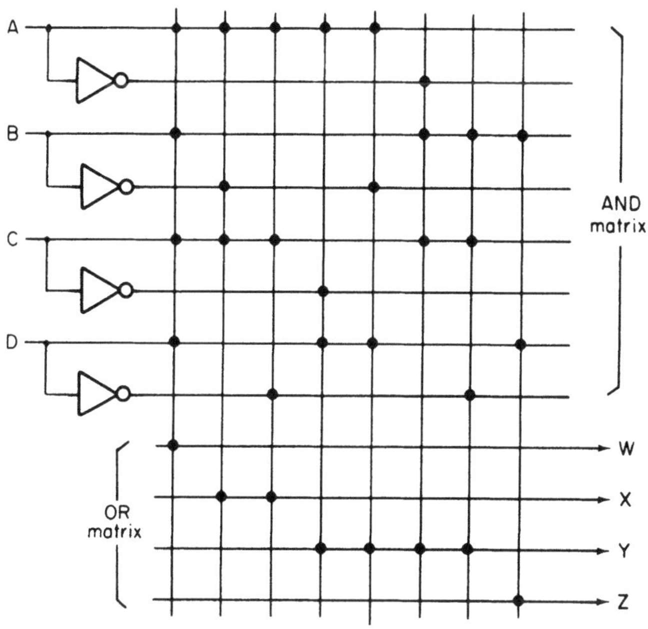

Figure 4.15 The programmed PLA

smaller, standard size PLA to be used there is nothing to be gained. If, however, the number of terms in the function does exceed the number of product terms provided in the chosen PLA, the first step should be to ensure that there is no duplication within the logic expressions. After that the possibility of combining several terms into a smaller term should be considered.

Example 4.3: Synthesize the functions

$$f_1 = ABC + A\overline{B}D + AC\overline{D}$$
$$f_2 = \overline{A}BC + AC\overline{D} + A\overline{B}CD$$
$$f_3 = A\overline{B}\overline{C}D + BCD$$

At first sight, eight product terms are required. However, it can be seen that both f_1 and f_2 contain the common term $AC\overline{D}$. Also, the terms $A\overline{B}CD$ and $A\overline{B}\overline{C}D$, which are required by f_2 and f_3 respectively, when ORed together provide the term $A\overline{B}D$, which is required by f_1. Thus the functions could be implemented using only six product terms, as in figure 4.16.

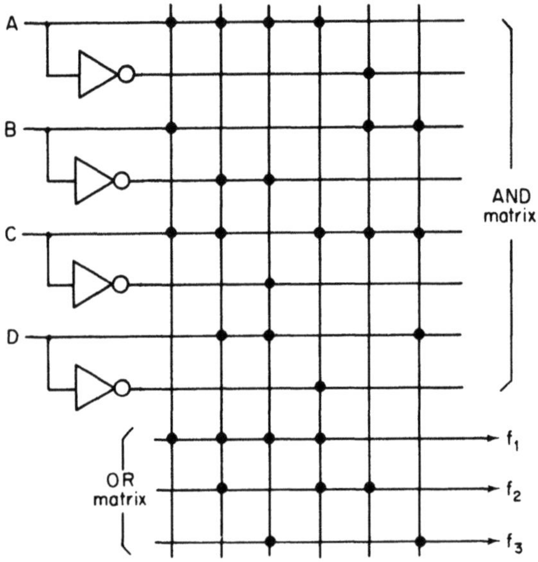

Figure 4.16 Combined terms reduce the size of PLA required

Another feature of some programmable logic arrays is the provision of a set of flipflops at the output of the sum array, to act as a data register or for use in providing internal feedback when generating state sequences. These devices are known as *field programmable logic sequencers*, FPLS, and again are available in different sizes and arrangements, the smallest catering for up to sixteen input variables and with six flipflops.

The programmable logic array is the most flexible form of device for implementing logic functions, since both the product matrix and the sum matrix are programmable. However, this means that they are relatively expensive devices and need special programming equipment. A slightly more limited variant has therefore become popular and is known as *programmable array logic*, PAL. In this arrangement fusible-link programming of the product matrix is retained, but the sum matrix is non-programmable. A range of devices is manufactured, giving the necessary choice of connections and gating in the sum matrix, and, as with the PLA, some are provided with sets of flipflops.

4.4 Multiplexers in design

A particularly useful example of general-purpose larger scale integration is the *multiplexer*. The device is available in several forms with more or fewer inputs, but one of the most popular is the four-line-to-one-line multiplexer, shown in figure 4.17. Any one of the four input lines, C0 to C3, can be connected to the output by the correct choice of code on the select inputs, A and B. Thus the value on input data line C0 is routed to the output if A = 0, B = 0; data line C1 is selected if A = 1, B = 0, etc. The additional control input, $\overline{\text{ENABLE}}$, allows the selected one of several multiplexers to be enabled as and when required. The negated signal, $\overline{\text{ENABLE}}$, indicates that, in this case, the enable input must be taken low to enable the circuit. Most large scale integrated circuits have at least one enable input, commonly called chip enable, CE, or device enable, DE, or device select, DS. Where more than one enable is provided they are logically interrelated so that, for example, all the enable inputs must be positive before the circuit is enabled.

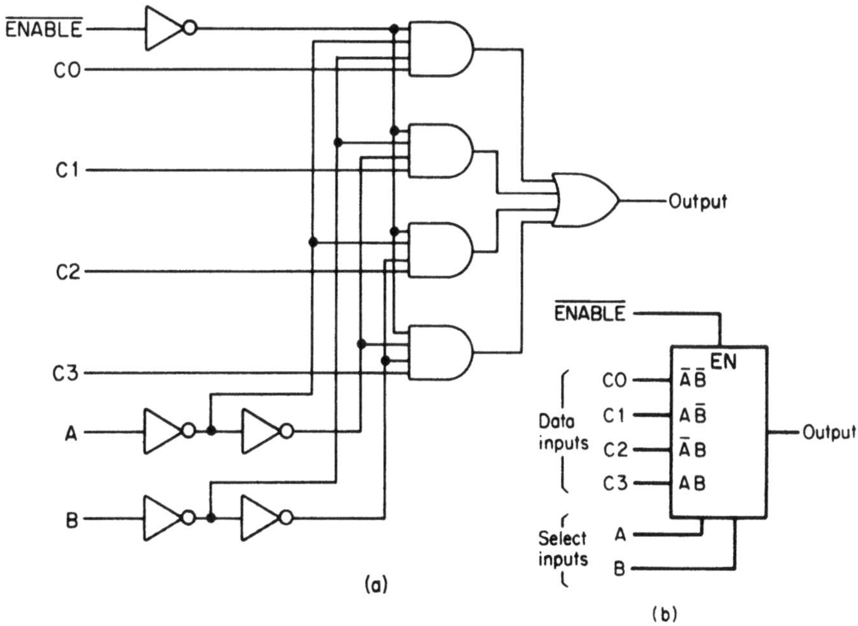

Figure 4.17 Four-line-to-one-line multiplexer: (a) logic circuit; (b) symbol

It is possible to use a multiplexer for the direct realization of a logical function expressed in the sum of products form. Consider the multiplexer shown in figure 4.17. Apart from the $\overline{\text{ENABLE}}$ signal, the output function

$$f = (\bar{A}\bar{B}C0 + A\bar{B}C1 + \bar{A}BC2 + ABC3)$$

is essentially a sum of products expression. Consider now the function

$$f = \bar{A}C + BC + \bar{A}B$$

By expansion of each term to the canonical form and omitting duplicated terms, we obtain

$$\begin{aligned} f &= \bar{A}\bar{B}C + \bar{A}BC + ABC + \bar{A}B\bar{C} \\ &= \bar{A}\bar{B}C + \bar{A}B(C + \bar{C}) + ABC \end{aligned}$$

Re-emphasizing the role of the different variables, we can say

$$f = \bar{A}\bar{B}(C) + \bar{A}B(1) + AB(C) + A\bar{B}(0)$$

since $(C + \bar{C}) = 1$ and $A\bar{B}(0) = 0$.

This function can now be realized at the multiplexer output if the input selected by $\bar{A}\bar{B}$ has variable C connected to it; the input selected by $A\bar{B}$ has logical zero connected; that selected by $\bar{A}B$ has logical one connected; and the final input, selected by AB, has variable C connected. This is shown in figure 4.18 together with the circuit required when using separate NAND gates.

Note that the conventional method necessitates at least two gate packages since 74 series packages are commonly available with either four 2-input gates or three 3-input gates. Furthermore, two four-line-to-one-line multiplexers are usually packaged together (Texas Instruments SN74LS153 for example). Though the cost of the more complex multiplexer package is higher than that of the simpler gates, package cost is often less important than package count since reliability, and hence overall cost, is more a function of the number of external connections that have to be made.

In the example considered, variables A and B were chosen to operate on the select inputs of the multiplexer and are, therefore, known as the *select* variables. The variable C operated on the data inputs and is therefore known as the *data* variable. The variables could equally well have been allocated differently, and in many cases one allocation gives a better circuit than alternatives. An optimum solution can be defined as that which leads to minimum loading of the data input source. In other words, more of the multiplexer data inputs are set to logical one or zero than for any other solution.

For a limited number of variables, the Karnaugh map is a convenient aid in determining which variable allocation is the best. Consider for example the function

$$f = \bar{A}B + BC + \bar{B}\bar{C}$$

Karnaugh map representations could be as shown in figure 4.19. The maps all

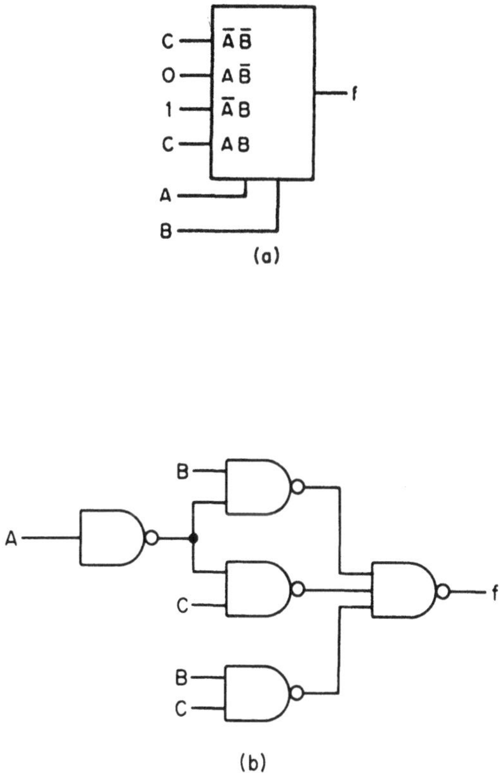

Figure 4.18 Realization of the function $f = \bar{A}C + BC + \bar{A}B$: (a) by use of a multiplexer; (b) by use of NAND gates

represent the same function but show differing patterns because the spatial relationship of the variables has been altered. This enables us to relate the variables to specific functions within the multiplexer.

In the case of figure 4.19a, variables A and B can be considered as the multiplexer select variables and variable C as the data variable. When the first pair, $\bar{A}\bar{B}$, is selected, the map indicates that the output value must be 1 if C is 0, and 0 otherwise. Thus for that selection the data input must have $\bar{C}$ connected. For the second pair, $\bar{A}B$, the output must be 1 whatever the value of C. In this case, the required data input is 1. For the third pair, AB, the output must be 1 if C is 1, and 0 otherwise, so the data input must have C connected. Where both entries are 0, the data input must have 0. Figure 4.19b and figure 4.19c enable

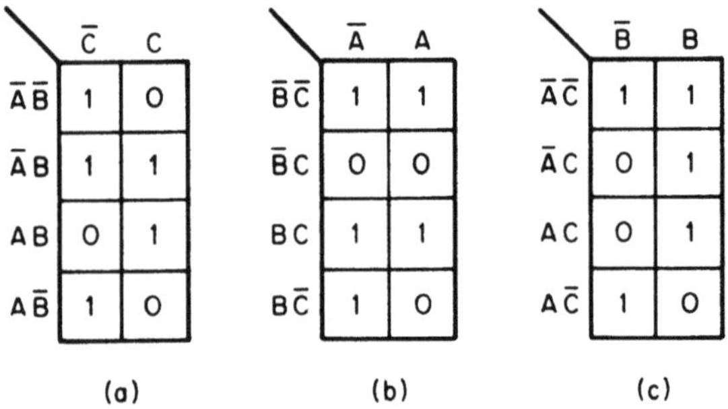

Figure 4.19 Maps of the function $f = \bar{A}B + BC + \bar{B}\bar{C}$

us to consider the cases of A or B as the data variables. It can be seen that the use of the A variable as the data variable leads to a realization giving minimal loading (figure 4.20b).

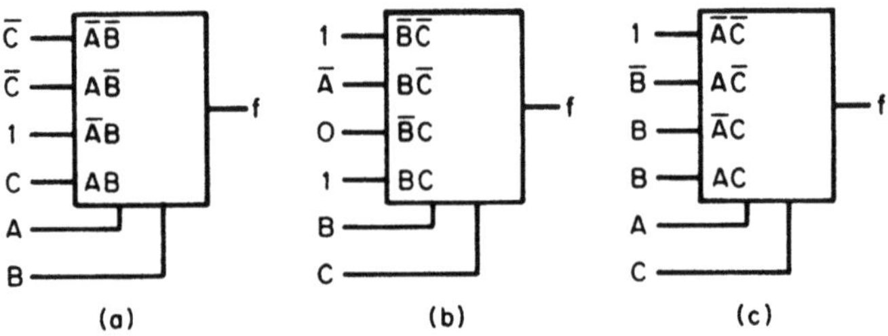

Figure 4.20 Multiplexer realization using: (a) C as the data variable; (b) A as the data variable; (c) B as the data variable

From a consideration of the logical function describing the output of a multiplexer, it is clear that a four-line-to-one-line multiplexer can realize any function of three variables. Similarly an eight-line-to-one-line can realize any function of four variables, and a sixteen-line-to-one-line multiplexer can realize any function of five variables. With certain simpler functions, however, it is possible to make use of a smaller multiplexer. For example

$$f = A\bar{B}\bar{C} + \bar{B}C\bar{D} + BD$$

can be rearranged as

$$f = \bar{B}\bar{C}(A) + \bar{B}C(\bar{D}) + B\bar{C}(D) + BC(D)$$

We can therefore use a four-line-to-one-line multiplexer with variables B and C connected to the select inputs and the data inputs connected to A, $\bar{D}$, D and D respectively. On certain multiplexers the inverted output is available (for example, Texas Instruments SN74LS151), and also, if necessary, the ENABLE inputs can be used in achieving extra logic functions.

Where functions of more variables are required it is possible to cascade multiplexers into a circuit of any desired size. The Karnaugh map and algebraic methods then become unwieldy and a numerical approach is preferable. Such a technique for synthesis using multiplexers has been proposed (Whitehead, 1977), which is essentially one of factorizing, using the boolean relationship $AB + A\overline{B} = A$, in which the variable allowing the maximum number of function terms to be so factored is sought. That variable is then used as the multiplexer data input variable. The method bears some resemblance to that used in the Quine-McCluskey (Bannister and Whitehead, 1973) minimization procedure. However, instead of attempting to reduce the product terms by checking for repeated adjacencies and removing variables progressively, each variable term is checked against all the product terms and possible adjacencies are noted. The variable yielding the greatest number of adjacencies is the one which will provide the most efficient solution when used as the multiplexer data variable.

The function to be realized is first expressed in its minterm form and the procedure is as follows:

(1) List the terms, in a single column, in accordance with the number of ones in the binary representation of those terms, separating the minterms into groups having the same number of ones. Include don't care terms, if any.

(2) Working from the top of the minterm column, compare terms in successive groups, listing those terms with unit differences. Repeat with a new list for those terms with a difference of two. Continue, making lists of pairs for all differences corresponding to the weights of each variable contained in the function.

(3) Compare the lists; the one having the greatest number of pairs corresponds to that variable which should be used as the multiplexer data variable.

Example 4.4: Use multiplexers to realize the function

$$f(ABCDE) = \Sigma m(0, 4, 5, 7, 8, 9, 12, 15, 21, 22, 26, 29, 30)$$

The table of figure 4.21 shows the listing in accordance with steps (1) and (2).

It can be seen that the greatest number of pairs is produced when a difference of eight is considered. Variable B, having the binary weight of eight, will thus be chosen as the multiplexer data variable.

To obtain the data input values for the multiplexer, all the 2^n minterms possible for an n-variable function (in this case, $n = 5$) are listed in pairs, each pair having a difference equal to the binary weight previously determined, starting with minterm zero. The minterms corresponding to those of the required function are ringed and, alongside each pair, values for the multiplexer data input are listed on the following basis:

(1) For pairs both ringed, or one ringed and the other a don't care, set the data input value to 1.

Minterm list	Variable weight E=1	D=2	C=4	B=8	A=16
0	4,5	5,7	0,4	0,8	5,21
4	8,9		8,12	4,12	
8			26,30	7,15	
5				21,29	
9				22,30	
12					
7					
21					
22					
26					
15					
29					
30					

Figure 4.21 Table for the selection of the data input variable

(2) For pairs both unringed, or one unringed and the other a don't care, set the data input value to 0.

(3) For pairs with one term ringed, set the data input equal to the data variable if the ringed term is in the right-hand column, and equal to the inverse data variable if the ringed term is in the left-hand column.

(4) For pairs both marked don't care, set the data input to the more convenient logic level.

Figure 4.22 illustrates this for the function under consideration.

The remaining variables, ACDE, are then assigned to the select inputs as determined by the choice of multiplexers. Further simplifications may be possible on an intuitive basis if the data-value column shows patterns of repeated groups following the binary weights of the selector variables. Figure 4.23 shows the final realization of the function.

4.5 Designing in silicon

It is now possible to realize logic circuits directly in silicon. This capability has been brought about by advances in integrated circuit technology and by the need to develop more and more complex systems. Instead of interconnecting a set of logic packages to implement a required design, with the attendant costs of printed circuit board manufacture, assembly, testing, etc., the designer is now able to implement designs either at the gate or device level in an integrated circuit (*gate array* or *semi-custom design*), or directly into the silicon by defining

A	C	D	E	$\bar{B}$	B	Data value
0	0	0	0	0	8	1
0	0	0	1	1	9	B
0	0	1	0	2	10	0
0	0	1	1	3	11	0
0	1	0	0	4	12	1
0	1	0	1	5	13	$\bar{B}$
0	1	1	0	6	14	0
0	1	1	1	7	15	1
1	0	0	0	16	24	0
1	0	0	1	17	25	0
1	0	1	0	18	26	B
1	0	1	1	19	27	0
1	1	0	0	20	28	0
1	1	0	1	21	29	1
1	1	1	0	22	30	1
1	1	1	1	23	31	0

Figure 4.22 Determination of data input value

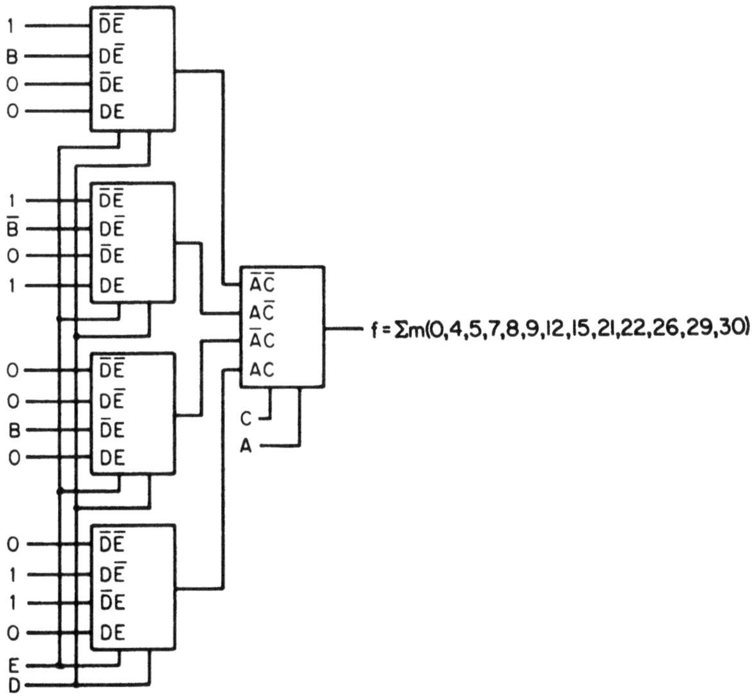

Figure 4.23 The final circuit

the diffusion masks at the processing stage (*VLSI* or *custom design*). Both processes are heavily dependent on computer-aided-design (CAD) techniques.

In semi-custom design the technique is to take as the starting point a standard array of basic logic circuits such as NOR gates, flipflops, counters and so on, or uncommitted transistors and resistors, fabricated in either MOS or bipolar technology. In some processes the circuits are already laid out on a silicon chip, but without the final layers of metallization. A unique interconnecting pattern is then defined by the circuit designer to be laid on to the chip by the semiconductor manufacturer before packaging the array. In other systems the designs of these logic circuits are provided as 'standard cells' held in a cell library to be called upon as 'building blocks' as and when required. Placement of the cells on the silicon then becomes part of the design process.

Once a design has been fabricated there is no way of correcting errors unless such a facility has been deliberately designed into the circuit. Consequently it is imperative that exhaustive checks are made at the design stage to ensure that the circuit really functions as expected. This is achieved by computer simulation. All semi-custom chip manufacturers provide extensive software facilities to enable their products to be designed reliably. A typical design procedure would be as follows:

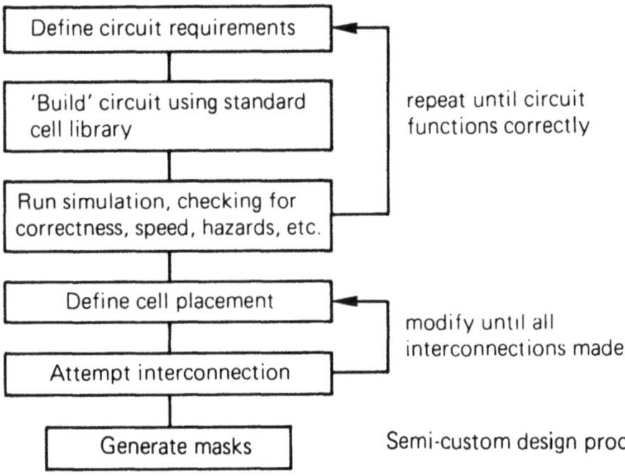

Semi-custom design process

Using MOS technology, it is also possible for the logic system designer to create a system from the basic transistor upwards. Such '*full custom design*' gives the advantage of flexibility, the ability to optimize the design for maximum packing density and/or performance and to create structures not readily available in a standard cell library.

As explained in the previous chapter, MOS devices are formed from layers of metal and polysilicon arranged in paths and areas over oxide-insulated *n*-type and *p*-type diffused silicon, a transistor switch being formed whenever a polysilicon path crosses over a diffusion area. The design goal therefore is to create a series of masks which, when used in the fabrication process, defines the structures and interconnecting paths which will perform the required logical function. Design rules which specify path widths, separations, overlaps, etc. are provided for the particular technology used.

To enable the designer to visualize the circuits in a form closely related to reality, *stick diagrams* are employed. Consider, for example, the ratioed NAND gate described in section 3.6. A stick diagram of the gate is shown in figure 4.24.

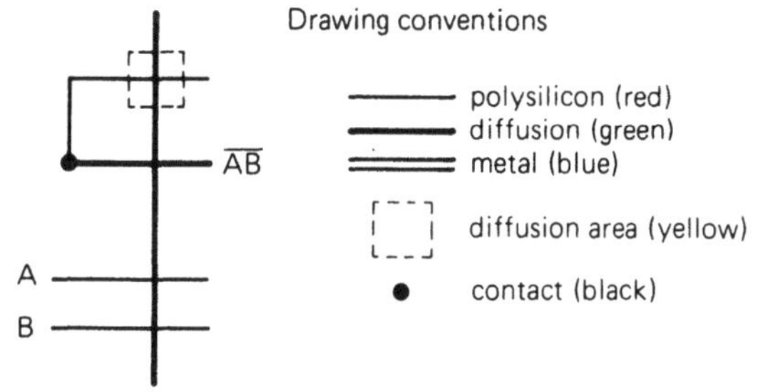

Figure 4.24 Stick diagram for a two-input NAND gate

Colours are normally used to highlight the different regions. It is possible to use such diagrams as graphical input to some CAD systems. Using a light-pen or 'mouse', the designer is able to create a circuit on a display screen and rely upon software to translate the design into process masks. The *silicon compiler* is the name often given to the software used for the design of such integrated circuits. The name arises because of the analogous nature to the software compilers used to translate high-level language into computer machine code. Silicon compilers allow designers to work with high-level functional descriptions of their logic designs, the conversion into the mask geometry for the MOS circuits and their interconnections being performed by the computer. In this way the tedium of ensuring error-free designs is left to the machine, freeing the designer to concentrate on more creative activity.

4.6 Design for testability

With the increasing complexity of integrated circuits, it is becoming more and more important to provide the means whereby the circuits can be tested [Lala (1985)]. This is a simple matter when considering a circuit such as a basic gate. For example, a three-input NAND gate presents the possibility of up to three faulty inputs and one faulty output. The object of the test is to determine which, if any, of these *nodes* is at fault. Testing the inputs involves establishing a *sensitive path* which in our simple case means setting two of the inputs to logical one.

The output is thus rendered sensitive to the logic state of the remaining input. By switching this input from logical one to logical zero we would expect the output to change correspondingly. The test can then be repeated on the other inputs. To test the output node the inputs must be set to such values that will *always* produce a known output. In this simple case, setting all three inputs equal to logical one and then to logical zero would establish the functionality of the output node. Of course, if the output was stuck at either logic level then all the input node tests could register faults. However, we assume with this type of test that only one fault will occur. Thus tests which indicate that every input node is faulty imply that the output is the faulty node. Naturally, for more complex circuits, the establishing of sensitive paths and the appropriate logic control states becomes a non-trivial matter.

Example 4.5: What test sequences are required to test a three-input NAND gate fully?

Let the inputs be A, B, C and the output be D. Setting $A = B = C = 1$ ensures that D must equal '0' This can be written as

$$A \cdot B \cdot C/\overline{D}$$

Setting $A = B = 1$ sensitizes the C input so that a transition on C should give rise to a change in D, thus:

$$A \cdot B \cdot \overline{C}/D \text{ (if } C \rightarrow 0 \text{ then } D \rightarrow 1)$$

Similarly for A and B.

Thus the complete sequence is

$$A \cdot B \cdot C/\overline{D};\ \overline{A} \cdot B \cdot C/D;\ A \cdot \overline{B} \cdot C/D;\ A \cdot B \cdot \overline{C}/D$$

If the last three all show a fault, then we can assume that the output D is stuck at '0', whereas a fault on the first would indicate that the output was stuck at '1'.

This one-dimensional path-sensitization technique suffers from the disadvantage that faults can be hidden if the circuit under test has a reconvergent fanout structure. An *n-dimensional* path sensitization approach, the *D-Algorithm* (Roth, 1980) overcomes this problem by enabling all possible sensitive paths to be identified and tested. This approach is powerful and is usually implemented on a computer (Bennetts, 1984).

Another useful method for detecting faults is by the use of *boolean differences*. In order to detect, at the output of a logical function block, if a fault exists in the path of a particular input variable x_i by applying some pattern to the inputs, then the output function F must be sensitive to changes in x_i. In other words, the output function F must change if we invert the sensitive input signal, x_i, and the output function values under the two conditions are not equivalent. Algebraically

$$F(x_1, x_2 \dots x_i \dots x_n) \oplus F(x_1, x_2 \dots \bar{x}_i \dots x_n) = 1$$

The expression $F(\dots x_i \dots) \oplus F(\dots \bar{x}_i \dots)$ is called the *boolean difference* of the function $F(\dots x_i \dots)$ and is often represented by the derivative function $dF(\dots x_i \dots)/dx_i$.

We know that a test for a fault on x_i exists if $d(F(x))/dx_i = 1$. If this function is zero then no test exists, since changing the state of the specified input does not affect that particular output.

We can further refine the process to test for specific 'stuck at' faults. If the logical AND of the boolean difference and $\bar{x}_i$ is used as the test function this checks for 'stuck at one' faults, since if the output x_i is able to switch to zero then the test function also changes to zero. Similarly, to test for a 'stuck at zero' fault, the test function is

$$T(x) = x_i \, dF(x)/dx_i$$

For example, consider the output function

$$F = (\bar{x}_2 + \bar{x}_3)x_4 + x_1 x_2$$

Suppose we wish to test for x_4 stuck at zero.
The boolean difference expression is

$$[(\bar{x}_2 + \bar{x}_3)(x_4 = 1) + x_1 x_2] \oplus [(\bar{x}_2 + \bar{x}_3)(x_4 = 0) + x_1 x_2]$$

$$= \bar{x}_2 + \bar{x}_1 \bar{x}_3$$

Thus for x_4 stuck at zero, the test function is given by

$$T(x) = x_4(\bar{x}_2 + \bar{x}_1 \bar{x}_3)$$
$$= x_4 \bar{x}_2 + x_4 \bar{x}_1 \bar{x}_3$$

so in applying the patterns

$$x_4 = 1;\ x_2 = 0$$
$$x_4 = 1;\ x_1 = 0;\ x_3 = 0$$

we obtain a conclusive test. If $F = 0$ for either of these patterns, then x_4 is stuck at zero. Note that the other inputs are don't care states for this particular test.

This procedure is not restricted to input variables; if we wish to test for a faulty internal connection, then the equations can be set up using a pseudo-input, which input can then be expressed in terms of x_2, x_3 and x_4.

Example 4.6: Determine the conditions necessary to test for an open circuit on interconnection ①-② in the circuit shown in figure 4.25. Assume open circuit gate inputs adopt the logical one state.

We establish a pseudo-input 'p' to gate ②. The function F is given by

$$F = abp + cd(\bar{c} + d)$$
$$= abp + cd$$

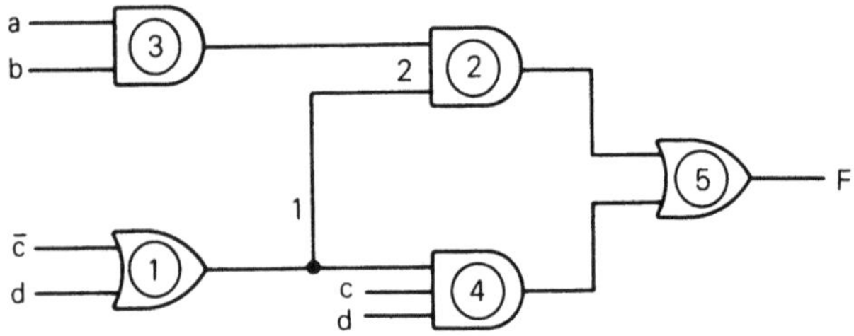

Figure 4.25 Example 4.6

so $dF/dp = F(a,b,c,d,p = 0) \oplus F(a,b,c,d,p = 1)$

$= cd \oplus (ab + cd)$

$= (\overline{cd})(ab + cd) + cd(\overline{ab + cd})$

$= ab\bar{c} + ab\bar{d} + \bar{a}cd + \bar{b}cd$

Now, $p = (\bar{c} + d)$ and $\bar{p} = c.\bar{d}$

and the set of all tests is given by $\bar{p}\,dF/dp$, which means that

$$T(x) = c\bar{d}(ab\bar{c} + ab\bar{d} + \bar{a}cd + \bar{b}cd) = abc\bar{d}$$

Thus the test pattern to check for line ①-② stuck at one is

a = 1; b = 1; c = 1; d = 0

If the line is open-circuit, when this pattern is applied F = 0 will *not* result.

It is becoming increasingly common for complex circuits to have test logic built in to the chip. One technique for such testing is by *signature analysis* [Wilkins (1986)], in which a linear feedback shift register (LFSR) is used to generate a unique code from a combination of its own feedback and signal responses from the circuit under test. This technique was originally developed by Hewlett-Packard for testing LSI systems. The advantage is one of signal compression. The N-bit code from such an LFSR of N bits may be part of a sequence of $(2^N - 1)$ bits in length and as such can be capable of indicating a fault in a circuit with many more than N different responses. Feedback shift registers are dealt with in chapter 5.

Problems

4.1 Realize the following functions in a minimal wired-OR NAND form:

$$(1)\ f = (\overline{A} + \overline{B})\overline{D} + \overline{A}BC$$
$$(2)\ f = A(\overline{B} + C) + \overline{A}(B + \overline{C})$$

4.2 A comparator circuit is required to indicate on two outputs Z_1, Z_2 whether the two-bit input number A_1, A_2 is greater than, equal to, or less than the two-bit input number B_1, B_2. Devise a suitable circuit using eight-line-to-one line multiplexers to give outputs of 01 if $A > B$, 10 if $A < B$, and 11 if $A = B$.

4.3 Realize the following function using only four-line-to-one-line multiplexers:

$$f(VWXYZ) = \Sigma m(1, 2, 3, 4, 5, 6, 7, 10, 14, 20, 22, 28)$$

4.4 A logic circuit is required to perform as a full-adder when the MODE signal, M, is at zero, and as a full subtractor when the MODE signal, M, is at one. Inputs A, B carry the digits to be added or subtracted, and C carries the CARRY/BORROW input. Output S carries the SUM or DIFFERENCE, and T the CARRY/BORROW output. Two types of multiplexer circuit are available: the 74150 is a sixteen-line-to-one-line multiplexer, and the 74151A is in eight-line-to-one-line multiplexer. Each type has a low active ENABLE input. Show how each type can be used in building the circuit.

4.5 By means of a diode matrix, show clearly how the following code could be generated from ten decimal input switches, 0–9.

A	B	C	D	Decimal
0	0	0	0	0
0	0	0	1	1
0	0	1	0	2
0	0	1	1	3
0	1	0	0	4
1	0	0	0	5
1	0	0	1	6
1	0	1	0	7
1	0	1	1	8
1	1	0	0	9

How could the array be modified to indicate when the switch pressed represents an even digit?

4.6 What is a programmable logic array (PLA) and how can it be considered as being a development of a read-only memory? A PLA has four inputs a, b, c, d, and two outputs Z_1 and Z_2. Using a lattice network to represent the internal structure of the PLA suggest a programming pattern necessary to give the logic functions

$$f_1 = AB + CD$$
$$f_2 = A(B + C) + B(A + D)$$

4.7 A logic network is required that will detect a single error in a received binary character, held in a register and consisting of four data bits and an odd parity check bit. Develop an expression for the network output function and draw the corresponding Karnaugh map. Hence, or otherwise, devise a suitable circuit using exclusive-OR gates. Show how the same result could be achieved by the use of a multiplexer.

4.8 Devise a test sequence for the complete testing of (a) a 3-input NOR gate, (b) an exclusive-OR gate.

References

Bannister, B. R. and Whitehead, D. G. (1973). *Fundamentals of Digital Systems*, McGraw-Hill, London

Bennetts, R. G. (1984). *Design of Testable Logic Circuits*, Addison-Wesley, Reading, Massachusetts

Evans, F. C. (1969). Use of inverse Karnaugh maps in realization of logic functions, *Electronics Letters*, **5**, No. 21, 670

Lala, P. K. (1985). *Fault Tolerant and Fault Testable Hardware Design*, Prentice-Hall, Englewood Cliffs, New Jersey

Roth, J. P. (1980). *Computer Logic, Testing and Verification*, Computer Science Press, Washington DC

Whitehead, D. G. (1977). Algorithm for logic circuit synthesis using multiplexers, *Electronics Letters*, **13**, No. 12, 355-6

Wilkins, B. R. (1986). *Testing Digital Circuits*, Van Nostrand Reinhold, New York

5 Sequential Logic Components

Earlier chapters have concentrated on the analysis and design of circuits in which the output signal value at any time is determined solely by the values of the input variables at that time. These are *combinational logic* circuits, though in practical terms we have to extend the definition to allow for the settling time necessitated by the unavoidable propagation delay through the logic gates.

Combinational logic design involves the interconnection of appropriate gates and subsystems, in many cases making use of multiplexers, programmable logic arrays and similar devices. The development of large scale integration techniques has now led to the possibility of building complete systems by the use of arrays of identical circuits (Edwards and Hurst, 1976). This has the advantage of simplifying the production processes since arrays can be mass-produced cheaply, with the individual characteristics of the system being defined by means of the interconnections laid down at a late stage in the manufacture. These arrays are known as *iterative arrays*, and each array produces an ordered pattern of bits, called the output *word* or *vector*, by operating in a predetermined manner on the input vector. In order to do this each cell of the array has to react to both the input signals and the information from its neighbour, as shown in the generalized circuit of figure 5.1. The parallel adder is one example of an iterative array: each cell, p (where $1 \leqslant p \leqslant n$), responds to the two input bits, a_{1p}, a_{2p} (where $k = 2$), which are to be added, and to the carry signal, b_{1p}, coming from its less significant neighbouring cell. The cell generates a sum output, Z_{1p}, and a new carry signal, Y_{1p}, to pass on to the next cell. The effect of each carry must

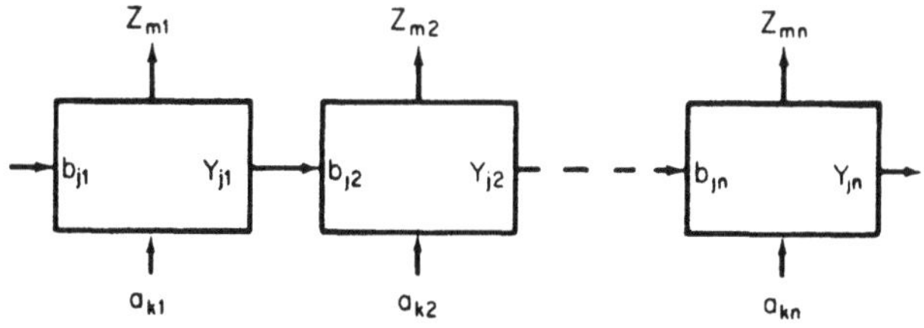

Figure 5.1 Generalized iterative array

propagate through the array, giving rise to the carry delay problems discussed in section 2.5.

The iterative array operates on an input vector in which the bits are provided simultaneously but are separated spatially along the array. An alternative method of achieving the same end result is to use time separation of the operations; the successive bits of the input data are applied to a single cell as sequences and the output sequence is extended at each time interval. Since a single cell is used to deal with successive operations, it is necessary to introduce some temporary storage capability, so that the information generated during one operation and necessary for the next operation is retained until required. In addition a feedback path is needed to route the stored information correctly. Such circuits are known as sequential circuits. The serial adder, for example, is the sequential form of the parallel adder. The outputs are now dependent not only on the values of the input variables at a given time, but also on the *state* of the circuit at that time, and the state is determined by previous input values. In order to take account of the time element we must relate present states and present inputs of the circuit to the next state which results from that combination. In generalized form the circuit contains two main sections, a combinational logic section containing both input and output gating, and a storage section, as in figure 5.2. The storage section 'remembers' the effects of previously applied input sequences and feeds back information to the combinational logic circuitry to be used in conjunction with the present inputs.

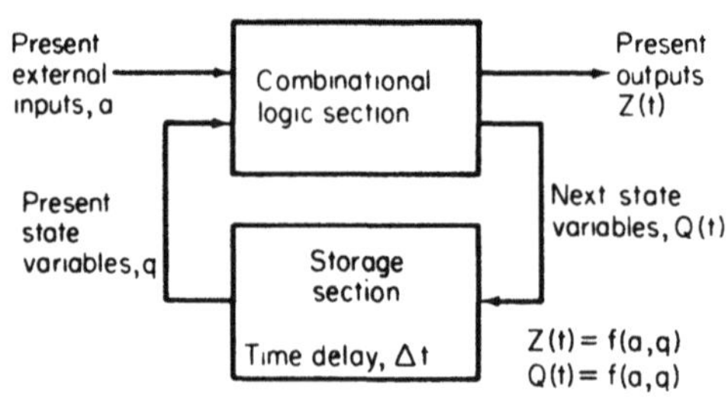

Figure 5.2 Generalized sequential circuit

5.1 Iterative arrays

The iterative array was shown in general form in figure 5.1. Each cell has one or more external inputs, a, and one or more internal inputs, b. Internal and external outputs, Y and Z respectively, are provided at each cell though in some arrays, circuit output arrays, the only external output is taken from the final cell of the array.

The outputs from any cell are determined by the inputs to that cell and all the possible output conditions may be listed on a modified truth table, known as the *flow matrix*. In drawing up the matrix, column headings list all the possible combinations of the external input variables, a, and row headings relate to the internal variables, b. The table entries are in two parts, the internal and external output values respectively.

Example 5.1: Draw the flow matrix for the circuit of figure 5.3.

The output expressions are $Y_1 = a\,\bar{b}_1 b_2$, $Y_2 = a\,\bar{b}_1 \bar{b}_2$, $Z = a\,\bar{b}_1 b_2$, and using these expressions the output values are tabulated to give

$b_1 b_2$ \ a	0	1
00	00,0	01,0
01	00,0	10,1
10	00,0	00,0
11	00,0	00,0
$b_1 b_2$	$Y_1 Y_2$,Z	

The rows of the table indicating the internal conditions of the cell are normally allocated letters, giving the matrix of figure 5.3b. As the fourth row entries merely repeat the third they are omitted.

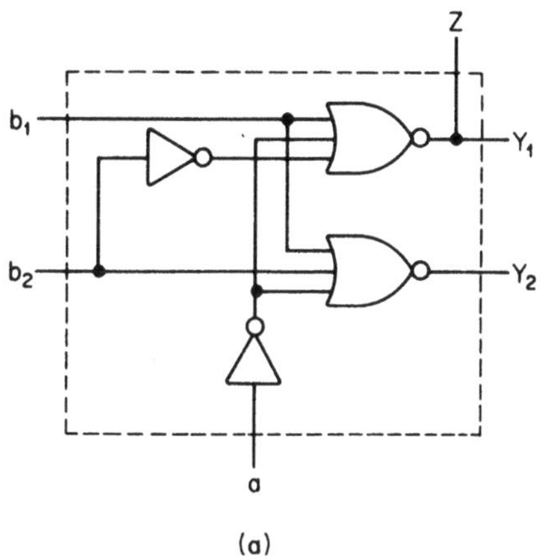

	0	1
A	A,0	B,0
B	A,0	C,1
C	A,0	A,0

(b)

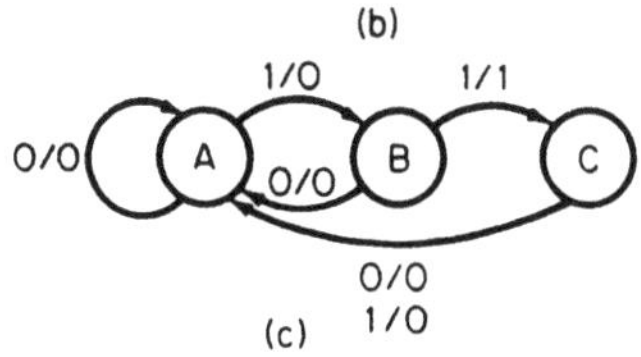

Figure 5.3 Analysis of cell action: (a) cell circuit; (b) flow matrix; (c) flow diagram

The flow matrix enables us to determine the output values of all the cells in an array under any given input conditions. Suppose five of the cells of example 5.1 are iterated, and the external inputs, a, are 10111. The leftmost cell internal inputs, b_1, b_2, are the only inputs whose values are not specified, since all other internal input values are governed by the internal outputs of the previous cell, and we must specify the *boundary conditions* for the array before we can determine the output values. In other words, we must indicate the starting point on the flow matrix. Unless otherwise stated the flow matrix is so arranged that the boundary condition is the heading of the first row. Thus, in this case $b_{11} = 0$, $b_{21} = 0$.

With the boundary condition specified and an external input a_1 of one the matrix indicates that condition B is fed to the second cell and an output $Z_1 = 0$ results from the first cell. The second cell, having condition B at the input and an external input a_2 of zero, feeds condition A to the third cell and gives an output $Z_2 = 0$. By stepping around the matrix in this way for each set of inputs at the five cells the external outputs are found to be 00010. In fact this array gives a one output only at any cell whose input is the second, fifth, $(3n + 2)$th member of any group of consecutive input ones.

An alternative form of flow matrix is the *flow diagram* (or state diagram in sequential circuit theory) in which the row headings occur as nodes which are interconnected by arrows carrying labels indicating the inputs leading to the transition and the output resulting from it. The matrix of figure 5.3b then becomes the flow diagram of figure 5.3c. The entry with each arrow or sling (an arrow terminating on its originating node) gives first the input and then the output value. The diagram is complete only when *closed*: that is when all input combinations at each node have been considered and assigned an arrow or sling. As with the truth table or Karnaugh map, the flow diagram is an invaluable first step in synthesizing an array since we are obliged to specify the output required for every input combination. We then proceed to the design of the cell.

Example 5.2: A one output is required from each cell in an array whose input is such that the total number of ones occurring in the input sequence up to that cell is an odd number. Design the cell circuit.

The array is to act in effect as an odd parity indicator, and the first cell, with boundary condition A, must give an indication of odd parity only if its input is a one. Thus a one input moves the operating point to condition B (figure 5.4a) whereas a zero input returns to condition A. At B, a zero input now retains odd parity, where a one destroys parity and operation returns to node A, closing the diagram. From the flow diagram we derive the flow matrix of figure 5.4b and assign logical values to the internal inputs. Our two-row matrix requires only a single bit, leading to the table of figure 5.4c. The equations describing the table are $Y = Z = \bar{a}b + a\bar{b}$ and the cell circuit is, therefore, an exclusive-OR circuit, one form of which, using distributed logic, is given in figure 5.4d.

The arrays considered so far are one-dimensional and are very useful in many applications, in particular in parallel adders and code converters. Many other applications, however, require more complex operators in more than one dimension. Multiplication and division, for example, use a range of operations and, if we are to retain the array approach, it is necessary to go to two or more dimensions. As we have seen, multiplication and division may be carried out by a suc-

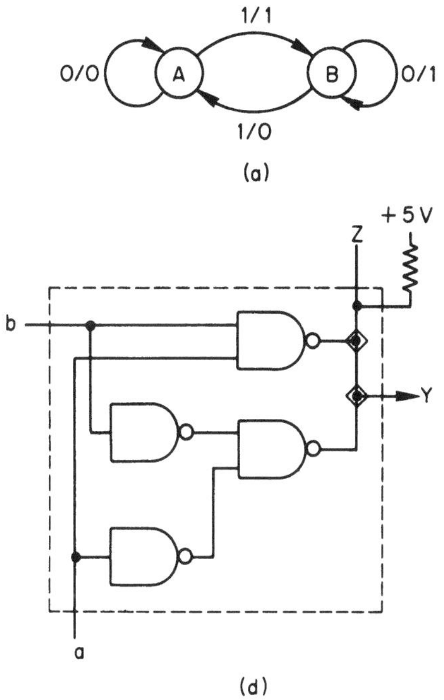

	0	1
A	A,0	B,1
B	B,1	A,0

(b)

b \ a	0	1
0	0,0	1,1
1	1,1	0,0
		Y,Z

(c)

Figure 5.4 Design of iterative array: (a) flow diagram; (b) flow matrix; (c) matrix with assigned variables; (d) cell circuit

cession of shift and add operations. That process is part serial, part parallel but the same operations can be carried out in a two-dimensional array.

Figure 5.5 shows an array which gives the product of two four-bit numbers, A0-3 and B0-3. Each cell contains a gated full-adder circuit which acts as a normal full-adder when the control signal, g, is at one. When g is at zero, however, the sum output, S, takes the value on input b_n. The multiplier bits, B, determine whether or not the multiplicand, A, is added to the shifted partial result at each stage of the array.

5.2 Flipflops

The fundamental difference between a sequential circuit and a combinational circuit is the ability of the former to store information over a period of time. The simplest forms of sequential circuit are flipflops which can be constructed from basic NAND/NOR circuits suitably interconnected to provide the necessary feedback. A flipflop has two stable states and may be switched from one to the other depending on the combination of input signals. Since the flipflop retains its state until switched to the other state, subject only to the power supply being

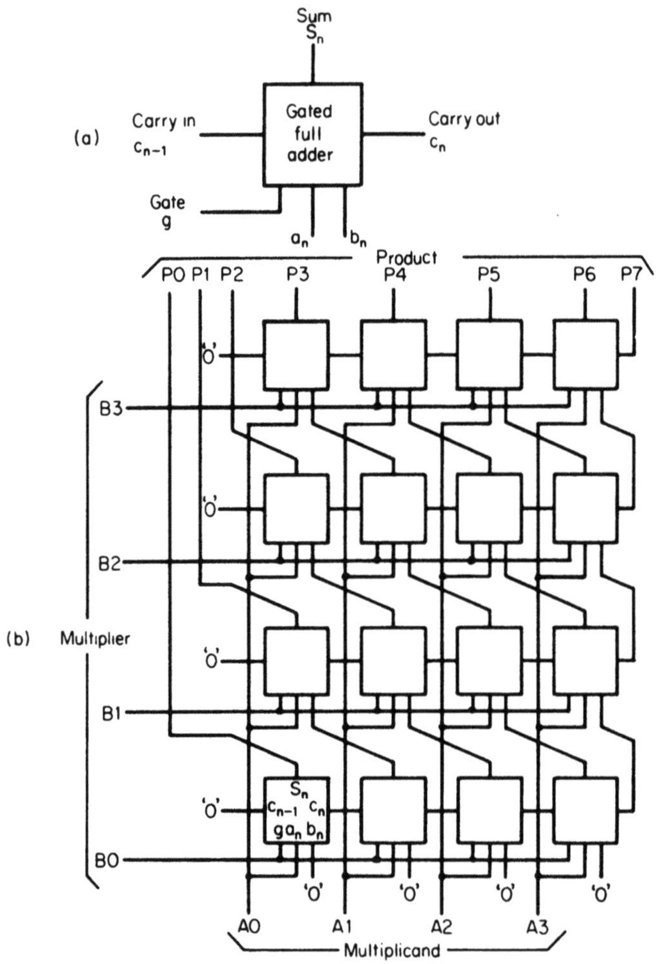

Figure 5.5 (a) Cell details. (b) Multiplier array

uninterrupted, it acts as a one-bit store or *memory*. Normally both *true*, Q, and *negated*, $\bar{Q}$, outputs are made available. Flipflops have been used in one form or another as storage elements throughout the development of digital systems and, as we shall see in the next chapter, are still major components in modern high-speed storage systems. In addition, by introducing extra circuitry to control the changing of state of arrays of flipflops it is possible to build up the complex circuits which are used to count pulses, shift binary codes and otherwise process digital information.

The simplest flipflop is the SR or *set–reset* flipflop shown in figure 5.6. Let us assume the Q output is 'high', that is logical one in positive logic, and both

S	R	Q^+
0	0	Q
1	0	1
0	1	0
1	1	–

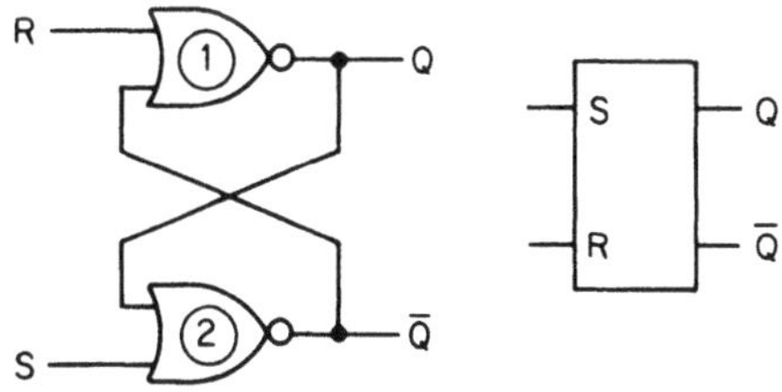

Figure 5.6 The NOR SR flipflop and truth table

inputs are 'low', logical zero. NOR gate 2 has one input high and the other low, so its output, $\bar{Q}$, is low. Both inputs to NOR gate 1, therefore, are low, and its output is high. The circuit is stable with Q at one and $\bar{Q}$ at zero. Equally, if we assume the Q output to be low we find the circuit to be stable with Q at zero and $\bar{Q}$ at one. The truth table included in figure 5.6 describes the operation of the flipflop, where Q^+ represents the new value of Q in response to the S and R input values. It is important to note that the input condition S = R = 1 leads to indefinable output states, and is usually avoided in logic design. In algebraic terms, taking S = R = 1 as a don't care condition, $Q^+ = S + \bar{R}Q$.

The flipflop can also be formed from cross-coupled NAND gates, but now the setting and resetting action is governed by low signals at the inputs. Both forms of the circuit are often referred to as *latches* and can be used to store a single bit of data of a transitory nature. Once switched, or *set*, by a signal on the set input, the flipflop remains in that state until a signal is received on the reset input. The latch is therefore widely used as a *flag* which detects the occurrence of some event and continues to indicate until appropriate action has been taken and the flag can be reset.

The same circuit is commonly used in overcoming the effect of contact bounce when a mechanical switch is operated. Most switch contacts bounce several times before settling, so a single operation of the switch may result in a series of pulses at intervals of several milliseconds. If a debouncing circuit is not used the digital circuitry may respond to each pulse giving completely false operation. The arrangement shown in figure 5.7 is usually sufficient provided the bounce is not so great that the contact returns to the original pole; the initial contact change sets the output of gate 2 high, and subsequent pulses have no effect. The flipflop resets only when the contact reverts to its original position.

A collection of latches forms a *register* as in figure 5.8. When data bits are present and the GATE DATA IN line is energized, flipflops with logical one on their input lines are set. When the GATE DATA IN signal is removed, the data set into the flipflops remains, irrespective of any change on the input lines, until the reset signal is received. Output gating is also shown.

The SR flipflop is *asynchronous*, meaning that the flipflop switches to the output state indicated by the input conditions as soon as those conditions are

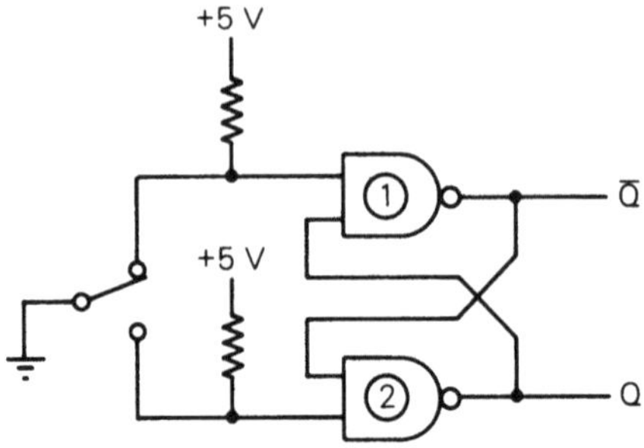

Figure 5.7 Switch debouncing circuit

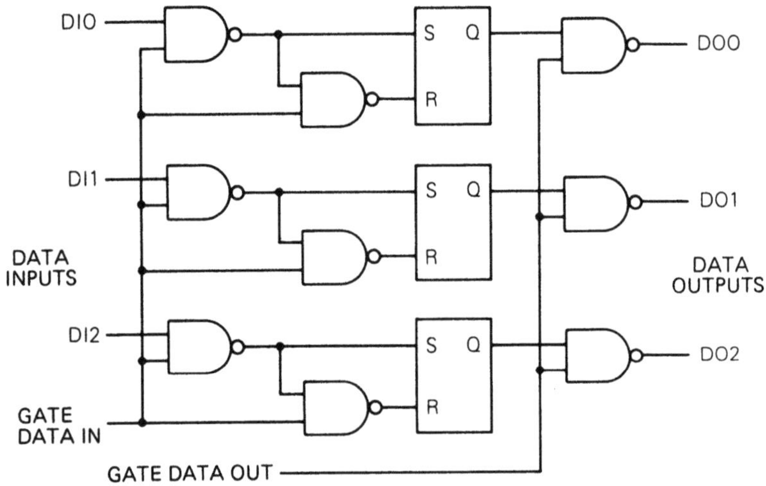

Figure 5.8 Simple register

recognized. The addition of the logic gating at the inputs allows the time of switching to be determined by an external control signal, shown in figure 5.8 as GATE DATA IN but commonly known as the *clock*. This clocking signal is so called because in many applications it is a series of pulses, often but not necessarily of fixed duration and constant repetition frequency, used to provide timing information throughout the system. It is also known as a *trigger*, *strobe* or *enable* signal. With its input gating each flipflop is now *synchronous* in operation; the input conditions indicate the next state but the actual time of switching is determined by the controlling signal. These devices are useful for storing data words and are provided in groups of four, six or eight in a single integrated circuit package. They are referred to as *transparent registers* or *latches* because, while the clock or enabling signal is at logical one, the outputs follow the values on the data lines, D; when the enable is taken to logical zero the outputs at that instant remain latched.

Another version of the circuit is the *D flipflop* shown in figure 5.9. In this version the data line, D, is allowed to determine the state of the output, Q, during the transition of the signal on the clock input from logical zero to logical one. This is achieved by means of the two additional *lockout flipflops* which form an integral part of the input gating. The flipflop is said to be *edge-triggered* since it switches on the positive-going edge of the clock signal, and the lockout flipflops prevent any further changes during that clock pulse period even though the logic condition at the D input changes. *Preset* and *clear* inputs are included to allow initial conditions to be set up if so desired but these inputs change the state of the flipflops asynchronously.

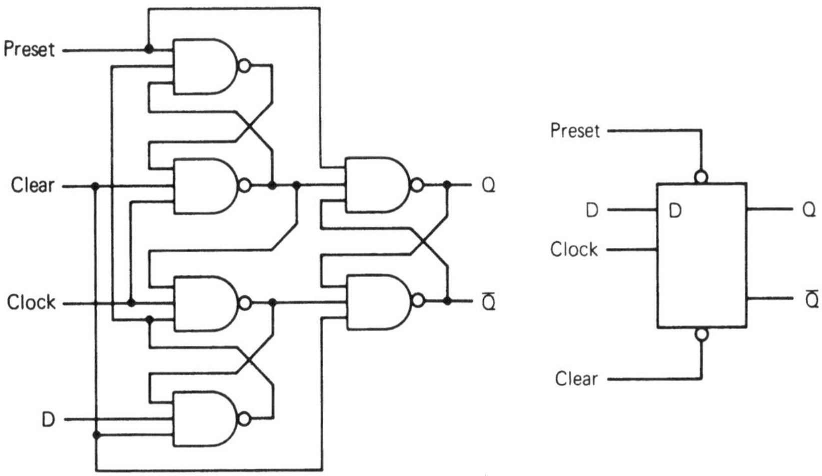

Figure 5.9 The D flipflop

If the Q̄ output of the D flipflop is connected to the D input the flipflop changes state each time the clock signal changes from zero to one. This type of circuit is called a *toggle*, and is capable of frequency division as shown in figure 5.10.

The D flipflop has a single data input, and appropriate set and reset signals are generated internally. An alternative form, the JK flipflop, has two inputs, J and K, which correspond to set and reset respectively, and again switching is controlled synchronously by the clock signal. JK flipflops are available in edge-triggered form, as described earlier, or in *master–slave* form, as in figure 5.11, which uses two flipflops in series. Information presented to the J and K inputs does not result in an output change until a pulse is received at the clock input. Any change of state of the output takes place on the back edge of this pulse, as it changes from high to low. Examination of the circuit shows that flipflop A

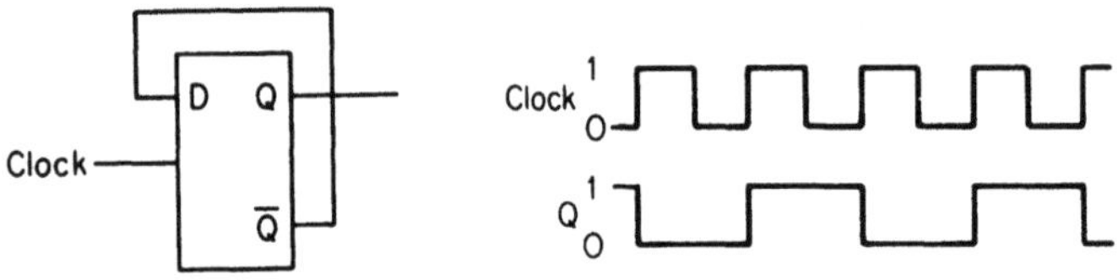

Figure 5.10 The D flipflop as a toggle

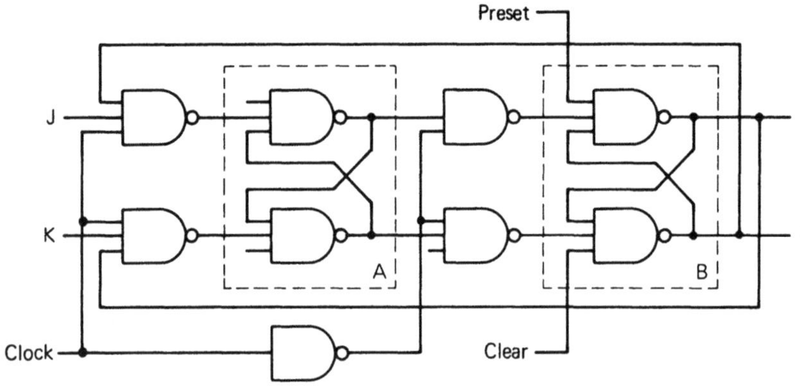

J	K	$Q^+ = J\overline{Q} + \overline{K}Q$
0	0	Q
0	1	0
1	0	1
1	1	$\overline{Q}$

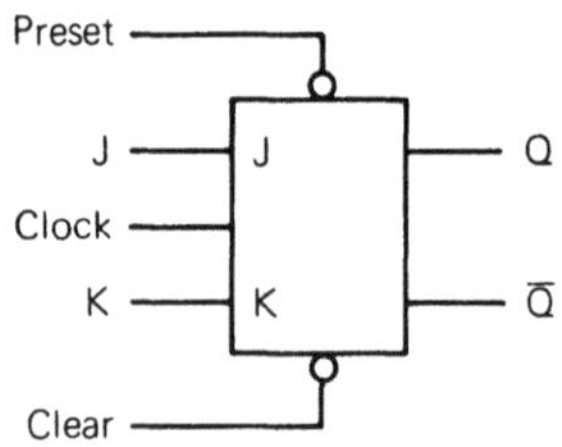

Figure 5.11 Truth table and logic circuit for the JK flipflop

receives information from the inputs when the clock signal is high, but during this period flipflop B is inhibited. When the clock signal returns to the low condition, flipflop B is enabled to adjust its state to correspond to the setting of flipflop A, which is in turn inhibited from responding to changes in J and K. J and K signals must be stable during the presence of the clock pulse. The feedback connections from the output of flipflop B to the input gates of flipflop A allow the J = K = 1 condition to be used and the flipflop then toggles.

The different types of flipflop are used in a variety of registers which act as high-speed temporary stores for data, or as *staticizers* or *latches* used in capturing or building up data words from randomly changing signals. The data words may be taken from the register in *parallel* or *serial* form. In parallel operation each data bit is routed, often via three-state or open-collector output gates, to a separate connection, and the total word is transferred to its destination along the interconnecting bus or highway. In serial operation a shift register is used to transfer the data bit by bit through a single output. Shift registers are widely used in communication systems and in arithmetic operations, since movement of a binary number one place left or right will change its value by a factor of two.

5.3 Shift registers

Figure 5.12a demonstrates the result of applying a series of pulses to the clock or shift input of a shift register into which a solitary one has been set. Each pulse causes the one, and of course the zeros, to move one place to the right. The construction of such a register using master–slave JK flipflops is shown in figure 5.12b. On the trailing edge of each pulse the flipflops adopt the state dictated by the current conditions on their J and K inputs, that is by the state of the preceding flipflop. Note that in this application D-type flipflops could also be used. Initial conditions may be preset into the register in parallel using the preset and clear inputs or additional gating. The register shown shifts right, but shift-left registers can be built simply by taking the inputs to flipflop n from the outputs of flipflop $(n + 1)$ rather than flipflop $(n - 1)$. In fact, bidirectional parallel-loading shift registers are available in integrated circuit form, such as the 74LS194 for example. The mode of operation is determined by the signals on two control inputs.

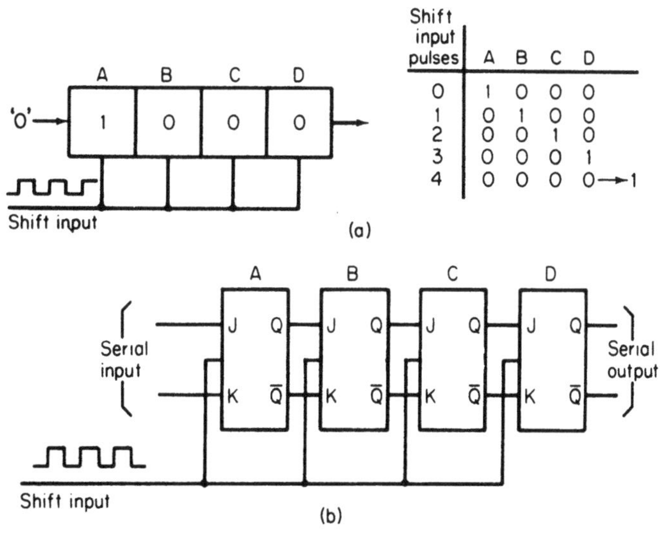

Figure 5.12 The shift register: (a) basic action; (b) circuit using JK flipflops

Large capacity shift registers make use of MOS devices in which data bits are stored *dynamically*, relying on the ability to store charge on the transistor gate. Using charge storage rather than bistable storage allows for a smaller circuit occupying less chip area, but the charge must be replenished periodically thus imposing a lower operating frequency limit. A simple shift register stage is shown in figure 5.13 using two antiphase clocks. Assuming initially that phase A clock is low, transistor T1 conducts, transferring data to the gate of T2. Transistors T3 and T4 are off. When phase A switches to high T1 turns off and, because phase B

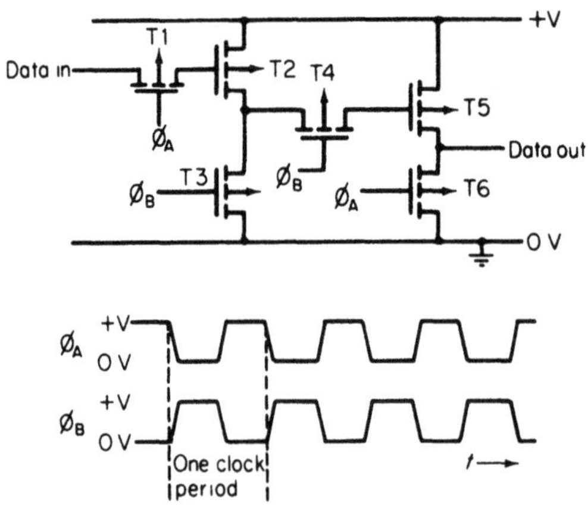

Figure 5.13 An MOS dynamic shift register stage

is now low, T3 and T4 turn on, with T3 acting as a switched load for T2. If the data signal transferred to the gate of T2 was low, T2 is still on and, behaving as an inverter, transfers the inverted data signal to the gate of T5. During the next clock period, as phase A goes low, T6 turns on and T5, acting as a further inverter, presents the data in the original form to the input of the next stage. In systems handling large amounts of serial data, such as image display systems, shift registers using charge coupled devices or magnetic bubble devices are often used because of their very small size and low power demands, and these are dealt with in chapter 7.

The action of a shift register means that, on each clock pulse, one bit of information is shifted out at one end of the register, and a zero or one must be provided as the input to the flipflop at the other end. An important group of circuits is formed when the value of the input bit provided is determined from the values already held in the register, rather than coming from an external source. Such circuits are *feedback shift registers*, FSR, and they are said to operate *autonomously*. In a *linear feedback shift register*, LFSR, the logic used in generating the input bit consists entirely of exclusive-OR functions, and, in general terms, the arrangement is as shown in figure 5.14. Each block represents a flipflop but

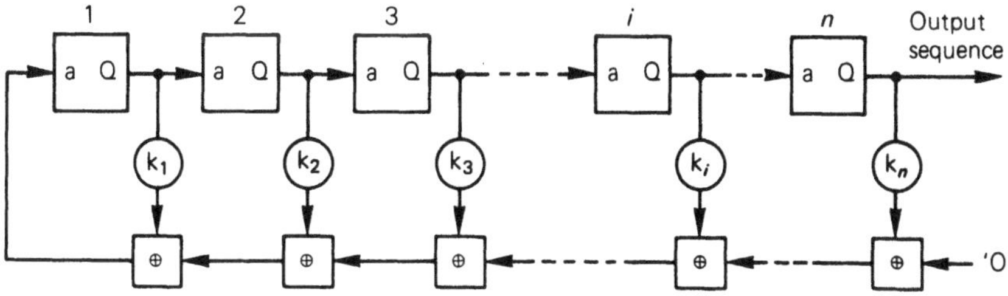

Figure 5.14 The linear feedback shift register

the clock signal is not shown explicitly. The value of k at each stage is one or zero to indicate whether that stage contributes to the feedback function or not.

We see that

$$a_i \mid_{i=2 \text{ to } n} = Q_{i-1}$$

whereas

$$a_1 = k_1 Q_1 \oplus k_2 Q_2 \oplus \ldots k_i Q_i \oplus \ldots k_n Q_n$$

where

$$k_i \mid_{i=1 \text{ to } n} = \text{'0' or '1'}$$

With any initial state of the register other than all zeros, the output sequence, as clock pulses are applied to the register, is an apparently random mixture of ones and zeros which repeats after a certain number of clock pulses. The all-zeros condition is excluded because the exclusive-OR functions used in the feed-back path would never generate any value other than zero at the input, and so the output sequence would also be zeros.

Certain feedback connections give longer sequences before repetition occurs, and a few give a *maximal length sequence*, or *m-sequence*. The length is determined by the number of flipflops, n, in the register, and the sequence consists of $2^n - 1$ bits, the -1 reflecting the absence of the all-zeros combination.

Example 5.3: What is the output sequence obtained from the LFSR of figure 5.15 when (a) $k_1 = k_4 =$ '1', $k_2 =$ '0'
and (b) $k_2 = k_4 =$ '1', $k_1 =$ '0'?

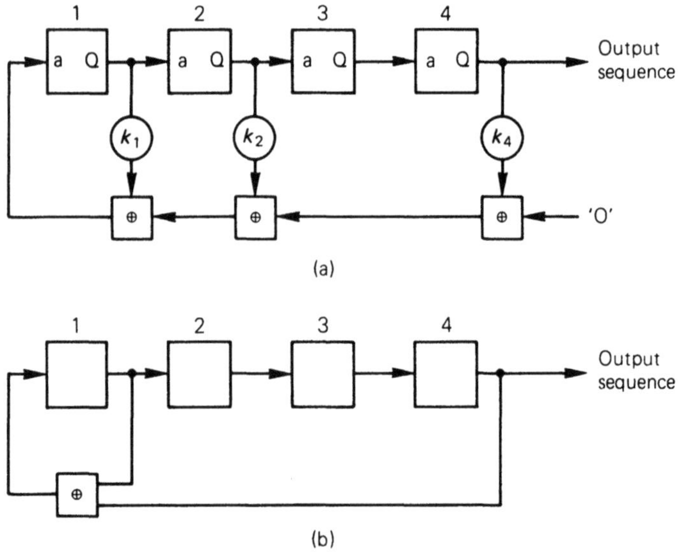

Figure 5.15 Example 5.3

(a) This connection produces a maximal length sequence. Start, for example, with 1000 in the register; then the successive combinations are

1	2	3	4
1	0	0	0
1	1	0	0
1	1	1	0
1	1	1	1
0	1	1	1
1	0	1	1
0	1	0	1
1	0	1	0
1	1	0	1
0	1	1	0
0	0	1	1
1	0	0	1
0	1	0	0
0	0	1	0
0	0	0	1

Thus the output sequence is 000111101011001.

(b) This arrangement produces any one of three short sequences dependent on the starting combination:

1	2	3	4
1	0	0	0
0	1	0	0
1	0	1	0
0	1	0	1
0	0	1	0
0	0	0	1

1	2	3	4
1	0	0	1
1	1	0	0
1	1	1	0
1	1	1	1
0	1	1	1
0	0	1	1

1	2	3	4
1	0	1	1
1	1	0	1
0	1	1	0

The number of m-sequences for a register of length n, and the feedback arrangements to achieve them, are not at all obvious, but have been worked out for a large number of cases [Messina (1972)]. A four-bit LFSR will produce only one m-sequence, but a 10-bit register can produce thirty distinct m-sequences, and a 30-bit register produces no less than 8 910 000 distinct sequences! Clearly, we have the ability to produce a large number of different sequences, but to see why they are so important we must look at their general properties.

The first important property is that the ones and zeros in the sequence occur randomly; there is no way of deducing from the sequence itself what the next bit will be, although the sequence always contains one more one than zeros. Strictly, of course, since the sequence ultimately repeats and we can recreate it whenever we wish, it is not mathematically random. Nonetheless, the sequences have random properties and are therefore also known as *pseudo-random binary sequences*, PRBS. Such sequences are very useful in producing repeatable 'white noise' test patterns when measuring the noise performance of communications, control and instrumentation systems.

Other properties are used in error-checking codes (Golomb, 1982). The m-sequence generated by an n-bit LFSR contains all combinations of n bits (except all zeros, of course), as shown in figure 5.16a for the sequence when n

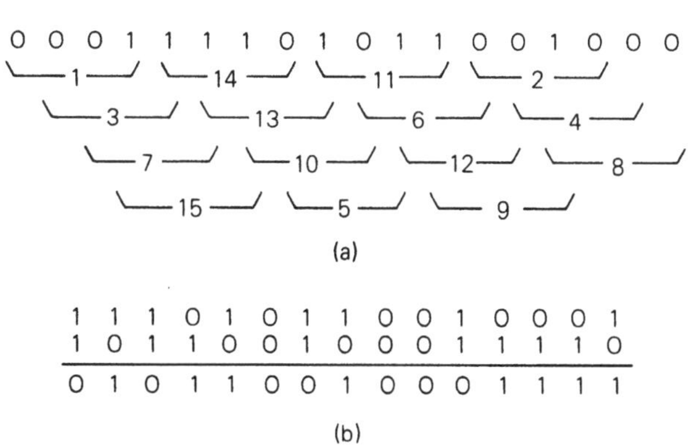

Figure 5.16 Properties of an m-sequence

equals four. Furthermore, if we take any shifted version of the m-sequence and add it, modulo-2, to the same m-sequence shifted a different number of bits, we get the same m-sequence yet again, figure 5.16b. In general, since modulo-2 addition gives a one whenever the two bits differ, and the resultant is always a shifted version of the same m-sequence, any m-sequence must differ from its shifted version in as many bits as there are ones in the sequence, that is $2^n/2$ or 2^{n-1}. In our four-bit example, for instance, the number of bits which differ is always eight. The basis of many error-checking codes is then to append a number of check bits to the data bits which are to be transmitted, in such a way as to form an m-sequence. These check bits are redundant as far as the data is concerned but ensure that each transmitted group of bits differs from any other group in a known number of bit positions. This is called the *Hamming distance* of the code (Hamming, 1950). If, for example, we build a four-bit data group into a 15-bit sequence, we know that each group differs from any other in eight places. We can then tolerate up to four errors in receiving the code group before it becomes impossible to determine what should have been from what was actually received.

5.4 Counters

A counter is a circuit which changes state each time a pulse is applied and gives, in a convenient form, an output code indicating the number of pulses received. The simplest counters give a binary code, but since decimal representation is often preferred, decade counters are also widely used. With suitable gating a counter for any radix can be constructed.

We have seen that a JK or D flipflop can be made to change state each time a pulse is applied, and this toggling can be used as the basis of the counter. If a signal alternating between logical zero and one, such as a square wave sequence, is applied to the clock input of flipflop A in figure 5.17a, toggling will occur. The signal at Q_A is itself a square wave, at half the original frequency, and causes flipflop B to toggle. Flipflop B in turn drives flipflop C, so that the state of the flipflops after each input pulse gives an indication, in binary, of the number of pulses that have occurred, as shown in figure 5.17b. With three flipflops the maximum binary count is seven and the sequence then repeats. In general, if there are n flipflops the maximum count is $2^n - 1$. A *reset* connection is provided on the counter to allow the flipflops to be preset to a known state, normally $Q_A = Q_B = Q_C = 0$ as shown. The counter in figure 5.17 is a binary up-counter, in that the count progresses from zero to seven. By taking the inverse signal from each flipflop as the clocking signal for the next flipflop, and initially setting the counter to 'all ones' instead of 'all zeros', a binary down-counter is obtained.

Counters relying on toggling of flipflops are known as *ripple* counters since, after an input pulse, the final state of the circuit is reached only when the flipflops have had time to change state successively. Because of the ripple delay, and the consequent incoherent edges in the output waveforms, this type of

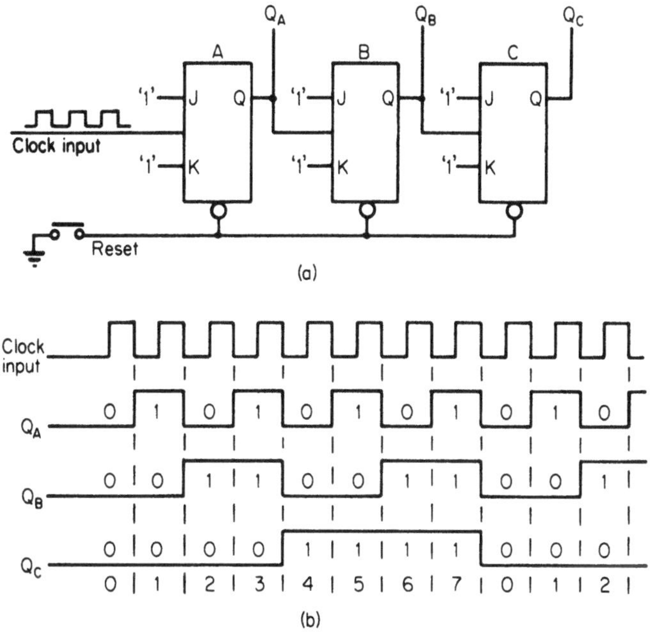

Figure 5.17 A three-stage binary counter: (a) circuit; (b) output waveforms obtained

circuit is unacceptable for large-capacity counters involving output decoding, or for work at high speeds. An alternative construction gives *synchronous* operation by feeding the input signal directly to each flipflop, and changes of state are determined by gating at the J and K inputs. JK flipflops with three-input AND gates at each J and K input are manufactured specifically for use in counters, and, to allow the flipflop to toggle, all the J inputs and all the K inputs must be at the one level. The design process involves the generation of the correct gating signals as the count proceeds.

Example 5.4: Design a four-stage synchronous binary up-counter using JK flipflops.

The required binary count is shown in figure 5.18a and from this is seen that the first stage is required to change each time an input pulse occurs. Thus flipflop A is to toggle, and $J_A = J_1 J_2 J_3 = 1$, $K_A = K_1 K_2 K_3 = 1$.

Also the second stage must change when the first stage is at one; thus $J_B = Q_A$, $K_B = Q_A$.

The third stage must change when both Q_A and Q_B are at one; hence $J_C = Q_A Q_B$, $K_C = Q_A Q_B$,
and finally the fourth stage must change when $Q_A Q_B$ and Q_C are all at one; $J_D = Q_A Q_B Q_C$, $K_D = Q_A Q_B Q_C$.

The complete circuit is shown in figure 5.18b.

Clock input pulse	Q_A	Q_B	Q_C	Q_D
0	0	0	0	0
1	1	0	0	0
2	0	1	0	0
3	1	1	0	0
4	0	0	1	0
5	1	0	1	0
6	0	1	1	0
7	1	1	1	0
8	0	0	0	1
9	1	0	0	1
10	0	1	0	1
11	1	1	0	1
12	0	0	1	1
13	1	0	1	1
14	0	1	1	1
15	1	1	1	1
16	0	0	0	0

(a)

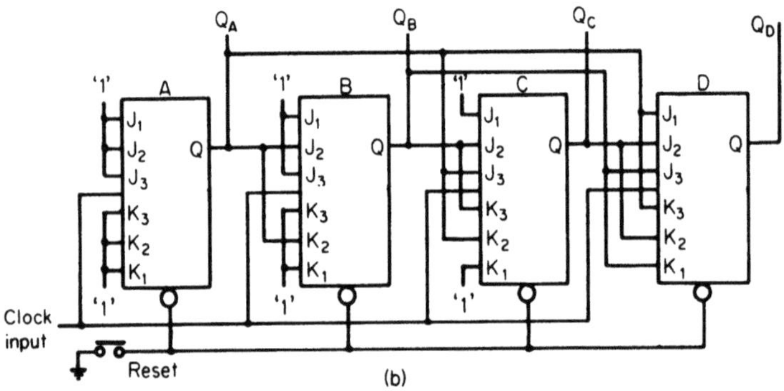

(b)

Figure 5.18 Synchronous binary counter: (a) binary count sequence; (b) circuit

The three-input flipflop has its limitations of course; the fourth stage of the counter in the example already has all its inputs in use, so if more than four stages are required it is necessary to change the procedure. In such a case it is possible to treat the fifth stage as being asynchronous, connecting its clock input directly to Q_D; then further stages, to a maximum of four, can be connected synchronously.

A decade counter uses a base of ten, and to give the ten different states at least four stages are necessary in the counter. On every tenth pulse the count must return to zero, and this is achieved by signals fed back from the final stage. Figure 5.19 shows a common decade counter, the 74LS90A, which is organized as a *divide-by-two* counter and a *divide-by-five* counter. The divide-by-two counter is simply a toggling flipflop, stage A, which can be used to feed a clocking signal to the divide-by-five counter, stages B, C and D. When stage D sets to one, at count 8, the feedback signals set the J input of stage B to zero and the K input of stage D to one so that when stage A switches from one to zero, that is as the count goes from 9 to 10, both stages B and D switch to zero as required. By connecting several of these counters together as in figure 5.20, BCD representation and counting can be achieved. Note that the counter as connected in figure 5.19 gives a standard binary count from zero to nine, so that the output waveform from stage D is asymmetrical. In some cases, however, it is more convenient to have a symmetrical waveform. This is very useful in frequency division systems, and a symmetrical waveform at one-tenth of the clocking frequency can be achieved by using stage A as the fourth stage of the counter: that is by connecting Q_D to input A, feeding the count input signal into input B, and taking the output from Q_A. The output code is not then the standard binary code of course.

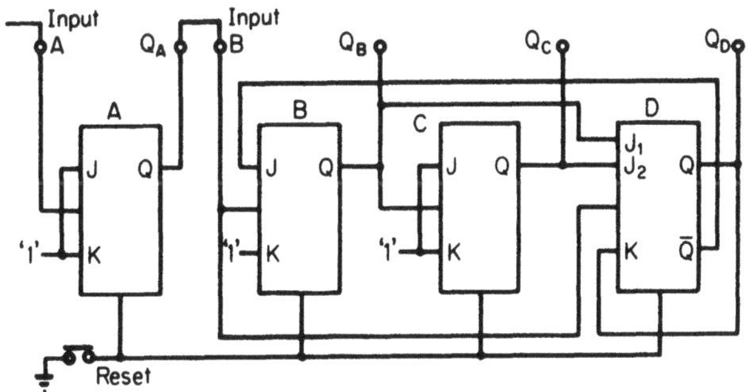

Q_D	Q_C	Q_B	Q_A	Input pulse
0	0	0	0	0
0	0	0	1	1
0	0	1	0	2
0	0	1	1	3
0	1	0	0	4
0	1	0	1	5
0	1	1	0	6
0	1	1	1	7
1	0	0	0	8
1	0	0	1	9
0	0	0	0	10

Flipflop reset to '0'

Flipflop prevented from setting to '1'

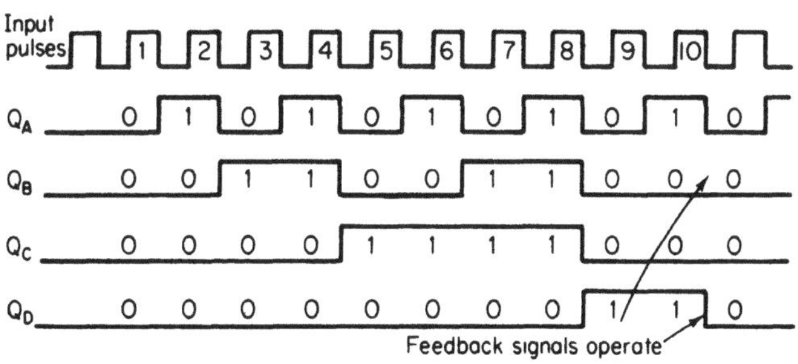

Figure 5.19 A decade counter, 74LS90A

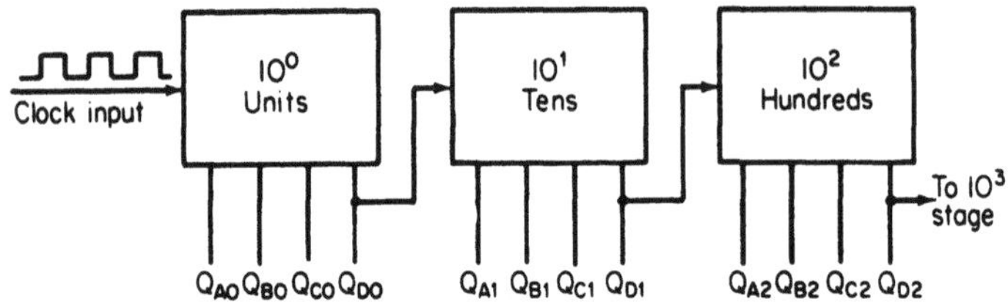

Figure 5.20 A BCD counter

A decade counter can be constructed by applying feedback connections to a five-stage shift register (figure 5.21a). This counter makes use of the Johnson or creeping code and, because of the twist in the feedback connections, is often called a *twisted ring* counter. This should not be confused with a *ring* counter which, as we shall see, is a counter having only one stage at '1' at any time. Although the output code is not binary, it is convenient where decoding to give a decimal representation is required, since no more than two inputs are necessary for each gate, as in figure 5.21c. In general terms an n-stage Johnson counter has a maximum count of $2n$, compared with the 2^n of a binary counter, and is

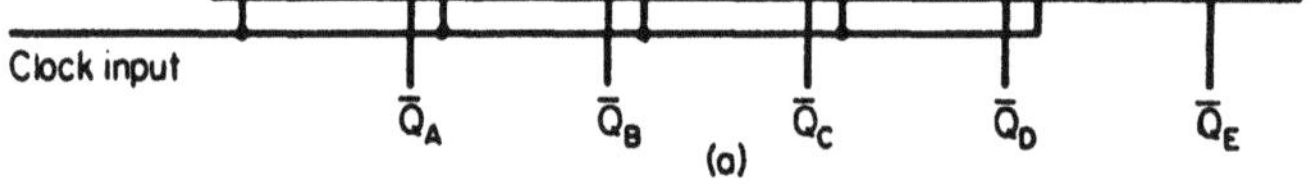

Q_A	Q_B	Q_C	Q_D	Q_E	Clock input pulse
0	0	0	0	0	0
1	0	0	0	0	1
1	1	0	0	0	2
1	1	1	0	0	3
1	1	1	1	0	4
1	1	1	1	1	5
0	1	1	1	1	6
0	0	1	1	1	7
0	0	0	1	1	8
0	(	0	0	1	9
0	0	0	0	0	10

(b)

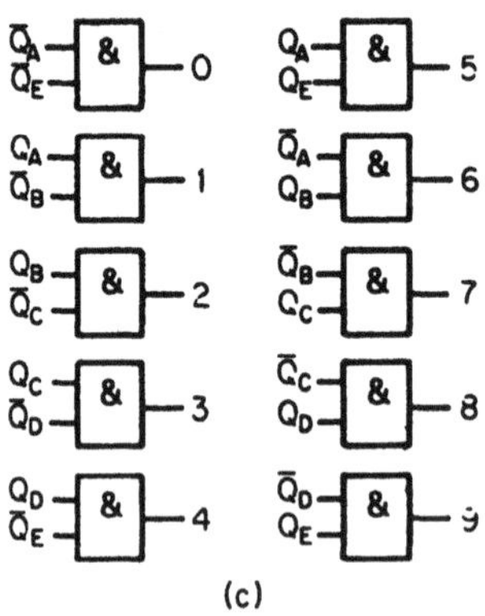

Figure 5.21 Five-stage Johnson counter with decimal decoding gates

practically limited, therefore, to relatively small bases. Johnson counters to any even base, $2n$, are achieved merely by adjusting the number of stages to n, and only minor changes are necessary in order to count to an odd base, $(2n - 1)$. The counter still requires n stages, but the feedback to the K input of the first stage must be taken from the penultimate stage rather than the last.

As we have seen, with suitable gating, counters to any base can be constructed, but in the majority of cases it is preferable to use one of the programmable counters available in integrated circuit form, rather than spend time devising a special circuit. A programmable counter is normally a four-stage synchronous counter with complex gating arrangements allowing it to count up or to count down dependent on which input is pulsed. In some cases a single input is used in conjunction with a count up/count down control input. Additional gating is included to allow the counter to be loaded with four data bits, as in a register, completely independent of the input pulses, when the LOAD signal is applied; see figure 5.22. Further pulses now cause the counter to count from

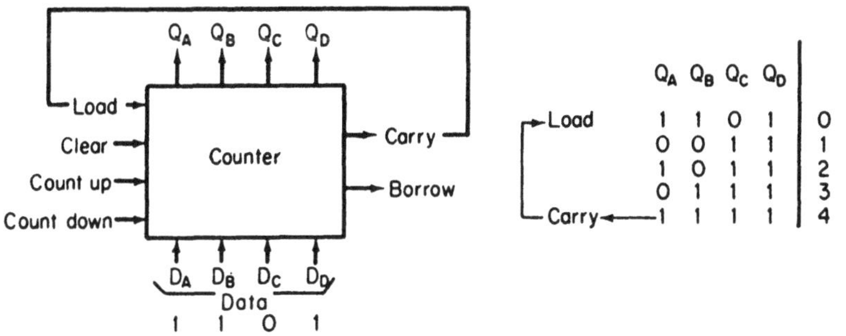

Figure 5.22 Programmable counter

that initial value. Yet more gates detect when the counter reaches its maximum value, and generate a CARRY output on the next pulse. Similar gates generate a BORROW signal when the minimum value is reached when counting down.

It is, therefore, very simple to cascade as many counters as necessary, but the most powerful feature of these counters is the ability to use the carry (or borrow) signal to reload the data value, thus allowing the count cycle length, or base, to be varied. Suppose the data word used is 1011, for example. Having loaded the counter with 1011 the count proceeds as pulses are applied, up to 1111. On the next pulse, the carry signal is generated and is used to reload 1011. Thus the sequence cycles every five pulses, giving a count of five.

A *ring counter* is formed when the outputs of the last stage of a shift register, such as in figure 5.12, are connected back to the inputs of the first stage, Q to J and $\overline{Q}$ to K. If all stages except one are initially set to zero, successive pulses to the ring counter will step the single one continuously around the ring. With

n stages the counter is a divide-by-n counter, but its main use is in generating sets of sequential timing waveforms from a free-running clock signal.

5.5 Timing circuits

The clock signal may be generated by a simple circuit such as is shown in figure 5.23a. Three open-collector gates are connected in a ring, and the frequency of operation is determined partly by the capacitor and resistor values and partly by the propagation performance of the gates. The frequency is not very stable and is limited to a range of a few megahertz. A much more stable oscillator can

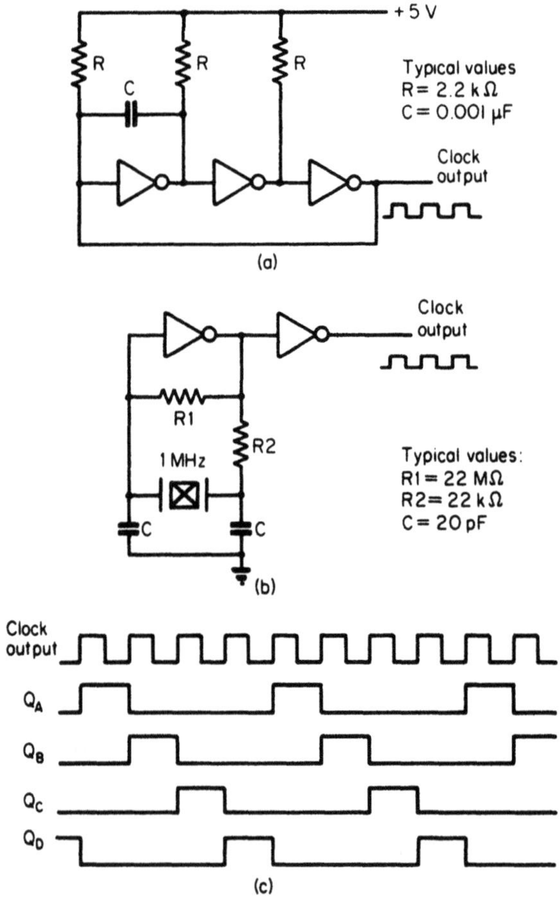

Figure 5.23 (a) Ring oscillator using open-collector TTL gates. (b) Crystal-controlled oscillator using CMOS gates. (c) Sequential timing waveforms generated by a ring counter

be constructed by the use of a CMOS gate biased into its active region, with a crystal to control the frequency very accurately, as in figure 5.23b. An inverter is included after the active gate to act as a high input impedance buffer preventing any capacitive loading effect of the subsequent circuitry and its wiring from affecting the frequency of operation. The ring counter is then used to produce the individual timing waveforms as illustrated in figure 5.23c, in which the number of flipflops is taken to be four.

It is sometimes necessary to control or generate a succession of signals from a single event, and the ability to create timing pulses of the correct duration and polarity is then very important. In many of these cases the required pulses can be derived by use of counters and gating from clocking waveforms already in use in the system. Wherever possible it is preferable to do it that way, but sometimes it is necessary to resort to special circuitry, and a monostable multivibrator may be the best answer. The monostable multivibrator, normally abbreviated to plain *monostable*, is a two-state circuit, but only one of the states is stable. When triggered into the quasi-stable state the circuit automatically switches back to its original stable condition after a time delay which is governed by specific timing components. This method of operation leads to the device also being known as a *one-shot*. Simple monostable circuits can be constructed using two CMOS gates and a resistor–capacitor network, but the performance is heavily dependent on the input characteristics of the gates chosen, and it is preferable to use one of the integrated circuit versions provided by the semiconductor manufacturers.

We saw in chapter 3 that races between two signals can lead to hazardous operation when both change in response to the same signal and ultimately feed through to the same gate. The race occurs as the result of differing delays in the separate paths taken by the signals, and, in certain cases, we can make use of these gate delays to generate pulses from simple signal transitions.

Example 5.5: Use a quad 2-input NOR gate package to generate a positive-going pulse whenever the input signal switches to logic '0'.

The circuit and timing waveform are shown in figure 5.24 (see page 158).

Problems

5.1 Modify the circuit of figure 5.17 to a down-counter, and verify its action by drawing the waveforms obtained from each flipflop.

5.2 Consider the two-dimensional array of figure 5.5 for multiplying two four-bit numbers, and show how the same result could be achieved using time sequenced operations rather than spatial.

5.3 If the counter of figure 5.19 is rearranged as suggested to give a symmetrical output from the fourth stage, what is the output code sequence from the counter?

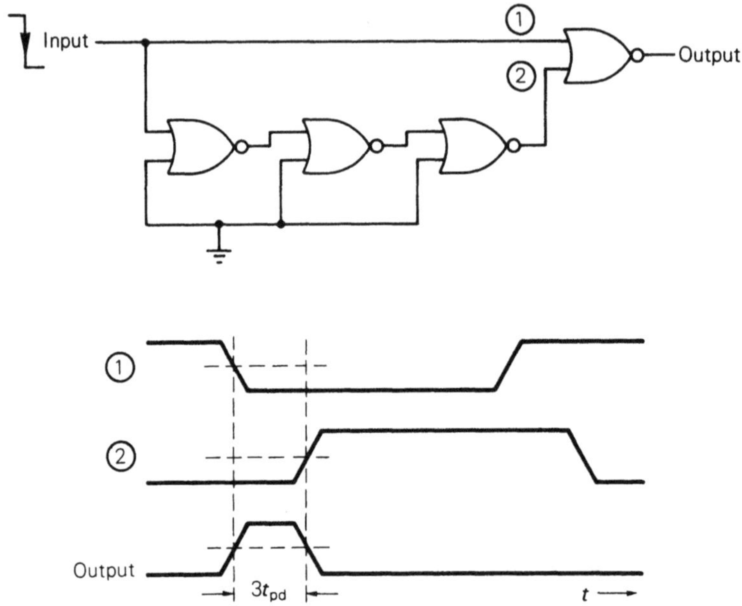

Figure 5.24 Use of gate delays in pulse generation

5.4 Design a ring-counter circuit to produce the timing waveforms of figure 5.23c using (a) D flipflops, (b) JK flipflops.

5.5 An iterative array is to consist of cells each with a single external input and a single external output. A one output is to occur from the cell at the left-hand end of alternate groups of input ones, starting with the leftmost group. Draw the cell matrix and hence design a cell circuit using NAND gates.

5.6 Develop the flow matrix for a parallel adder and use it to check the addition of the numbers 00100 and 01101.

5.7 The all-zeros state is not included in a maximal length sequence, but in testing logic circuits it is often necessary to carry out exhaustive testing which requires the application of all 2^n possible input combinations. Modify the circuit used in example 5.3a, by the addition of a single AND gate, to include the all-zeros state. What then is the output sequence?

5.8 The 4-bit LFSR of example 5.3 produces only one maximal length sequence, but it can produce the same sequence in reverse if the feedback connections are taken from flipflops 3 and 4. Confirm that this is so.

5.9 Use 2-input NAND gates to generate a negative-going pulse whenever the input signal goes to logic '1' (in the same way as in example 5.5).

5.10 We can easily halve the frequency of a clock waveform by use of a toggling flipflop, but it is much more difficult to double the frequency. Devise an arrangement using delays in exclusive-OR gates to produce two pulses for each cycle of the input clock waveform.

References

Edwards, C. R. and Hurst, S. L. (1976). An analysis of universal logic modules. *Int. J. Electron.*, **41**, 625-8

Golomb, S. W. (1982). *Shift Register Sequences*, Aegean Press, California

Hamming, R. W. (1950). Error-detecting and error-correcting codes, *Bell Syst. Tech. J.*, **29**, 147-60 (April)

Messina, A. (1972). Considerations for non-binary counter applications, *Computer Design*, **11**, No. 11 (November)

6 Sequential Logic Analysis and Design

The counters and registers we considered in the previous chapter are standard sequential circuits, but many applications require other specially designed circuits, and we must now look to the general theory behind all sequential circuits in order to formulate design methods which are broadly applicable.

There are many different ways in which the inherent behaviour of a sequential logic circuit can be described and summarized. The concept of successive states, as discrete intervals of time pass, necessitates some form of description which includes state transition information related to the allied changes in input and output variables.

State diagrams are ideal for small systems, but the rapid increase in complexity as the size of the system increases has led to the development of more sophisticated chart description methods. We will introduce the basic concepts of state operation by analyzing simple circuits using state tables, and then show how they can be used in the design of circuits to meet given specifications. Finally we will look at a popular chart method, the *algorithmic state machine* chart, which presents the same information in an alternative and often more comprehensible form.

6.1 Analysis using state tables

The output signals from a sequential circuit are dependent on both the *internal state* of the circuit and the input variables, and to determine the reaction of the circuit to any input changes we must construct the *state table*. This is the equivalent in sequential circuit theory of the flow matrix in iterative array theory. As a first step, with the circuit of figure 6.1, our combinational logic knowledge allows us to develop expressions for the excitation signals applied to the storage elements (JK flipflops in this case), and for the circuit output values in terms of the flipflop settings at that time. Thus the *excitation equations* are $J_A = a\bar{b}$, $K_A = \bar{a}$, $J_B = \bar{a}b$ and $K_B = ab$. The output equations are $Z_1 = q_A c$,

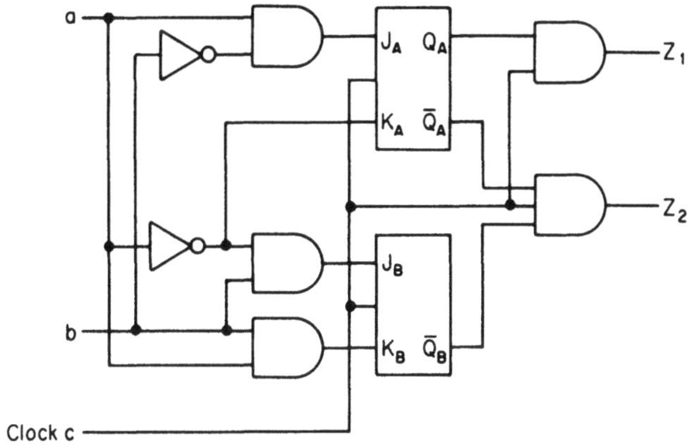

(a)

Present inputs J	Present inputs K	Next state Q
0	0	q
0	1	0
1	0	1
1	1	$\bar{q}$

(b)

Figure 6.1 Analysis of a sequential circuit: (a) circuit for analysis; (b) truth table

$Z_2 = \bar{q}_A\bar{q}_B c$, where $q_A q_B$ represent the present states of the flipflops, so that $q = 1$ when the flipflop is set, and $\bar{q} = 1$ when it is reset.

However, the relationship between the flipflop state and its excitation signals is governed not by combinational logic but by the time-dependent transition behaviour of that type of flipflop. The state to which the flipflop will next set, under given excitation conditions, is represented by the *next state variable*, Q, and for a JK flipflip the *transition equation* is $Q = J\bar{q} + \bar{K}q$. We first use the excitation equations to plot the values of the signals reaching the flipflops A and B under all input and present state conditions, figure 6.2a. The clock signal, c, controls the timing of the state changes but plays no part in determining what the next state will be so is not included in the excitation equations. The table of figure 6.2a is the *excitation table*, in which the column headings relate to the input variable values and the row headings to the present state of the flipflops. In this case the table has identical rows indicating that the signals to the flipflops do not depend on the state of the flipflops.

The transition equations allow us next to replace each entry of the excitation table with the setting of the flipflops resulting from the specified excitation conditions, and the *transition table* of figure 6.2b results. All the excitation entries for $\bar{c}$ $(c = 0)$, which corresponds to the absence of the clock pulse, lead to no change in the flipflop settings, as shown by the row entries merely repeating the row headings. When the clock pulse is present $(c = 1)$, however, the row entries vary, indicating that the next state of the flipflops is governed by the present states and the action of the excitation signals upon them. Thus with input 00, for instance, the excitation signals 0100 ensure that flipflop A will change to the reset condition and flipflop B will remain in its previous state.

The *output table* of figure 6.2c contains entries giving the output signal values and, by superimposing the transition and output tables and allocating

(a)

q_Aq_B \ ab	00	01	11	10
00	0100	0110	0001	1000
01	0100	0110	0001	1000
11	0100	0110	0001	1000
10	0100	0110	0001	1000

$J_A\ K_A\ J_B\ K_B$

(b)

q_Aq_B \ ab	c=0 00	01	11	10	c=1 00	01	11	10
00	00	00	00	00	00	01	00	10
01	01	01	01	01	01	01	00	11
11	11	11	11	11	01	01	10	11
10	10	10	10	10	00	01	10	10

$Q_A\ Q_B$

(c)

q_Aq_B \ ab	c=0 00	01	11	10	c=1 00	01	11	10
00	00	00	00	00	01	01	01	01
01	00	00	00	00	00	00	00	00
11	00	00	00	00	10	10	10	10
10	00	00	00	00	10	10	10	10

$Z_1\ Z_2$

Figure 6.2 Tables for the sequential circuit: (a) excitation table; (b) transition table; (c) output table

letters to the various state codes, we arrive at the state table, figure 6.3. Each entry gives the next state of the circuit and the present output values.

Our analysis of the circuit is now complete and we can predict, with the aid of the state table, the precise sequence of state changes and output values resulting from any sequence of inputs.

Let us assume, for instance, that the circuit is initially in state A with the input variables at 00 and no clock pulse present. We shall represent the total state information as A-00, referring to the present state and present input values respectively. Also $(\text{A-00})\bar{c}$ will mean that the circuit is in total state A-00 with

(a) Present inputs

Present state \ ab	c=0 00	01	11	10	c=1 00	01	11	10
A	A,00	A,00	A,00	A,00	A,01	B,01	A,01	D,01
B	B,00	B,00	B,00	B,00	B,00	B,00	A,00	C,00
C	C,00	C,00	C,00	C,00	B,10	B,10	D,10	C,10
D	D,00	D,00	D,00	D,00	A,10	B,10	D,10	D,10

Next state / Present outputs

(b)

	Null	c=1 ab 00	01	11	10
A	00	A,01	B,01	A,01	D,01
B	00	B,00	B,00	A,00	C,00
C	00	B,10	B,10	D,10	C,10
D	00	A,10	B,10	D,10	D,10

Figure 6.3 State tables for circuit of figure 6.1: (a) state table; (b) simplified format

no clock pulse present. The next state entry for (A-00)$\bar{c}$ is A, from figure 6.3, indicating that no state transition will occur. In other words the circuit is in a stable state.

Examination of the left-hand half of the state table shows that the next state entries are all identical with the heading of the row in which they occur, and the circuit is stable when the clock pulse is not present. This is a property of all clocked, or synchronous, sequential circuits. Also, for our circuit, the output value for all states is 00 when the clock pulse is absent, since both output gates are controlled by the clock signal.

When the clock signal becomes '1' the 'operating point' moves from (A-00)$\bar{c}$ to (A-00)c. We will assume that the flipflops are of the master-slave type so that any change of state occurs on the back edge of the clock signal. However, the entry under (A-00)c is A-01, indicating that the output signals have the value 01 as long as the clock pulse is present, and that the next state of the circuit is still A. At the end of the clock pulse the operating point moves back to (A-00)$\bar{c}$ and the output value reverts to 00. Let the input signals now change to 01. Our total state is (A-01)$\bar{c}$, which is still stable and gives an output value 00. The next clock pulse, however, takes us to (A-01)c, again giving outputs 01, but now indicating a transition to state B. Thus, when the clock pulse is removed, the operating point moves to (B-01)$\bar{c}$. A further change in input signals to 10 leads to state (B-10)$\bar{c}$ followed, when the clock pulse is applied, by (B-10)c. The entry for (B-10)c gives the next state as C, and the operating point moves to (C-10)$\bar{c}$ when the clock pulse is removed.

The use of the table is now clear; with this type of circuit, input variable changes lead to the operating point moving *within* a row but state changes leading to *diagonal* movement are controlled by the clock pulses. To ensure correct action of the circuit it is necessary to restrict input variable changes to the inactive periods between clock pulses, and it is arranged that the delay within any synchronous circuit is such that only one transition can occur during each clock pulse.

It is apparent from the entries in the left-hand half of the state table that the information there is independent of input values and a single *null* column is sufficient to present the same information, as in figure 6.3b.

The type of circuit considered so far has output gating which includes the clock signal and so gives pulse-type output signals. The circuit with this arrangement is often referred to as a PP circuit (Cadden, 1959) meaning that it uses pulse-type inputs and gives pulse outputs. An alternative form of output gating is possible, which is independent of the clock signal and therefore gives a constant output signal level governed only by the present state and present inputs of the circuit. Such a circuit is described as a PL circuit, indicating pulse-in, level-out. The state table for the circuit should strictly retain all the columns in the null region since the output values may differ under different input signal conditions. Movement around the table still follows the action of the clock pulses, causing the operating point to move alternately between the null region and active region of the table.

A third version of the table arises when the circuit is asynchronous. The circuit is now an LL circuit, meaning level-in, level-out. Since there is no clock signal, movement of the operating point around the table is governed solely by the changes of the input variables, and there is no null region at all. This operation is also referred to as *fundamental mode operation*.

The circuit of figure 6.4 operates in fundamental mode and uses a read-only memory which is programmed with the data listed. The list contains all the state and input and output information we need to plot the state table in figure 6.4b. The circuit will reach a stable condition only when the indicated next state is the same as the present state. Stable states are shown circled on the table, and it should be noted that each row contains at least one stable state. Uncircled entries represent transient states.

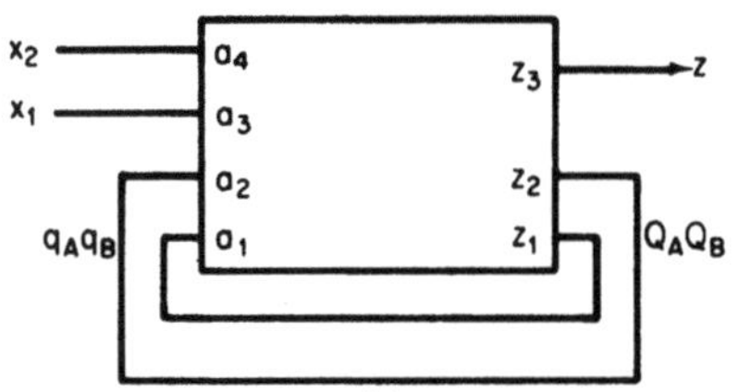

(a)

	x_1x_2 00	01	11	10
(00) A	A,1	A,1	B,1	A,1
(01) B	A,1	C,1	B,1	C,1
(11) C	D,0	C,0	C,0	C,0
(10) D	D,0	A,0	C,0	A,0

(b)

Present state; Present input; Next state; Present output

q_A	q_B	x_1	x_2	Q_A	Q_B	z
0	0	0	0	0	0	1
		0	1	0	0	1
		1	1	0	1	1
		1	0	0	0	1
0	1	0	0	0	0	1
		0	1	1	1	1
		1	1	0	1	1
		1	0	1	1	1
1	1	0	0	1	0	0
		0	1	1	1	0
		1	1	1	1	0
		1	0	1	1	0
1	0	0	0	1	0	0
		0	1	0	0	0
		1	1	1	1	0
		1	0	0	0	0

(c)

Figure 6.4 Fundamental mode circuit: (a) sequential circuit; (b) state table; (c) ROM program

As an illustration of the use of the table let us suppose that the circuit is in state B-11. Since the next state entry is B, the circuit is stable, and the output value is 1. The input variables now change to 10 and the entry for state B-10 gives the next state as C. In state B-10 the circuit is no longer stable and a transition takes place, involving vertical movement on the table, and the circuit reaches the new stable state C-10. Note that during the transition the output value changes from 1 to 0. A change of input variables now to 11 would not disturb the circuit since C-11 is a stable state but a change to 00 would induce a transition to D, and so on.

In practice, with all types of circuit the null information, if any, is often omitted and must be inferred from the row headings of the table.

We have seen that the analysis of a sequential circuit requires knowledge of the state variables. When bistable, delay or memory elements are used in the circuit it is clear that the state variables correspond to the output signals of those elements, but where a circuit relies on simple feedback loops for its storage properties it is sometimes difficult to decide which signals act as state variables.

In order to discover the state variables it is necessary to locate the minimum number of points of the circuit which, if the circuit were broken at those points, would result in all closed loops being interrupted. The decision on the number of state variables is not critical as long as all loops are catered for; the resulting state tables will differ only in the transient indications.

Example 6.1: Derive the state tables for the asynchronous circuit of figure 6.5.

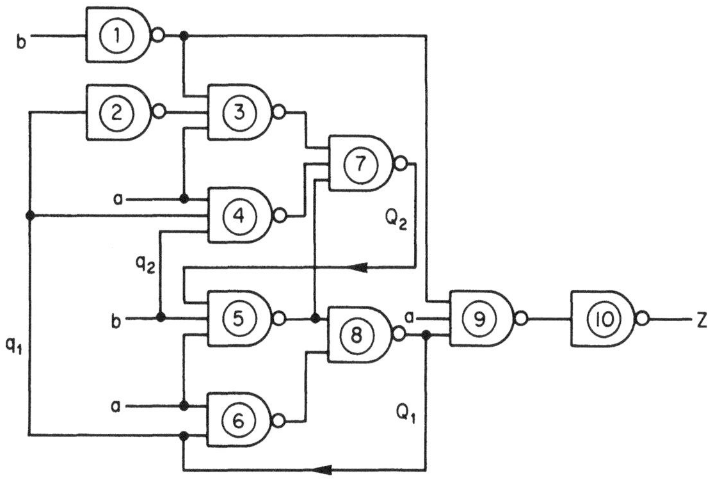

(a)

q_1q_2 \ ab	00	01	11	10
00	00,0	00,0	00,0	01,0
01	00,0	00,0	11,0	01,0
11	00,0	00,0	11,0	10,1
10	00,0	00,0	11,0	10,1

Q_1Q_2 Z

(b)

ab	00	01	11	10
A	A,0	A,0	A,0	B,0
B	A,0	A,0	C,0	B,0
C	A,0	A,0	C,0	D,1
D	A,0	A,0	C,0	D,1

(c)

Figure 6.5 Analysis of sequential circuit: (a) sequential circuit; (b) transition table; (c) state table

In order to locate the state variables we note that the output of gate 8 is part of the loops which include gates (8, 6, 8), (8, 4, 7, 5, 8) and (8, 2, 3, 7, 5, 8). The output signal of gate 8 is therefore selected as the first state variable, Q_1. The loop including gates (7, 5, 7), however, has not been covered and a second variable, Q_2, must be allotted at the output of gate 7. The analysis now proceeds as before

$$
\begin{aligned}
Q_1 &= aq_1 + abq_2 \\
Q_2 &= abq_1 + a\bar{b}\bar{q}_1 + abq_2 \\
Z &= a\bar{b}q_1
\end{aligned}
$$

These equations lead to the tables of figure 6.5, and by inspection it is apparent that this circuit detects the input sequence 10, 11, 10.

6.2 Synthesis using state tables

The synthesis of a sequential circuit involves the reversal of the analytical procedure and takes the state table as the starting point in working towards the final circuit.

In common with combinational logic circuits, the initial word-form of the specification is often ill-defined and the ambiguities are best resolved by drawing up the state table. If the final circuit is to be synchronous a convenient half-way stage to the state table is the state diagram, which is the sequential circuit version of the flow diagram discussed in connection with iterative arrays. In fact the flow diagram that is used in the early stages of an iterative array design could equally well be used in the design of a sequential circuit. Recall that in example 5.2 we developed the flow diagram and matrix for an array giving an output from each cell whose input is such that the total number of ones occurring in the input sequence, to that point, is an odd number. In sequential circuit terms the equivalent operation requires a one output during all periods in which the total number of ones in the time-separated sequence applied to the single input is an odd number. This is a modulo-2 counter. The state diagram and corresponding state table are shown in figure 6.6. Each present state is represented by a node of the diagram. The arrows carry an indication of the present input causing the transition and the present output, respectively.

The state diagram shown is in the form of a Mealy model (Mealy, 1955). An alternative representation is the Moore model (Moore, 1956), in which the output values are considered to be dependent solely on the internal state of the circuit. Each node therefore carries the output value occurring when the circuit is in that state, and the transition arrows indicate the input values only. Thus the Moore model is the representation of an *internal state* circuit and normally refers to a PP (pulse-in, pulse-out) circuit. The Mealy model is the more general and includes the internal state circuits but, in design, the Moore model often tends to a simpler final circuit. The Moore model of the modulo-2 counter is given in figure 6.7.

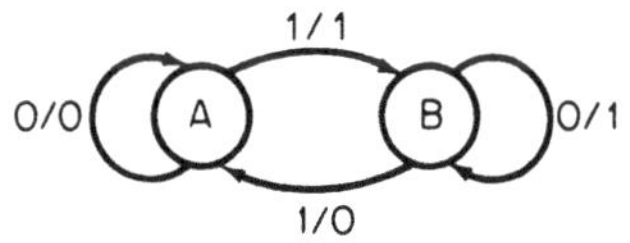

(a)

	N	0	1
A	A,0	A,0	B,1
B	B,0	B,1	A,0

(b)

	0	1
A	A,0	B,1
B	B,1	A,0

(c)

Figure 6.6 Modulo-2 counter: (a) state diagram; (b) Mealy model state table; (c) abbreviated form of table

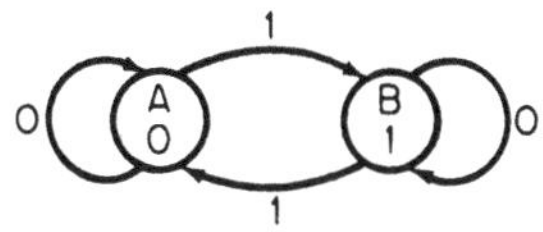

(a)

	0	1	Output Z
A	A	B	0
B	B	A	1

(b)

Figure 6.7 Alternative model of the modulo-2 counter: (a) state diagram; (b) Moore model state table

In drawing the state diagram it is not vital to use the lowest possible number of nodes, or states, as redundant states can be recognized and removed later. The primary objective at this stage is to get the diagram to 'close', ensuring that the response to all possible input values has been considered at every node.

As a further example, figure 6.8 shows the state diagram for a circuit which responds to the input sequence 1101. This type of circuit is sometimes used in detecting marker codes on bidirectional magnetic tape systems. We see that the successive stages of the recognition of the sequence step the circuit on, and the output signal indicates when the final '1' of the sequence is received. Any input breaking the sequence 'resets' the circuit to the starting condition. Note that the timing of the sequence 1101 is from left to right. Also, though the final '1' in the sequence could act as the first bit of the next sequence, a completely self-contained sequence is required each time in this application; the '1' input at D therefore leads back to A, not to B.

The state diagram can also be used in connection with asynchronous circuits but it is important to note that each node now has a sling, so that the circuit is

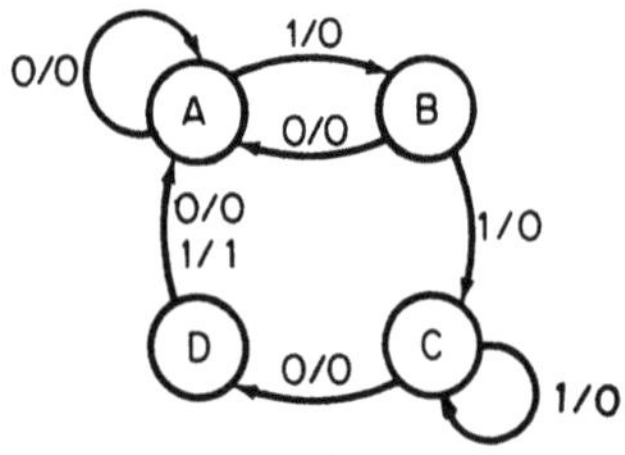

(a)

	0	1
A	A,0	B,0
B	A,0	C,0
C	D,0	C,0
D	A,0	A,1

(b)

Figure 6.8 1101 sequence detector: (a) state diagram; (b) state table

correctly shown as remaining in that state until the input variables change. It is also of interest to note that a single-input asynchronous circuit is virtually useless since the only input sequence possible is alternate ones and noughts.

An alternative approach, often preferable when drawing up the state table for an asynchronous circuit, is to develop the *primitive flow table* for the circuit. The primitive flow table differs from the previously described state table in that every row of the table has one and only one stable state. The table necessarily has a large number of rows but can be reduced, by minimizing and merging processes, to the usual form which is more convenient for circuit implementation.

One of the most convenient ways of developing the primitive flow table is to work from the output waveforms which can be expected to arise under all the different operating conditions in the circuit.

Example 6.2 : A parcel-sorting system is to direct each of the parcels travelling on a conveyor belt into one of three traps according to length; those parcels less than one metre in length go straight to the first trap, those between one and two metres are diverted to the second trap, and any longer than two metres are diverted to the third. Three photocells spaced one metre apart provide input signals to the trap control circuits which must deduce the length of each parcel and route it accordingly. Derive the primitive flow table.

The waveforms from the photocell circuitry are shown in figure 6.9 together with the controlling output signals, Z_1, Z_2, needed to activate the trap mechanisms.

We know that our table must have eight columns, since there are eight possible combinations of the photocell outputs. We indicate that the circuit is initially in the quiescent state, no parcels having been sensed, and enter state A as a stable condition (figure 6.10). The outputs are 00. A parcel now passes photocell 1, and the input conditions become 100, so the table must indicate a transition to B-100. Unless something radically wrong develops in the system, the input 100 is the only possible input to follow 000; that is, a parcel of any length must pass photocell 1 first. All other entries in row A, therefore, are don't cares. Now, assuming that the parcel detected is less than

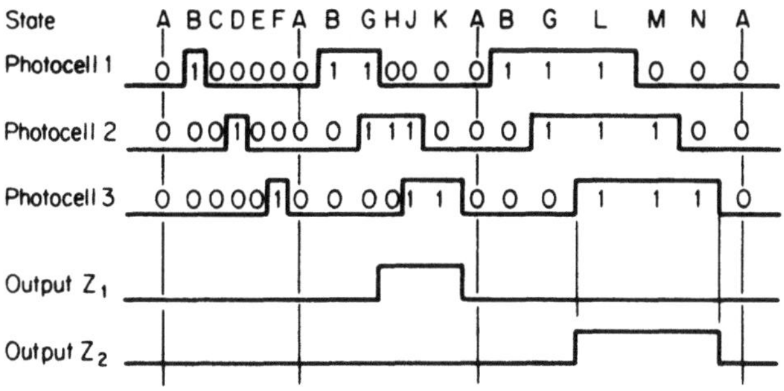

Figure 6.9 Waveforms for the parcel-sorting system

	000	001	010	011	100	101	110	111
A	(A,00)	–	–	–	B,00	–	–	–
B	C,00	–	–	–	(B,00)	–	G,00	–
C	(C,00)	–	D,00	–	–	–	–	–
D	E,00	–	(D,00)	–	–	–	–	–
E	(E,00)	F,00	–	–	–	–	–	–
F	A,00	(F,00)	–	–	–	–	–	–
G	–	–	H,Ø0	–	–	–	(G,00)	L,0Ø
H	–	–	(H,10)	J,1Ø0	–	–	–	–
J	–	K,10	–	(J,10)	–	–	–	–
K	A,Ø0	(K,10)	–	–	–	–	–	–
L	–	–	–	M,01	–	–	–	(L,01)
M	–	N,01	–	(M,01)	–	–	–	–
N	A,0Ø	(N,01)	–	–	–	–	–	–

Figure 6.10 Primitive flow table for the parcel-sorting system

one metre long, it will clear the first cell before passing the second, and the input reverts to 000. The circuit does not return to state A, however, since a parcel has already been detected, and operation must instead move to a new state C. Every time a new state arises we start a new row of the table, with that stable state directly below. As the parcel passes cells 2 and 3 the circuit switches through states D, E and F, and back to state A as the parcel clears the third cell. Longer parcels give rise to the later waveforms of figure 6.9 and new states are introduced as required, so that the full table is as shown.

The reader is advised to follow the process right through. Note that one of the outputs during transient states G-010, G-111, K-000 and N-000 is shown as don't care. The don't care entry indicates that, at this stage, we cannot be sure whether it is better to let the output change while the circuit is still in the original, now unstable, state, or to wait until the new stable condition is established. The decision is taken during the simplification of the table.

The table of figure 6.10 has assumed that the parcels are widely separated so that each parcel clears the third photocell before the next parcel is detected at the first. If we remove that limitation further entries are necessary on the table.

In some cases it is easier to develop the total state table from tables representing subsections of the final circuit. For example, the state table for a circuit which recognizes every second occurrence of a particular sequence can be built up from two sections, the first recognizing the given sequence and the second indicating every second occurrence of a signal from the first section.

Two circuits whose operation is defined by separate state tables may be connected logically either in series or in parallel, and a single state table developed defining the overall action of the composite circuit.

In series connection the output of the first circuit becomes the input to the second.

Example 6.3: Derive the state table for the composite circuit formed when the sequence detector, K1, figure 6.11, is combined with the modulo-2 counter, K2.

By connecting the circuits in series the overall circuit action results in an output of one on completion of every second occurrence of the input sequence 0101. The initial states of K1 and K2 are A and E respectively. The first entry

	0	1
A	B,0	A,0
B	B,0	C,0
C	D,0	A,0
D	B,0	A,1

K1. 0101 sequence detector

	0	1
E	E,0	F,0
F	F,0	E,1

K2. Modulo-2 counter

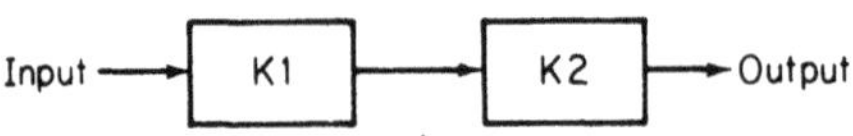

(a)

	0	1
AE	BE,0	AE,0
BE	BE,0	CE,0
CE		

(b)

	0	1
AE	BE,0	AE,0
BE	BE,0	CE,0
CE	DE,0	AE,0
DE	BE,0	AF,0
AF	BF,0	AF,0
BF	BF,0	CF,0
CF	DF,0	AF,0
DF	BF,0	AE,1

(c)

Figure 6.11 Composite circuit, series connection: (a) circuits in series; (b) preparation of composite table; (c) completed table

on our composite table, therefore, is state AE (figure 6.11b). A zero input to K1 gives a zero output and a change of state to B. The zero output causes K2 to give a zero output and a reset to state E. Thus the composite entry on our new table is BE,0.

A one input to K1, when in state A, gives a reset to state A and output zero. The zero as an input to K2 leads again to E,0. The composite entry under AE-1 is therefore AE,0. In completing this row of the table a new state, BE, has been introduced, so we list BE under AE and turn our attention to the effect of each input value on this new state. The entry under B-0 is B,0, and that for E-0 is E,0, so the entry on the new table becomes BE,0. B-1 has an entry C,0 and again the zero entry at E gives E,0, so we enter CE,0 and list the new state CE in the column of row headings. Work continues in this way until all input values have been covered for all composite states introduced, as shown in figure 6.11c.

In developing the composite table for two circuits connected in parallel we assume that the input variables are applied to the two sections simultaneously. The output values are determined by some boolean relationship between the outputs of the separate sections.

Example 6.4: Derive the state table for the composite circuit formed when the two detector circuits, K3, K4 of figure 6.12, are connected in parallel in conjunction with an AND gate.

The state table, K3, defines the operation of a two-input circuit which gives an output value of '1' on the first, fourth, seventh, and $(3n + 1)$th consecutive occasion that both inputs have been at '1'. State table K4 relates to a circuit giving an output '1' when the total number of ones applied to its inputs is an even number. If we operate the two circuits in parallel, with a simple AND function at the output, we arrive at a circuit which gives an output '1' when (a) the inputs have both been at '1', once, four, seven, and $(3n + 1)$ times consecutively, *and* (b) the total number of ones applied to the circuit is an even number.

Again we start our composite table (figure 6.12b) at the initial states of the constituent tables, AD. State A-00 gives the entry A,0 and D-00 gives D,0, so the composite state resulting is AD. Since the two outputs are zero the AND output circuit will also present a zero. State A-01 gives A,0 and D-01 gives E,0 leading to the composite state AE with an output again at zero. The A-11 entry is B,1 and D-11 entry is F,1, so the composite state is BF, and since both circuits give a one output, the overall output is one. The final entry on the first row is AE,0.

The two new states introduced in arriving at the first row are listed in the row headings, and we continue with a row headed AE. On completion of the process the full composite table is as shown in figure 6.12c.

	00	01	11	10
A	A,0	A,0	B,1	A,0
B	A,0	A,0	C,0	A,0
C	A,0	A,0	A,0	A,0

K3. Double-one detector

	00	01	11	10
D	D,0	E,0	F,1	E,0
E	E,0	F,1	E,0	F,1
F	F,1	E,0	F,1	E,0

K4. Even number detector

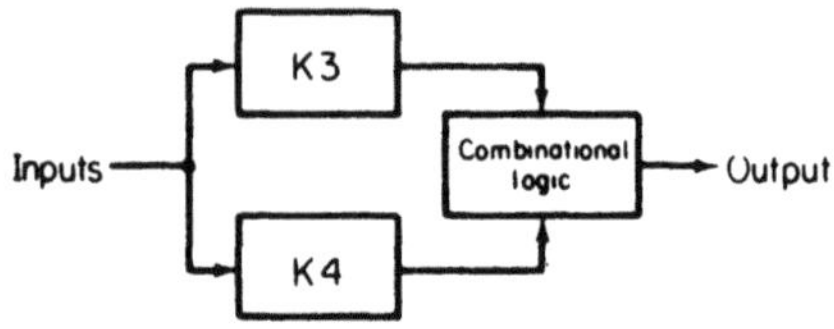

(a)

	00	01	11	10
AD	AD,0	AE,0	BF,1	AE,0
AE	AE,0	AF,0	BE,0	AF,0
BF				

(b)

	00	01	11	10
AD	AD,0	AE,0	BF,1	AE,0
AE	AE,0	AF,0	BE,0	AF,0
BF	AF,0	AE,0	CF,0	AE,0
AF	AF,0	AE,0	BF,1	AE,0
BE	AE,0	AF,0	CE,0	AF,0
CF	AF,0	AE,0	AF,0	AE,0
CE	AE,0	AF,0	AE,0	AF,0

(c)

Figure 6.12 Composite circuit, parallel connection: (a) circuits in parallel; (b) preparation of composite table; (c) completed table

In drawing up the state table for a circuit we have made no effort to limit the number of states to a minimum. Having obtained a state table we must be able to recognize equivalent and redundant states in order to arrive at the best form of the table for circuit implementation. Unfortunately it is not usually possible to define the best form of the table since a sequential circuit is a complex mixture of storage elements and combinational logic; over-simplification of the table may well reduce the amount of storage needed but will simultaneously increase the complexity of the input and output logic circuits. It is usually good policy to minimize the table as far as possible and then to consider the addition of appropriate redundant states which lead to a simpler circuit. This approach is often necessary anyway in order to remove hazards.

Synchronous circuit state tables are normally fully, or very nearly fully, specified; that is, there are few, if any, don't care entries either for next state or for present output. A don't care or unspecified state entry suggests that the total condition will not arise, and we can infer that all subsequent states must also be don't cares.

Two states can be said to be equivalent if all possible input sequences result in the same output sequences, regardless of which of the two states is used as the starting point. The first requirement for equivalence, therefore, is that the states have the same output values. If so they are said to be *output compatible*, and we know that an input sequence of one bit, in either state, gives the same output. Nevertheless, this is not a sufficient condition for equivalence, and it is necessary to check that sequences of any greater length give the same outputs. As the next step we check that input sequences of two bits give the same output sequences, then three bits, four bits, and so on. This checking is most readily achieved by use of a *partitioning* process.

Example 6.5: Minimize the state table of figure 6.13a.

	00	01	11	10
A	A,0	B,0	C,1	B,0
B	B,0	D,0	E,0	D,0
C	D,0	B,0	F,0	B,0
D	D,0	B,0	C,1	B,0
E	B,0	D,0	G,0	D,0
F	D,0	B,0	D,0	B,0
G	B,0	D,0	B,0	D,0

(a)

	00	01	11	10
(AD) A	A,0	B,0	C,1	B,0
(BEG) B	B,0	A,0	B,0	A,0
(C) C	A,0	B,0	F,0	B,0
(F) F	A,0	B,0	A,0	B,0

(b)

Figure 6.13 (a) State table for partitioning. (b) Reduced state table

The total set of states of the table is partitioned into subsets of states having the same output values. The first partition of the table, for example, gives

$$P_1 = (AD)(BCEFG)$$

Now, taking the two states, AD, of the first subset, we check the successor states for each input combination. State A has next state entries A, B, C, B for different inputs, and these, in terms of the subset 1 and 2 of P_1, refer to 1, 2, 2, 2. State D has next state entries D, B, C, B, or 1, 2, 2, 2. As far as the first subset is concerned, therefore, the two states are equivalent for any input sequence of length two bits.

The states of subset 2 of P_1 give next state entries respectively of 2, 1, 2, 1; 1,2, 2, 2; 2, 1, 2, 1; 1, 2, 1, 2; and 2, 1, 2, 1. Clearly, these states do not all give the same response to two-bit input sequences, though B, E, and G are equivalent. Thus our second partition is

$$P_2 = (AD)(BEG)(C)(F)$$

In subdividing the second subset of P_1 we may have upset the equivalence we detected in subset (AD) and it is necessary to continue the partitioning process and checking until no further splitting is necessary. In the worst case, if no two states are equivalent, our set of states will subdivide until there are as many subsets, each of one element, as there are states in the original set.

Returning to the table, the third partition is

$$P_3 = (AD)\,(BEG)\,(C)\,(F)$$

since, in terms of the subsets of P_2, A and D both give 1, 2, 3, 2 as successors, and B, E and G all give 2, 1, 2, 1. P_3 is therefore a repeat of P_2 and the partitioning process is complete. We have shown that the table can be reduced to four rows, and the reduced table is drawn up using the states listed first in each subset (figure 6.13b). Any reference to equivalent states is replaced by the first state in that subset.

Incompletely specified tables may sometimes be dealt with in the same way, but the process rapidly becomes unworkable since all possibilities with any unspecified states must be considered.

Example 6.6: Minimize the state table of figure 6.14a.

	00	01	11	10
A	B,0	-	C,Ø	B,0
B	C,0	E,1	B,0	-
C	C,0	D,1	-	E,0
D	-	A,1	B,Ø	-
E	-	-	A,1	-

(a)

		00	01	11	10
(AB)	A′	B′,0	D,1	B′,0	A′/B′,0
(BC)	B′	B′,0	D,1	A′/B′,0	D,0
(DE)	D	-	A′,1	A′,1	-

(b)

Figure 6.14 Minimization of incompletely specified state table: (a) incompletely specified table; (b) reduced table

The steps in the partitioning of the table are:

$$P_1 = (ABCD)\,(ACDE)$$
$$P_2 = (AB)\,(BC)\,(CD) \quad (ACDE)$$
$$P_3 = (AB)\,(BC)\,(CD) \quad (A)\,(C)\,(DE)$$
$$P_4 = (AB)\,(BC)\,(CD) \quad (A)\,(C)\,(DE)$$

P_4 contains a large number of subsets since all possibilities have been included, and a minimal table is made up by taking a selection from the subsets to include all the original states. The subsets chosen must be such as to give a *closed* partition. This means that where the equivalence of two states depends on the equivalence of two other states, the latter states must be together in a subset if the former states are, in fact, to be equivalent. It would appear that we could select (AB) (C) (DE) as our subsets, but on checking it is found that (AB) indicates a successor state (BC) with inputs 00 or 11, so (BC) must be one of our states: (C) alone is not sufficient. Thus the minimal table has three states (AB) (BC) (DE) and is shown in figure 6.14b. The choice between the two possible next states shown under A′ and B′ – 10 and 11 respectively – is best left to a later stage in the design of the circuit.

We have seen that two states can be equivalent if they are output-compatible, even though the next states are not the same. This is only possible if the next states are equivalent, or compatible. However, the next states themselves may be compatible only if some other states are compatible. These implied compatibility conditions arise when the state diagram has branches giving alternative but equivalent paths, as in figure 6.15.

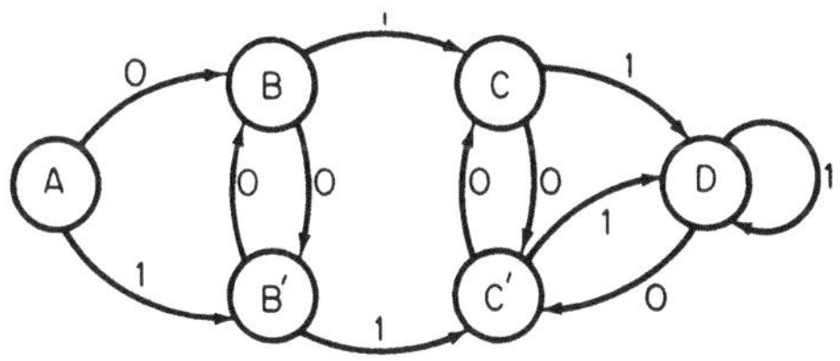

Figure 6.15 State diagram with equivalent paths

In a loosely specified table, such as the flow table for an asynchronous sequential circuit, the implied compatibles can become so involved and their number so unwieldy that a tabular method is the only practical method to consider. Paull and Unger (1959) developed the *implication chart* for such cases. The chart allows us to consider, in turn, the compatibility of all possible pairings of the states. The entry in each cell has one of three forms: (1) a cross if the two states are incompatible (this initially is because of output incompatibility, but extends later to pairs shown to be incompatible for other reasons); (2) a tick if the two states are unconditionally compatible; (3) the pair (or pairs) of states on the compatibility of which depends the compatibility of the pair under consideration. These are the implied compatibles. We shall use the chart to minimize a fully specified table and then progress to less fully specified tables.

Example 6.7: Minimize the table of figure 6.16.

In completing the chart of figure 6.17a we first consider state A with each other state in turn. As the outputs for state A(0011) do not occur with any other states, A cannot be compatible with any other state and we enter crosses in all the A column. Continuing with column B, we come eventually to BG. These two stages are compatible since the output requirement is

	00	01	11	10
A	A,0	G,0	B,1	B,1
B	K,0	B,0	C,1	B,0
C	A,0	D,0	D,0	J,1
D	A,0	H,0	H,0	C,1
E	E,0	F,0	B,0	D,0
F	E,0	D,0	A,0	A,0
G	K,0	G,0	C,1	G,0
H	A,0	J,0	C,0	J,1
J	A,0	J,0	D,0	C,1
K	K,0	F,0	G,0	J,0

Figure 6.16 State table to be minimized

satisfied and the only implied compatible is BG itself. We therefore place a tick in cell BG.

In the next column, states C and D are output compatible, but we cannot say that they are fully compatible until we know whether D and H, and C and J are compatible. The pairs DH and CJ become the entry for cell CD. The chart is completed as shown and we are now able to use the entries to decide which of the implied compatibles are not in fact compatible. By working through the chart systematically, we cross out the entry under EF. By the time we have worked right through the chart, all 'cross' entries should be underlined, as in figure 6.17b. The remaining entries indicate compatible pairs, which in this case are

(BG) (CD) (CH) (CJ) (DH) (DJ) (EK) (HJ)

Note that C ≡ D ≡ H ≡ J, since C ≡ D, C ≡ H, C ≡ J, D ≡ H, D ≡ J, and H ≡ J. (C, D, H, J) is said to be a *maximal compatible*, and it is convenient to

	A	B	C	D	E	F	G	H	J
B	x	-	-	-	-	-	-	-	-
C	x	x	-	-	-	-	-	-	-
D	x	x	DH CJ	-	-	-	-	-	-
E	x	x	x	x	-	-	-	-	-
F	x	x	x	x	DF AB AD	-	-	-	-
G	x	✓	x	x	x	x	-	-	-
H	x	x	DJ CD	HJ CH CJ	x	x	x	-	-
J	x	x	DJ	HJ DH	x	x	x	CD CJ	-
K	x	x	x	x	BG DJ	EK DF AG AJ	x	x	x

(a)

	A	B	C	D	E	F	G	H	J
B	x	-	-	-	-	-	-	-	-
C	x	x	-	-	-	-	-	-	-
D	x	x	DH CJ	-	-	-	-	-	-
E	x	x	x	x	-	-	-	-	-
F	x	x	x	x	~~DF AB AD~~	-	-	-	-
G	x	✓	x	x	x	x	-	-	-
H	x	x	DJ CD	HJ CH CJ	x	x	x	-	-
J	x	x	DJ	HJ DH	x	x	x	CD CJ	-
K	x	x	x	x	BG DJ	~~EK DF AG AJ~~	x	x	x

(b)

Figure 6.17 Implication chart: (a) first pass through the chart; (b) second pass

present the total states as a set of maximal compatibles. Since the set of states must include all the states, some of the maximal compatibles may contain only one state. In effect that state is compatible with itself but not with any other. In this example the maximal compatibles are

(A) (BG) (CDHJ) (EK) (F)

The maximal compatibles indicate the number of states required in the minimized table and, as before, it is necessary to ensure that the indicated state subsets do give a closed partition of the total set. In this case they do, and our minimized table can be drawn up as in figure 6.18.

		00	01	11	10
(A)	A	A,0	B,0	B,1	B,1
(BG)	B	E,0	B,0	C,1	B,0
(CDHJ)	C	A,0	C,0	C,0	C,1
(EK)	E	E,0	F,0	B,0	C,0
(F)	F	E,0	C,0	A,0	A,0

Figure 6.18 Minimized state table

The implication chart for an incompletely specified table is drawn up in exactly the same way but, because of the don't care entries, more cells of the chart contain ticks. Not all the implied compatibles can exist at the same time, since some arise from one choice of don't care state or output value, while others arise from mutually exclusive choices of don't care values. The final choice of don't care values must be made when selecting the maximal compatibles, by choosing the smallest possible set of maximal compatibles to give a covering of the total set and still retain the closure property.

Example 6.8: Minimize the incompletely specified table of figure 6.19a.
By use of the chart, figure 6.19, we arrive at the set of compatibles

(AD) (AH) (BC) (BD) (CD) (CF) (DE) (EG) (EH) (GH)

In terms of maximal compatibles, the set is

(AD) (AH) (BCD) (CF) (DE) (EGH)

Our first choice of maximal compatibles could be (A) (BCD) (EGH) (F), since all eight states are included. In checking for closure we find that B, C and D give successor states CD and BC for inputs 11 and 10 respectively, and these are both contained within the selected maximal compatibles. E, G and H, however, give AD and CF, which are not contained within the selected compatibles, and we must amend our choice to include AD and CF.

	00	01	11	10
A	A,0	-	B,0	B,1
B	-	E,0	D,0	C,0
C	F,0	-	-	B,0
D	-	E,0	C,0	-
E	A,0	-	B,0	G,1
F	F,0	C,0	-	D,0
G	A,0	C,0	-	H,1
H	D,0	F,0	B,0	-

(a)

	A	B	C	D	E	F	G
B	x	-	-	-	-	-	-
C	x	✓	-	-	-	-	-
D	BC	CD	✓	-	-	-	-
E	~~BG~~	x	x	BC	-	-	-
F	x	~~CE CD~~	BD	~~DE~~	x	-	-
G	~~BH~~	x	x	~~CE~~	GH	x	-
H	AD	~~EF BD~~	~~DE~~	~~EF BG~~	AD	~~DF CE~~	AD CF

(b)

	00	01	11	10
(AD) A	A,0	E,0	B,0	B,1
(BCD) B	C,0	E,0	B,0	B,0
(CF) C	C,0	C,0	-	B,0
(EGH) E	A,0	C,0	B,0	E,1

(c)

Figure 6.19 Minimization of incompletely specified table: (a) state table; (b) implication chart; (c) minimized table

We therefore try (AD) (B) (EGH) (CF). Checking through again, A and D give BC, and C and F give BD, neither of which we have included with our maximal compatibles. Our next try is (AD) (BCD) (EGH) (CF) which we find closes correctly and allows us to proceed to the minimal table.

There is one further restriction to bear in mind when seeking compatibles in the table for an asynchronous circuit. In an asynchronous circuit state changes occur immediately the inputs change, and two states can be equivalent only if the stable conditions exist for the same input values. This additional constraint means that the loosely specified tables, such as the primitive flow tables, will in fact give far fewer implied compatibles than for a synchronous case. The table of figure 6.10 for example gives only one non-trivial maximal compatible, (ACE), leading to the simplified table of figure 6.20a.

As we see, a large part of the table is still unspecified and we can take the minimizing a stage further by making use of the don't care possibilities. By suitable choice of entry we can merge two or more rows of the table to form a new composite state. For example, the rows A and B of the table could give the row A/B of figure 6.20b. Row A would also merge with rows D, F or L. There may be many alternatives available in the choice of rows to merge and a *merger diagram* gives an indication of the possibilities with distinctive patterns showing the groups of rows which can be merged. The table of figure 6.20 leads to the merger diagram of figure 6.21a and the minimal table is found by selecting the groups shown. Thus the table of figure 6.21b is formed by merging rows ABDF, GHJK and LMN respectively, though other equivalent choices are available. Note that

	000	001	010	011	100	101	110	111
A	A,00	F,00	D,00	–	B,00	–	–	–
B	A,00	–	–	–	B,00	–	G,00	–
D	A,00	–	D,00	–	–	–	–	–
F	A,00	F,00	–	–	–	–	–	–
G	–	–	H,Ø0	–	–	–	G,00	L,0Ø
H	–	–	H,10	J,10	–	–	–	–
J	–	K,10	–	J,10	–	–	–	–
K	A,Ø0	K,10	–	–	–	–	–	–
L	–	–	–	M,01	–	–	–	L,01
M	–	N,01	–	M,01	–	–	–	–
N	A,0Ø	N,01	–	–	–	–	–	–

(a)

	000	001	010	011	100	101	110	111
A/B	A,00	F,00	D,00	–	B,00	–	G,00	–

(b)

Figure 6.20 (a) Reduced primitive flow table. (b) Merged rows A and B

the number of stable states on the merged table is the same as on the unmerged table.

It will be recalled that in analyzing a sequential circuit we reached the stage at which all the next state values, in binary form, were listed on a transition table for all input and present state values. The state table was then derived by superimposing the output table and the transition table, with the various state codes replaced by letters. In our efforts to synthesize a circuit, the reverse process is now necessary, and is referred to as *state assignment* or, sometimes, secondary state assignment.

The minimum number of state variables n in an assignment is governed by the number of rows r of the state table, since $2^{n-1} < r \leqslant 2^n$. Thus a four-row table indicates a need for at least two state variables, whereas a five-state table needs at least three. In general, the total number of apparently different state assignments for n state variables, with r states, is

$$P = 2^n!/(2^n - r)!$$

and, with nine rows and four state variables, the number of distinct codes is already in excess of 10^7. In practice, however, many of these state assignments are repeats of others with the state variables relabelled and it has been shown (McCluskey and Unger, 1959) that the number of unique codes in in fact only

$$P' = (2^n - 1)!/(2^n - r)!n!$$

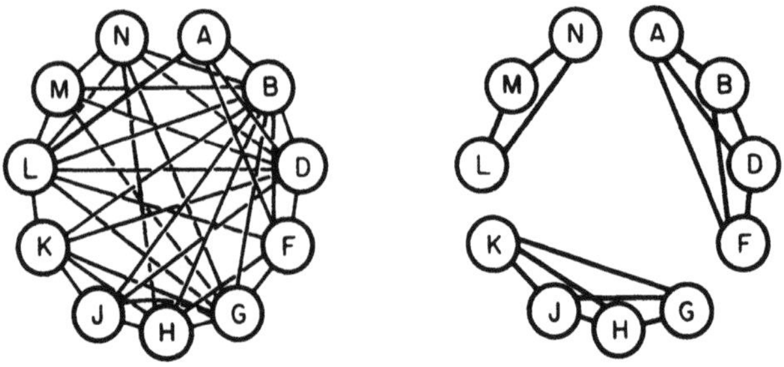

(a)

		000	001	010	011	100	101	110	111
(ABDF)	P	P,00	P,00	P,00	–	P,00	–	Q,00	–
(GHJK)	Q	P,Ø0	Q,10	Q,10	Q,10	–	–	Q,00	R,0Ø
(LMN)	R	P,0Ø	R,01	–	R,01	–	–	–	R,01

(b)

Figure 6.21 (a) Merger diagram. (b) Merged table

The assignment of state variables to ensure optimum performance when using individual storage elements, such as flipflops, is not easy and a great deal of work has been carried out in trying to devise systematic allocation methods (Hill and Peterson, 1974; Muroga, 1979). However, the availability of cheap LSI circuits allows us to go almost directly to a reliable implementation by means of a technique which makes use of the fact that any sequential circuit can be considered as a combinational logic section operating in conjunction with a storage section. We deal with each section as a separate entity, using a read-only memory to provide the necessary combinational logic, and a register to act as the storage section. If we assign codes to the internal states of the required circuit as convenient, we can use the total state information, comprising the present state and the present inputs, as the address of a location in the ROM. The data entry at that location is programmed partly to specify the next state and partly to provide the present output signals. At each clock pulse the next state values are latched into the register and the ROM address is adjusted to select a new location. The design problem is now one of deciding how best to program the ROM using the data contained in the state table or diagram.

In order to illustrate the method let us consider again the 1101 sequence detector circuit defined in figure 6.8. The state diagram is repeated in figure 6.22 and, as there are four internal states and two possible values of the single input in each state, there are eight possible total states. These are listed in figure 6.22b as the ROM addresses together with the code which each addressed location

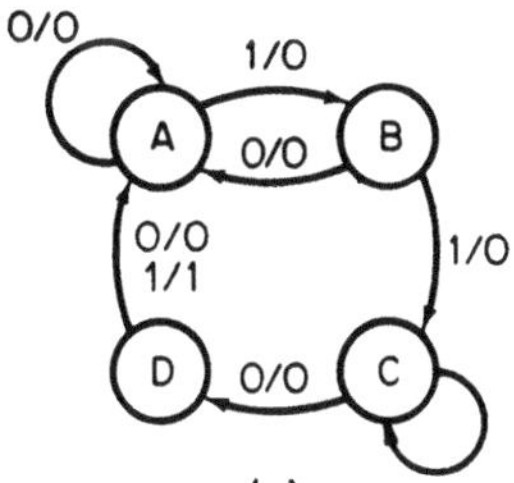

(a)

	Total state: Internal state	Total state: Input	Next state	Output
(A)	00	0	00	0
	00	1	01	0
(B)	01	0	00	0
	01	1	10	0
(C)	10	0	11	0
	10	1	10	0
(D)	11	0	00	0
	11	1	00	1
	$q_A q_B$	x	$Q_A Q_B$	Z
	ROM address		ROM data	

(b)

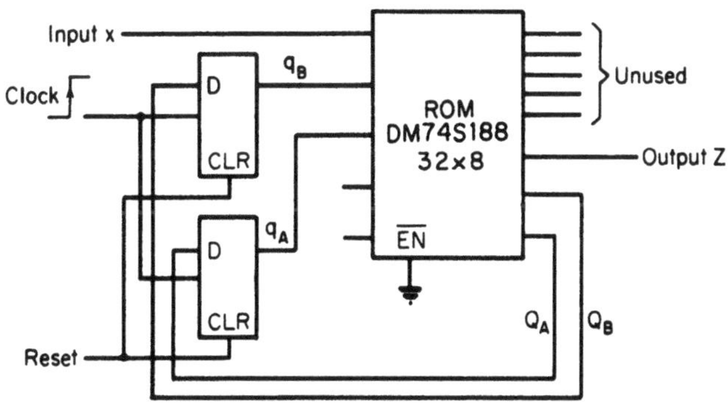

(c)

Figure 6.22 1101 sequence detector: (a) state diagram; (b) ROM program; (c) circuit

must hold to satisfy the state diagram. Thus, for example, total state 00–1, or A-1, must give a zero output and lead to state 01, or B. A ROM such as the DM74S188, containing 32 words of eight bits, would be suitable for this application, and edge-triggered flipflops, such as the 74LS74 dual D-type, are used to ensure there is no race condition resulting from the fast response of the ROM to address changes. The circuit is then as shown in figure 6.22c.

Counters can be designed using this technique by assuming no external inputs. The total state is the same as the internal state.

Example 6.9: Design a modulo-7 counter.

The counter must count to six and then reset to zero. The necessary sequence is shown in figure 6.23 with a circuit which makes use of a 256 eight-bit word ROM (82LS135) and a quad D-type latch (SN74LS175).

Note that this circuit uses only three of the eight ROM outputs for state information and the remaining outputs can be used to carry the corresponding

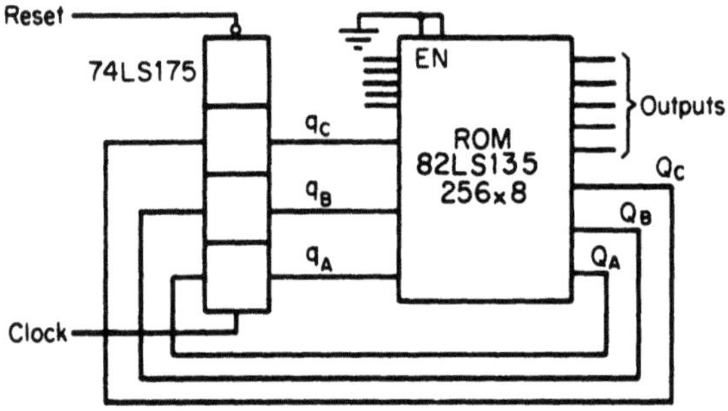

(a)

	Internal state	Next state	Outputs
(A)	0 0 0	0 0 1	× × × × ×
(B)	0 0 1	0 1 0	× × × × ×
(C)	0 1 0	0 1 1	× × × × ×
(D)	0 1 1	1 0 0	× × × × ×
(E)	1 0 0	1 0 1	× × × × ×
(F)	1 0 1	1 1 0	× × × × ×
(G)	1 1 0	0 0 0	× × × × ×
	q_A q_B q_C	Q_A Q_B Q_C	

(b)

Figure 6.23 Modulo-7 counter: (a) circuit; (b) ROM program

output code in any format we care to program. Also the circuit uses only a few of the total number of states, or addresses, available in the ROM. By making use of some of these spare locations we can dispense with the external register, so that the circuit becomes effectively asynchronous. The clock input is now considered as merely one of the inputs to the ROM as shown in the modified counter of figure 6.24. This method of operation utilises two locations in the ROM for each state of the counter, providing controlled transitions at both the beginning and end of the clock input. One disadvantage of removing the external register from the circuit is that we now have no easy method of resetting the circuit – that is, of setting the ROM address to zero. It is necessary, therefore, either to include the quad 2-input AND gates (74LS08) as shown, or to connect the Reset input to one of the unused address lines and to program the ROM appropriately. In the latter case, assuming the Reset is low active as before, we must program the sixth address bit as '1' for all the code combinations shown in figure 6.24b, and then for all codes where bit six is '0' we program the outputs as all zeros.

As another example of an asynchronous circuit we return to the parcel-sorting system of example 6.2. The state table eventually minimized to the three rows shown in figure 6.21 (and reproduced for convenience in figure 6.25a). Only two state variables are required, and two outputs, so the 74S188 32 eight-bit word ROM is more than adequate.

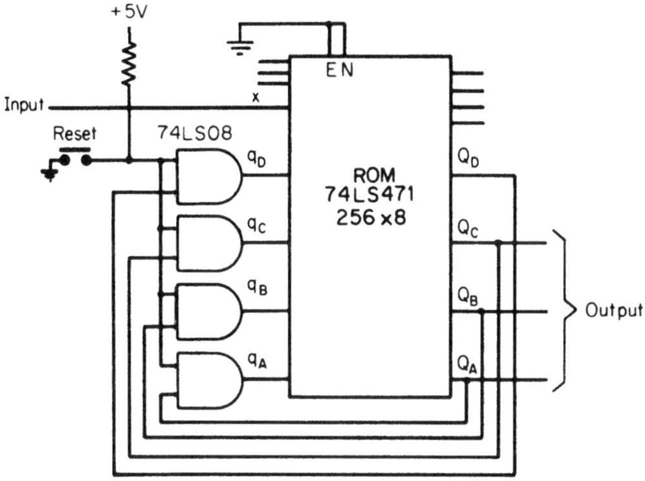

(a)

	Internal state	Input	Next state	Output
(0)	0000	0	0000	000
	0000	1	0001	
	0001	1	0001	
	0001	0	0010	
(1)	0010	0	0010	001
	0010	1	0011	
	0011	1	0011	
	0011	0	0100	
(2)	0100	0	0100	010
	0100	1	0101	
	0101	1	0101	
	0101	0	0110	
(3)	0110	0	0110	011
	0110	1	0111	
	0111	1	0111	
	0111	0	1000	
(4)	1000	0	1000	100
	1000	1	1001	
	1001	1	1001	
	1001	0	1010	
(5)	1010	0	1010	101
	1010	1	1011	
	1011	1	1011	
	1011	0	1100	
(6)	1100	0	1100	110
	1100	1	1101	
	1101	1	1101	
	1101	0	0000	

(b)

Figure 6.24 Asynchronous modulo-7 counter: (a) circuit; (b) ROM program

Allocating P as 00, Q as 01 and R as 10 we draw up the program of figure 6.25b, using the next states and present outputs specified in the state table. In this case we find that only one ROM location is required for each total state of the circuit, as asynchronous operation leads automatically to a stable state each time an input changes. In arriving at the final circuit of figure 6.25c we have again had to include the AND gates to allow resetting. Finally we see that certain entries on the state table are don't care entries. It may well be worthwhile including ROM entries under these conditions to set the circuit into an error state if one of these disallowed conditions does arise. Alternatively, use could be made of some of the unused outputs to indicate errors.

In chapter 3 we saw that combinational logic circuits are susceptible to static and dynamic hazards caused by races between signals within the circuit. These glitches can be troublesome when the combinational circuitry is working in conjunction with sequential circuits, which may interpret the spike as a genuine

	000	001	010	011	100	101	110	111
P	(P,00)	(P,00)	(P,00)	–	(P,00)	–	Q,00	–
Q	P,Ø0	(Q,10)	(Q,10)	(Q,10)	–	–	(Q,00)	R,0Ø
R	P,0Ø	(R,01)	–	(R,01)	–	–	–	(R,01)

(a)

	Internal state	Inputs	Next state	Outputs
(P)	00	000	00	00
	00	001	00	00
	00	010	00	00
	00	011	–	–
	00	100	00	00
	00	101	–	–
	00	110	01	00
	00	111	–	–
(Q)	01	000	00	00
	01	001	01	10
	01	010	01	10
	01	011	01	10
	01	100	–	–
	01	101	–	–
	01	110	01	00
	01	111	10	01
(R)	10	000	00	00
	10	001	10	01
	10	010	–	–
	10	011	10	01
	10	100	–	–
	10	101	–	–
	10	110	–	–
	10	111	10	01
	$q_A q_B$	$P_1 P_2 P_3$	$Q_A Q_B$	$Z_1 Z_2$
	ROM address		ROM data	

(b)

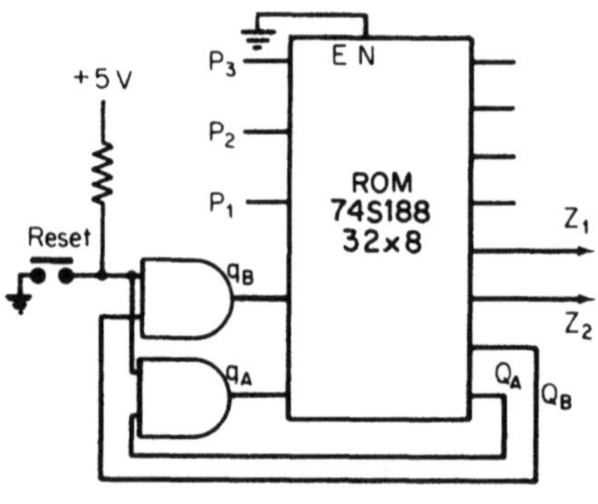

(c)

Figure 6.25 Parcel-sorting system: (a) minimal state table; (b) ROM program; (c) circuit

signal and change state accordingly. A third hazard can arise in an asynchronous sequential circuit caused by a race between a changing input variable and a resulting change in state variable. Unger (1959) has shown that this *essential* hazard exists if three consecutive changes in an input variable lead to a state which is different from the state reached after the first of the changes.

The circuit of figure 6.5 contains an essential hazard as can be seen from the state table, part of which is reproduced in figure 6.26. Assuming an initial state A-11, a change in the b variable leads to state B-10: a second change in b leads to C-11, and a third change to D-10. The third change has not brought us

		ab 11	10
(00)	A	A,0	B,0
(01)	B	C,0	B,0
(11)	C	C,0	D,1
(10)	D	C,0	D,1
q_1q_2			

Figure 6.26 State table containing essential hazard

back to B-10. Incorrect action occurs in this circuit when in state A-11 if the change in the b variable leads to q_2 changing to one before the input variable change is registered at all the gates. The operating point on the table moves to B-11 rather than to A-10. State B-11 is unstable and leads on to C-11. When the change in b is fully registered, operation moves to C-10, which again is unstable, and the circuit finally rests in state D-10. In order to eliminate the hazard we must either rearrange the state table, possibly by adding states, or include delays to ensure that the switching signals work through the circuit in step. Unger has shown that a single delay, correctly placed, is sufficient to remove an essential hazard from any circuit.

The design methods outlined above, using read-only memories, have the advantage of minimizing the occurrence of hazards, and where greater flexibility is required the same techniques can be used with a programmable logic array, PLA, or programmable array logic, PAL, in place of the ROM.

6.3 Algorithmic state machine chart representation

The ASM chart is an alternative form of graphical representation of the state transitions and output functions resulting from changes in the inputs to a sequential circuit or system. It is particularly useful in the early stages of design, as it can present all the fundamental system operation requirements without undue concern with the actual method of implementation. The chart is built up as an array of interconnected blocks, each having a single entry point but with the possibility of several different output routes. It has a marked similarity in appearance to a computer program flowchart but, unlike the flowchart, time is discontinuous and each ASM block relates to a particular state of the circuit. The operations indicated within a block, and any branching decisions associated with them, occur simultaneously.

There are three main symbols, which are usually referred to as the *state box*, the *decision box*, and the *conditional output box* (Green, 1986). Every ASM block has one state box, with the general form shown in figure 6.27a. The state name is shown to the left of the box and the box contains the list of state outputs. These can indicate *immediate outputs*, as in a Moore type of circuit where the outputs are determined by the state variables alone, or *delayed outputs* in which the state transitions are delayed. The ASM block may contain one or more decision boxes, in which a logical relationship between the specified input values at that time is tested and an exit path indicated for true and false conditions, figure 6.27b. The conditional output box, figure 6.27c, indicates outputs which are dependent on both the present state and the present input values, as in the Mealy representation of a circuit. These outputs become active only if the conditions specified are true. Individual ASM blocks are interconnected to form a chart of the total behaviour of the sequential system, as in the example of figure 6.28 which represents the 1101 detector circuit previously encountered in figure 6.8. The box outlines shown dotted on the diagram are normally omitted.

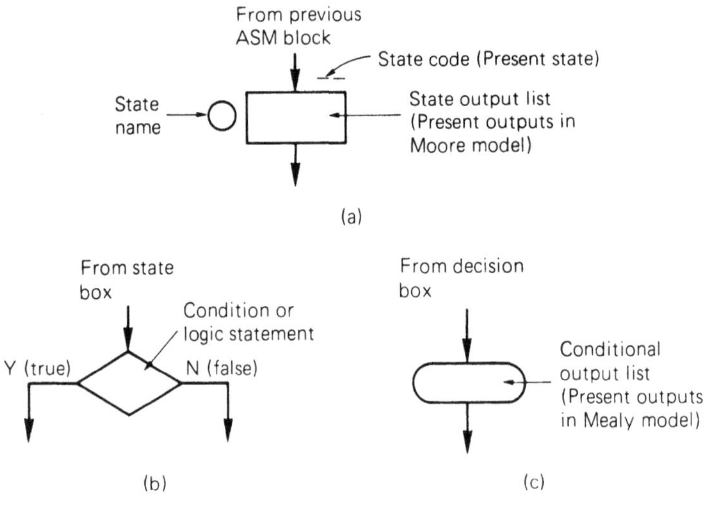

Figure 6.27 ASM chart symbols: (a) state box; (b) decision box: (c) conditional output box

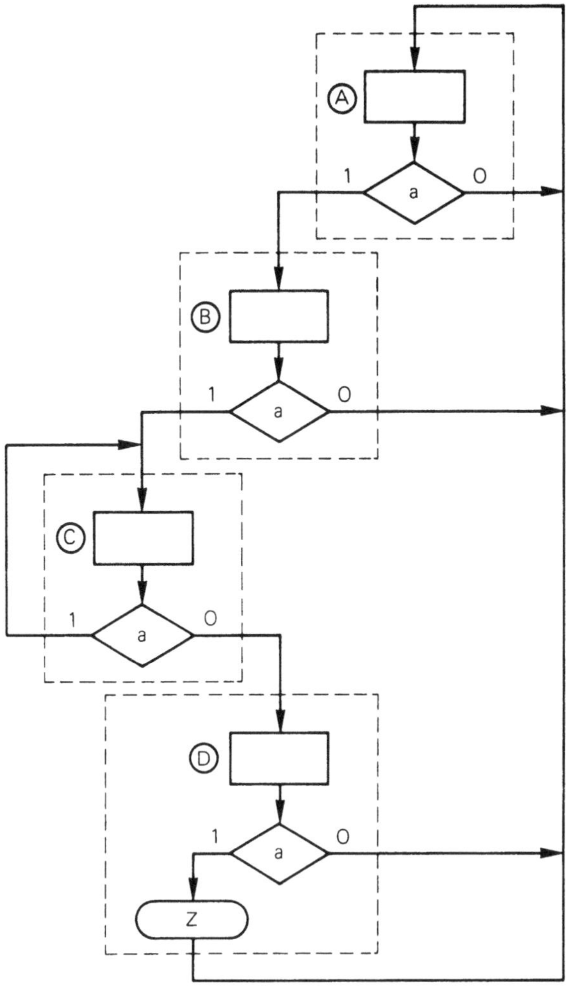

Figure 6.28 ASM chart of 1101 sequence detector

Sequential circuits and systems come in a seemingly infinite range of complexities, but it is possible to categorize them into five main types (Clare, 1973). We can illustrate these in terms of our generalized model introduced in figure 5.2, but we must now subdivide the combinational logic section into two, figure 6.29a, to emphasize that some of the circuitry is concerned with the generation of output signals and some with the state variables. The first category, A, is of purely combinational logic circuits, figure 6.29b, which are included for completeness though the state concept has no significance in this case. The corresponding ASM chart is essentially a set of decision boxes. Any path from one state to the next is called a *link path*, and decision boxes along the path indicate

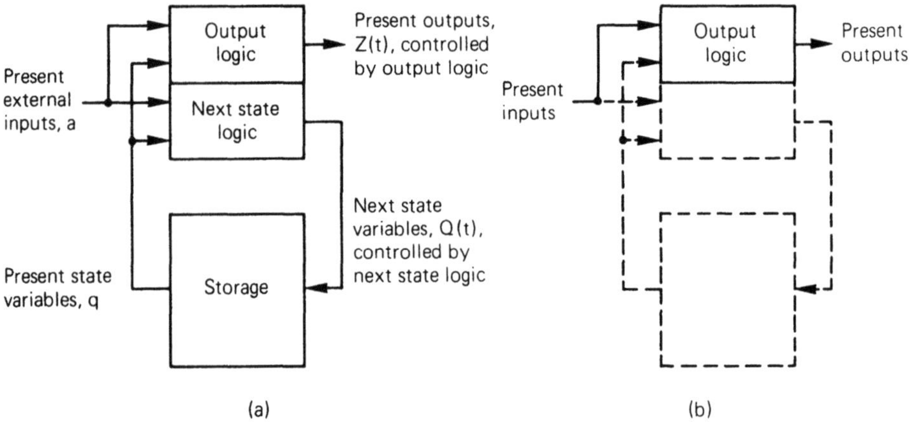

Figure 6.29 (a) Generalized sequential circuit. (b) Combinational logic circuit

an AND function whereas alternative paths indicate an OR, in much the same way as switches in series and parallel. When desirable, the box structure can be simplified by entering more complex logic functions in fewer boxes.

Example 6.10: Draw the ASM chart for the cell circuit used in the iterative array of figure 5.4.

The cell outputs are given by $Z_i = Y_i = \bar{a}b + a\bar{b}$, and the chart is therefore as shown in figure 6.30.

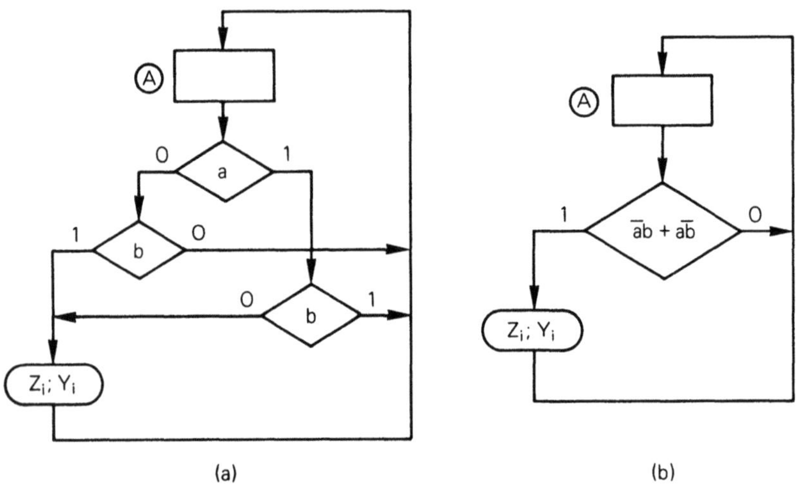

Figure 6.30 ASM chart for $Z_i = Y_i = \bar{a}b + a\bar{b}$ using: (a) multiple decision boxes; (b) simplified structure

The simplest form of sequential circuit is a delay element, and it is this type of circuit which is placed in the second category, B. It is, in effect, a combinational circuit followed by the delay, figure 6.31a. Because there is no internal feedback the behaviour of the circuit is defined entirely by the present inputs and the decision boxes are therefore common to all states.

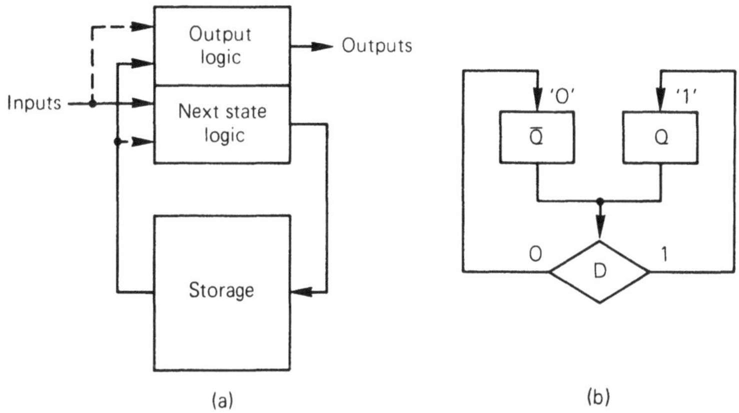

Figure 6.31 (a) Delay circuit. (b) ASM chart for D flipflop

Example 6.11: Draw the ASM chart for a D-type flipflop.

The flipflop has two states and its setting at any time is determined by the value on the D input when the flipflop is clocked, so a single decision box is sufficient, figure 6.31b.

Category C contains those circuits in which the next state and the present outputs are determined solely by the present states, figure 6.32a. This group comprises *autonomous circuits*, such as counters and linear feedback shift registers, in which there are no external inputs other than the clocking pulses. No decision boxes are involved as the circuit steps through a predetermined sequence of states, as, for example, in the modulo-7 counter considered in example 6.9 and shown in ASM chart form in figure 6.32b.

The final two categories, D and E, are the most general and correspond respectively to the Moore and Mealy models of a sequential circuit. In both cases the present states and present inputs determine the next state, but in the former the present outputs are dependent only on the present state, figure 6.33a, whereas in the latter the outputs depend on the present inputs as well as the present state, figure 6.33b. Thus, apart from the decision boxes, category D charts consist only of state boxes, each of which contains an output list, but category E charts also have conditional output boxes which specify when the

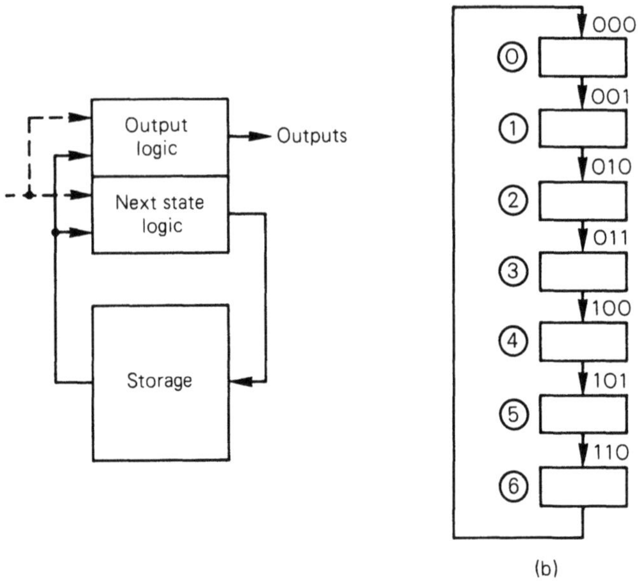

Figure 6.32 (a) Autonomous circuit. (b) Modulo-7 counter

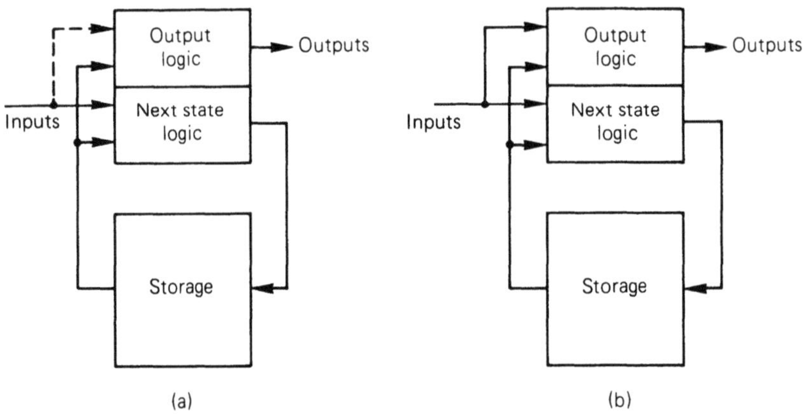

Figure 6.33 (a) The Moore model. (b) The Mealy model

output signals are to be generated. Category E represents the most general form of circuit, and all other categories can be considered as subsets of that category.

We complete this chapter with an example of design using the ASM chart as our starting point.

Example 6.12: A communications system is to include a variable delay unit which has a single data input line, a, and a single data output line, Z. Two control bits, $d_1 d_2$, are to indicate what delay is required, in the following way:

> d_1 and d_2 both at zero indicates no delay;
> either d_1 or d_2, but not both, at one indicates a delay of one clock interval;
> both d_1 and d_2 at one indicates a delay of two clock intervals.

Design a suitable circuit using (a) JK flipflops, (b) a PROM and (c) programmable array logic, PAL.

The first stage in the design is to decide exactly what the specification requires, and this is best done by drawing the ASM chart. In order to satisfy the different delay requirements we need at all times to know the last two input bit values, so that the appropriate bit can be routed to the circuit output. Four states are therefore sufficient to define the circuit, corresponding to previous input bit combinations of 00, 01, 10 and 11. Associated with each state box on the chart, shown in figure 6.34, is the decision logic which determines the output required and the next state, according to the value of the incoming data bit.

The information on the chart is now transferred to the state table of figure 6.35a and we must assign values to the state variables. One method of deciding on an assignment is to try to minimize the number of state transitions which would involve both state variables changing. This often leads to a simpler circuit and one that is easier to test. We can show on a form of Karnaugh map the successor states for each of the four states, under the different input conditions, and we find that the assignment which leads to the least number of diagonal transitions, and therefore the fewest occasions on which both state variables must change, is A = 00, B = 01, C = 10 and D = 11, figure 6.35b. As we have seen, optimum state assignment is a difficult area, but in many cases the number of states and the complexity of output gating can be traded off against each other. Modern computer-aided design tools allow the different balances to be explored. The design of test sequences is easier for combinational circuits, and, now that integrated circuit technology has removed the necessity to minimize the overall circuitry as far as possible, it is often preferable to simplify the state structure at the expense of the gating circuitry. Using the assignment suggested gives the transition table of figure 6.36a.

Part (a) of the specification asks us to use JK flipflops, so, using the known transition behaviour of the JK flipflop, the next stage is to produce an excitation table from the information on the transition table. The operation of

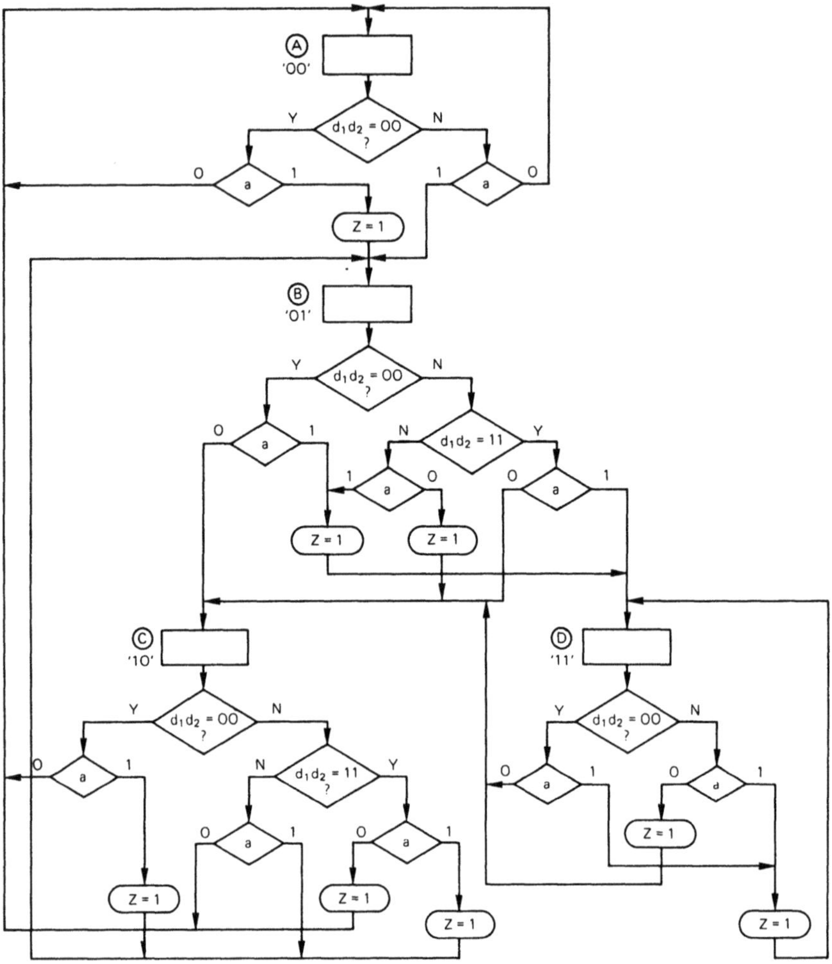

Figure 6.34 ASM chart for the variable delay unit of example 6.12

	d_1d_2 000	001	010	011	100	101	110	111
A	A, 0	B, 1	A, 0	B, 0	A, 0	B, 0	A, 0	B, 0
B	C, 0	D, 1	C, 1	D, 1	C, 1	D, 1	C, 0	D, 0
C	A, 0	B, 1	A, 0	B, 0	A, 0	B, 0	A, 1	B, 1
D	C, 0	D, 1	C, 1	D 1	C, 1	D, 1	C, 1	D, 1

(a)

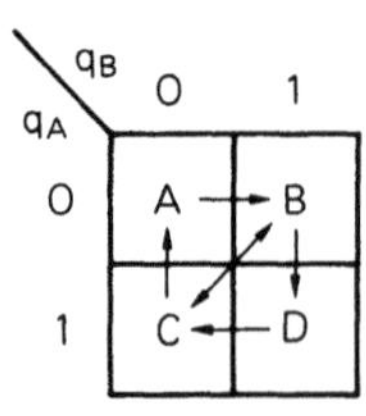

(b)

Figure 6.35 (a) State table. (b) State assignment

	d_1d_2 000	001	010	011	1001	101	110	111
(A) 00	00, 0	01, 1	00, 0	01, 0	00, 0	01, 0	00, 0	01, 0
(B) 01	10, 0	11, 1	10, 1	11, 1	10, 1	11, 1	10, 0	11, 0
(C) 10	00, 0	01, 1	00, 0	01, 0	00, 0	01, 0	00, 1	01, 1
(D) 11	10, 0	11, 1	10, 1	11, 1	10, 1	11, 1	10, 1	11, 1

(entries: Q_A Q_B, Z)

(a)

	d_1d_2a 000	001	010	011	100	101	110	111
00	0∅, 0∅, 0	0∅, 1∅, 1	0∅, 0∅, 0	0∅, 1∅, 0	0∅, 0∅, 0	0∅, 1∅, 0	0∅, 0∅, 0	0∅, 1∅, 0
01	1∅, ∅1, 0	1∅, ∅0, 1	1∅, ∅1, 1	1∅, ∅0, 1	1∅, ∅1, 1	1∅ ∅0, 1	1∅, ∅1, 0	1∅, ∅0, 0
10	∅1, 0∅, 0	∅1, 1∅, 1	∅1, 0∅, 0	∅1, 1∅, 0	∅1, 0∅, 0	∅1, 1∅, 0	∅1, 0∅, 1	∅1, 1∅, 1
11	∅0, ∅0, 0	∅0, ∅0, 1	∅0, ∅1, 1	∅0, ∅0, 1	∅0, ∅1, 1	∅0, ∅0, 1	∅0, ∅1, 1	∅0, ∅0, 1

(entries: J_A K_A, J_B K_B, Z)

(b)

Figure 6.36 (a) Transition table. (b) Excitation table using JK flipflops

:he JK flipflop is conventionally summarized in the form

J	K	Q^+
0	0	Q
0	1	0
1	0	1
1	1	$\overline{Q}$

But we have to know what JK values are needed to ensure a particular transition, and the same information can be presented in a more useful form as

Q	Q^+	J	K
0	→ 0	0	∅
0	→ 1	1	∅
1	→ 0	∅	1
1	→ 1	∅	0

where ∅ indicates a don't care value. These don't care conditions arise because of the JK flipflop's ability to toggle when J = K = 1, and are included to give maximum flexibility in later stages of the design. We proceed to build up the excitation table of figure 6.36b, which is now, in effect, the five maps of J_A, K_A, J_B, K_B and Z all superimposed, though the axis labelling is not in the correct order. Rearranging to standard Karnaugh map form gives the maps of figure 6.37a, from which, and by judicious use of the don't care possibilities, we obtain the functions:

$$J_A = q_B, \quad K_A = \overline{q}_B, \quad J_B = a, \quad K_B = \overline{a}$$

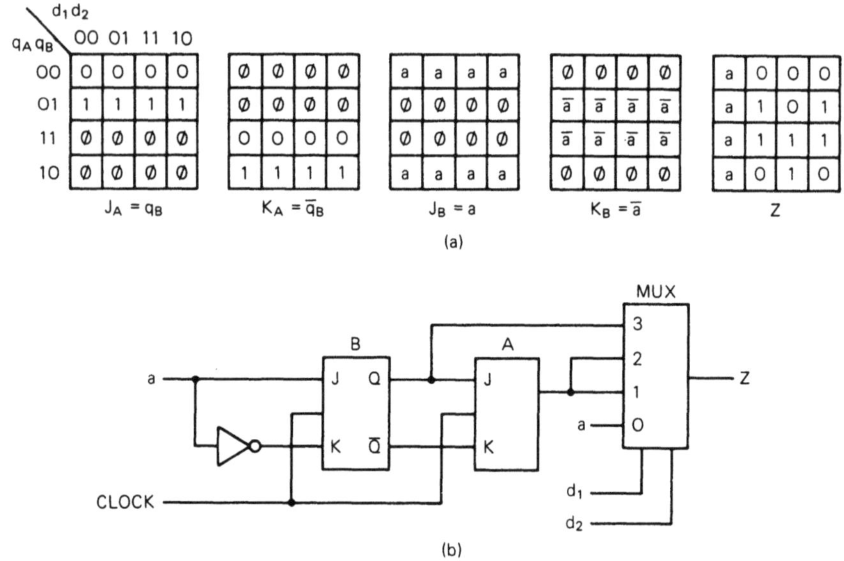

Figure 6.37

and

$$Z = \overline{d}_1\overline{d}_2 a + \overline{d}_1 d_2 q_B + d_1\overline{d}_2 q_B + d_1 d_2 q_A$$

The use of a multiplexer gives the simple circuit of figure 6.37b.

If we are to use a PROM as our main circuit element, as required in part (b), we can take the PROM program details directly from the transition table, and this is shown, with the circuit, in figure 6.38.

Finally, we are asked to use programmable array logic to build the delay unit. Our circuit is on a very small scale and does not make good use of the capabilities of a PAL device, but it will serve to illustrate the principles involved. Let us choose one of the smallest of the devices available, the 16R4. The 16 tells us that the programmable AND array can handle sixteen variables, eight external and eight fed back from the output gates, and the R4 indicates that there are four registered outputs using D-type flipflops. The logic diagram of the device is shown in figure 6.40. Each vertical line on the matrix is associated with one polarity of one of the variables, and horizontal lines feed the fixed AND-OR gating to the right of the diagram. The AND array is manufactured with an intact fusible link at each crossover point in the matrix, so, initially, all variables are connected to all AND gates. Programming the PAL involves blowing open those fuses which we do not want. The accepted convention when designing with these devices is to mark with a cross the intersections where we wish to retain a fuse link, so that the associated variable appears in the AND term at that point.

$q_A q_B$	d_1	d_2	a	Q_A	Q_B	Z
0 0	0	0	0	0	0	0
	0	0	1	0	1	1
	0	1	0	0	0	0
	0	1	1	0	1	0
	1	0	0	0	0	0
	1	0	1	0	1	0
	1	1	0	0	0	0
	1	1	1	0	1	0
0 1	0	0	0	1	0	0
	0	0	1	1	1	1
	0	1	0	1	0	1
	0	1	1	1	1	1
	1	0	0	1	0	1
	1	0	1	1	1	1
	1	1	0	1	0	0
	1	1	1	1	1	0
1 0	0	0	0	0	0	0
	0	0	1	0	1	1
	0	1	0	0	0	0
	0	1	1	0	1	0
	1	0	0	0	0	0
	1	0	1	0	1	0
	1	1	0	0	0	1
	1	1	1	0	1	1
1 1	0	0	0	1	0	0
	0	0	1	1	1	1
	0	1	0	1	0	1
	0	1	1	1	1	1
	1	0	0	1	0	1
	1	0	1	1	1	1
	1	1	0	1	0	1
	1	1	1	1	1	1

(a)

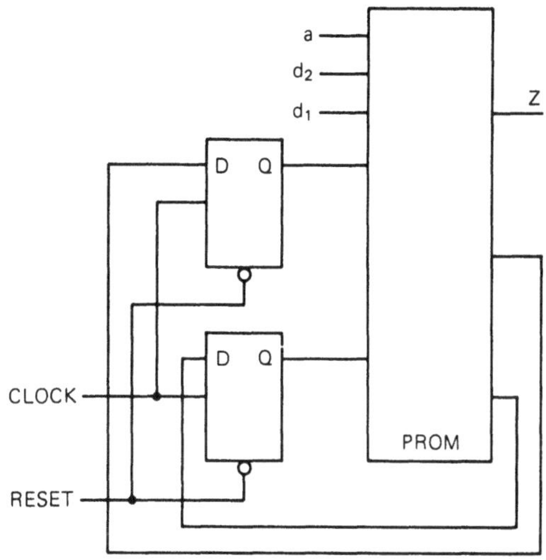

(b)

Figure 6.38

Our circuit requires two state variables so we use two of the D-type flip-flops provided. The excitation signals to the flipflops can be read directly from the transition table of figure 6.36a, but again the labelling of the axes is not in standard Karnaugh map form, and when rearranged we get the maps of figure 6.39. From the maps, $Q_A = q_B$, $Q_B = a$ and $Z = \overline{d}_1 \overline{d}_2 a + \overline{d}_1 d_2 q_B + d_1 \overline{d}_2 q_B + d_1 d_2 q_A$. This information is then transferred directly to the PAL logic diagram, as shown on figure 6.40.

$q_A q_B$ \ $d_1 d_2$	00	01	11	10
00	0	0	0	0
01	1	1	1	1
11	1	1	1	1
10	0	0	0	0

$Q_A = q_B$

	00	01	11	10
00	a	a	a	a
01	a	a	a	a
11	a	a	a	a
10	a	a	a	a

$Q_B = a$

	00	01	11	10
00	a	0	0	0
01	a	1	0	1
11	a	1	1	1
10	a	0	1	0

Z

Figure 6.39 Functions for the programmable array logic

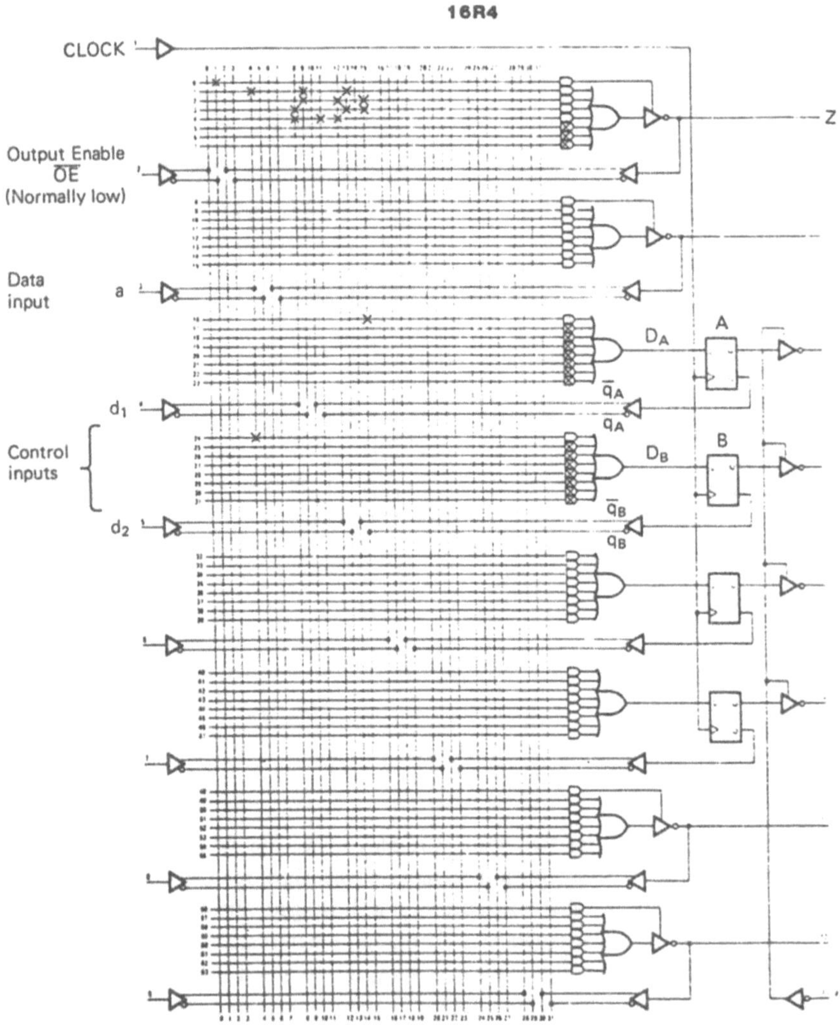

Figure 6.40 Programmable Array Logic device 16R4 (*courtesy of Monolithic Memories Inc.*)

Problems

6.1 Modify the counter circuit of figure 6.23 to count to base 9.

6.2 Complete the table of example 6.2 if it cannot be assumed that the parcels are widely separated.

6.3 Does the state table of figure 6.4 indicate the presence of any essential hazards? If so, are they critical?

6.4 A four-stage Johnson counter is assembled on a printed circuit board but testing indicates that the output sequence obtained is as follows

0, 8, 12, 14, 15, 6, 11, 4, 10, 13, 6, 11, 4, 10, 13, 6, 11, etc.

By drawing the transition and excitation tables corresponding to the output sequence obtained under test, deduce what single fault is present on the board.

6.5 Use an implication chart to show that the operation of the circuit, specified by the table below, requires a minimum of two state variables.

	00	01	11	10
A	A,0	–	E,1	B,1
B	E,0	C,1	B,0	–
C	–	B,0	C,1	D,0
D	A,0	–	F,1	B,1
E	B,0	–	B,0	–
F	–	C,1	B,0	G,1
G	D,1	D,1	–	G,0

6.6 Draw the ASM chart for a JK flipflop.

6.7 Modify the ASM chart of the 1101 detector, figure 6.28, to accept the final '1' of the 1101 sequence as the first '1' of a following sequence.

6.8 Modify the circuit and program of the PROM-based circuit in example 6.12 to remove the necessity for the dual D-type flipflop.

6.9 If it can be assumed that the sequences in example 6.12 can be provided as spatially separated sequences, design an iterative array cell to give the required delay properties.

6.10 List the ROM program required by the modulo-7 counter of figure 6.24 when the reset input acts as an address bit rather than controlling the AND gates.

References

Cadden, W. J. (1959). Equivalent sequential circuits, *IRE Trans. Circuit Theory*, **CT6** (March), 30-34

Clare, C. R. (1973). *Designing Logic Systems Using State Machines*, McGraw-Hill, New York

Green, D. H. (1986). *Modern Logic Design*, Addison-Wesley, Reading, Massachusetts

Hill, F. J. and Peterson, G. R. (1974). *Introduction to Switching Theory and Logic Design*, 2nd edn, John Wiley, New York

McCluskey, E. J. and Unger, S. H. (1959). A note on the number of internal variable assignments for sequential switching circuits, *IRE Trans. Electronic Computers*, **EC-8**, No. 4 (December), 439-40

Mealy, G. H. (1955). A method for synthesising sequential circuits, *Bell Syst. Tech. J.*, **34**, 1045-80

Moore, E. F. (1956). Gedanken-experiments on sequential machines, *(Automata Studies) Ann. Math. Stud.*, **34**, 129-53

Muroga, S. M. (1979). *Logic Design and Switching Theory*, John Wiley, New York

Paull, M. C. and Unger, S. H. (1959). Minimizing the number of states in incompletely specified sequential switching functions, *IRE Trans. Electronic Computers*, **EC-8**, No. 3 (September), 356-67

Unger, S. H. (1959). Hazards and delays in asynchronous sequential switching circuits, *IRE Trans. Circuit Theory*, **CT6** (March), 12-25

7 Storage Systems

All organizations require some means of retaining information for later reference, and digital systems are no exception.

For small digital systems registers alone may suffice, but for more sophisticated applications the cost of registers becomes prohibitive and alternative methods have been devised. These have ranged from the Williams' cathode-ray tube store of 1949, in which charge patterns on small areas of a conventional cathode-ray tube screen were used to store data at a density up to ten bits per square centimetre, through magnetic core and semiconductor memories to the video disc in which a laser is used to scan minute pits impressed in the surface of a metal-coated plastic disc, giving bit densities in excess of ten million bits per square centimetre. The popular compact disc, or CD-ROM, only 120 mm in diameter, provides storage in excess of 600 Mbytes.

By far the most popular form of mass storage in early years was the square loop ferrite core store (Rajchman, 1952), first introduced in the MIT Whirlwind computer in the early 1950s, but developed considerably since then. For more than two decades, until the advent of large scale semiconductor memories, the ferrite 'core store' was the only reliable and economic random-access read-write memory available to computer designers (Bannister and Whitehead, 1983). Though now almost entirely eclipsed by the semiconductor memory, the core store is still to be found in applications where the inherent non-volatility of its data storage mechanism and a total resistance to radiation damage are of prime importance (Ford, 1982).

Most memory systems use *position addressing*; that is, each stored word occupies a known position in the memory and, by selecting the coordinates of that position, the desired word is *accessed*. An alternative approach, which until recently has been hampered by the lack of suitable storage elements but is now finding increased application, is the *content addressed* memory (or *associative* memory) in which the information is identified not by its location but by recognition of key parts of the word itself. It has been suggested that this is a more 'natural' organization, since it is thought possible that the human brain uses a similar technique. However, at the present state of our knowledge it is unwise to carry the analogy too far.

A position addressed memory unit is accessed by presenting a binary code or *address*. A memory containing 4096 words, for example, will require a binary code of twelve bits, since 2^{12} equals 4096. The unique codes, from all zeros through all ones, correspond to the 4096 unique locations in the memory where the data bytes or words are held. Because binary codes are used to define the addresses, memory capacities are almost always in powers of two; that is, the number of words held in a given memory could be 1024 (termed 1K), or 2048 (2K), or 16374 (16K), and so on. The binary address specified by the user is translated by an address decoder, within the memory unit, into the signals necessary to select the indicated location. How this translation is achieved depends on the internal organization of the memory.

All position addressed memory systems have an addressing arrangement which falls in one of the following groups:

(a) word organized or linear select, known as 2D (two dimensional);
(b) bit organized, known as 3D;
(c) a compromise between word organized and bit organized systems, aptly known as $2\frac{1}{2}$D (Gilligan, 1966).

The simplest system is the 2D or word organized memory in which a complete data word is accessed by a uniquely selected *word* line. This line is selected by means of the *address*, presented as a binary number to a word selection decode matrix. The data word is obtained in parallel over the *digit* lines and, when required, new bit values are written into that chosen location through the same digit lines. Figure 7.1 shows such a memory of N bits total, with M bits per word. The number of words that can be stored is therefore N/M. A two-state storage device is situated at the intersection of each word line (W) and digit or sense line (D). In order to select one of the N/M words the number of address bits required is $\log_2$ (N/M). 3D and $2\frac{1}{2}$D methods were developed for use in core memories whereas semiconductor memories use two-dimensional selection and we will examine the 2D memory in relation to such devices.

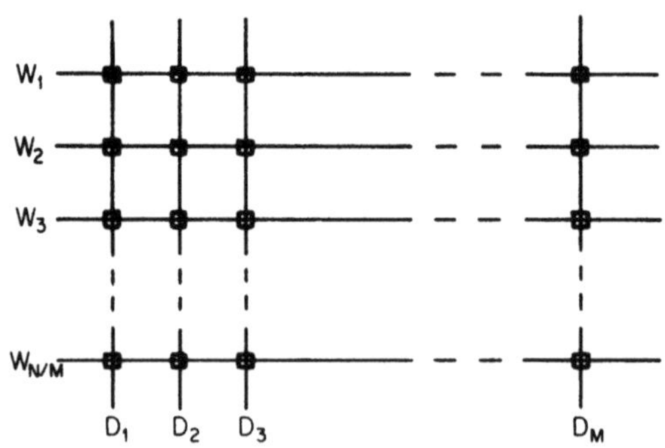

Figure 7.1 Layout of 2D memory

7.1 The random access memory

Semiconductor memories are broadly classed as either random access memories (RAM) in which data words can be written into or read out of any location at will, or read-only memories (ROM) in which the data words, once written in, are stored permanently and can only be accessed. Note that the ROM is also randomly accessible. At this stage we will consider the random access memory, and a typical storage cell for an MOS memory is shown in figure 7.2.

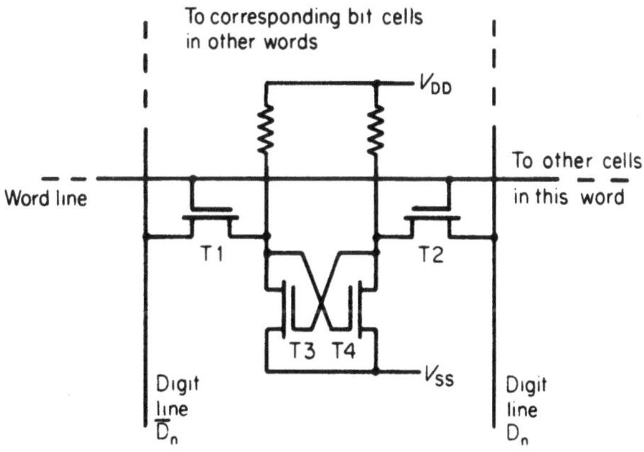

Figure 7.2 MOS memory cell

To store information a signal is applied to a selected word line. This signal brings into conduction the transistors T1 and T2 in every cell associated with that word line. To write a one where required, the appropriate signal is impressed on the digit line. Thus in the cell shown, a one would be stored by forcing D_n to logical one and $\bar{D}_n$ to logical zero. The flipflop formed by T3 and T4 will then take up the appropriate state. When the word line selection signal is removed, turning off T1 and T2, the flipflop remains in its impressed state.

Since the flipflop, once set, retains the data, the memory is said to be static but it is important to note that the data is *volatile* in nature; that is, if the power supply is removed, the information stored is lost. Some semiconductor memories are available with a 'power-down' facility in which all peripheral circuitry can be switched off, leaving only the memory array to be powered. Under these conditions power demand is very low and can easily be supplied by batteries. MOS memories are generally categorized by their relatively low power consumption; bipolar memory arrays tend to have greater dissipation but usually a faster speed of access.

A typical bipolar memory cell is shown in figure 7.3 using multi-emitter transistors. With T1 on and T2 off, for example, current can flow through T1 using either of the emitters. However, in its quiescent state, the sense/digit lines

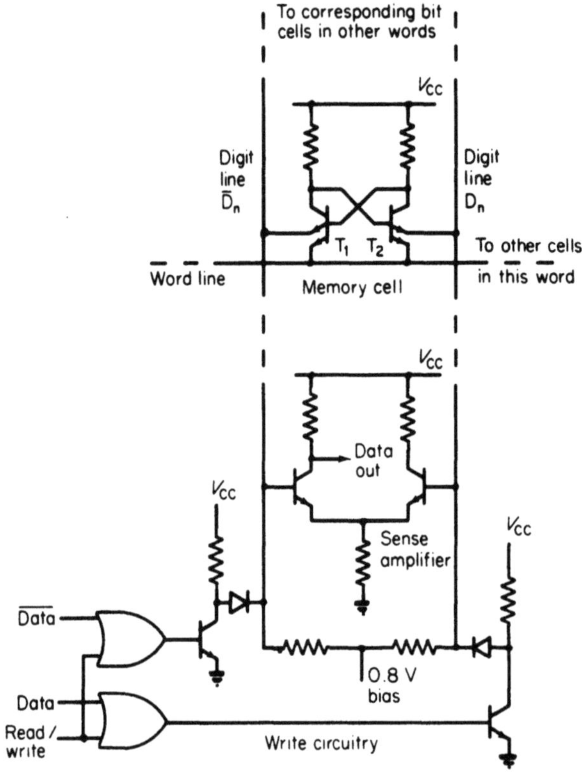

Figure 7.3 Bipolar memory cell and associated circuitry

D_n, $\bar{D}_n$, are held 0.8 V higher than the word line, reverse-biasing the emitters associated with them. To read, the word line voltage is raised, diverting current to the appropriate digit/sense line emitter. To write, a signal impressed upon the digit/sense lines can force the flipflop into the corresponding state. The *access time* (figure 7.4) of a typical bipolar memory is 30 ns, compared with 200 ns for the MOS memory, but the power dissipation for a 1 K x 1-bit bipolar array is around 475 mW.

To read out information already stored, a word line is selected and, instead of forcing the flipflops into a particular state, the logical levels on the digit/sense lines are monitored, thus providing an indication of the state of the flip-flops associated with the selected word line. The typical arrangement for a 2D system is shown in figure 7.5 and usually the selection matrix and buffer latches are fabricated with the memory array on a single silicon chip. The external organization often appears in a different form. For example, a 4 Kbit memory can appear to the user as 4096 separate one-bit 'words' as shown in figure 7.6. By decoding the word lines (rows) and the digit/sense lines (columns) in this manner, a flexible unit suitable for building larger memories is obtained.

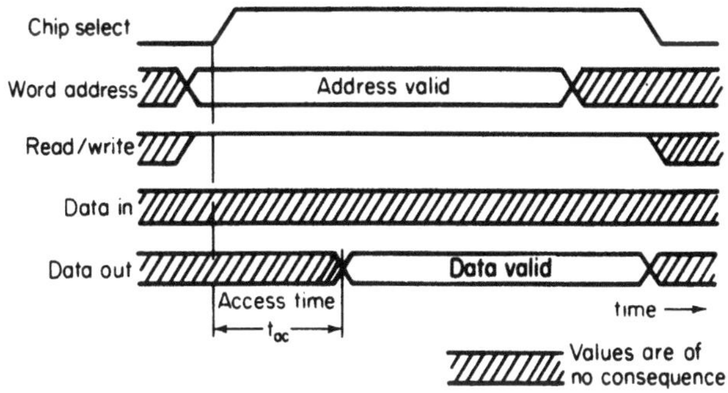

Figure 7.4 Typical read cycle timing showing access time

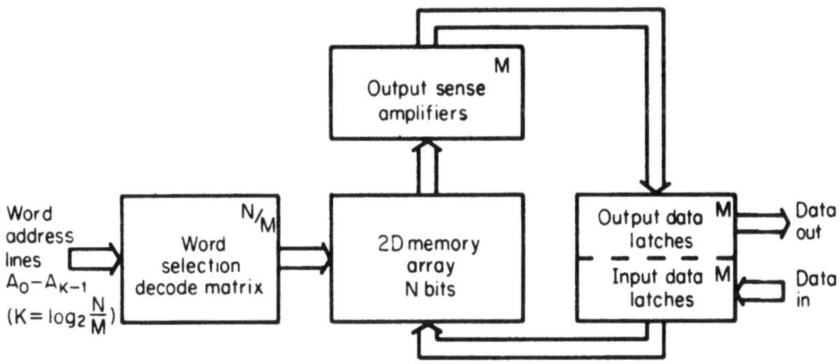

Figure 7.5 Typical arrangement of 2D system with N/M words

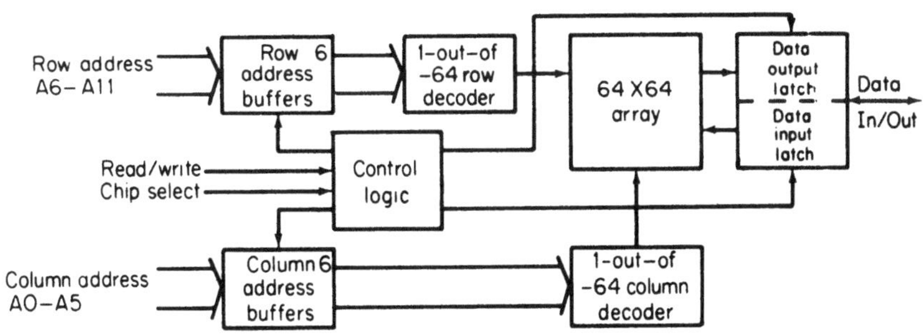

Figure 7.6 4096 word x 1 bit memory chip organization

To construct larger memories use must be made of the chip select control lines. In addition to the input buffers, the chip selection controls the three-state output buffers on the chip and, until activated, maintains that device effectively disconnected. It is thus possible to parallel several memory devices using the same low-order address lines and common data lines by suitably connecting the higher order address bits to the chip select pins.

Memory design continues to evolve very rapidly and RAM organized on a *byte-wide* basis is now very popular. It is possible to obtain an 8K x 8 bit CMOS RAM in a single package (the Hitachi HM6264, for example), though thirteen address lines, A_0 through A_{12}, and eight bidirectional data lines, D_0 through D_7, plus several control signals and power supplies, necessitate a 28-pin package. The introduction of CMOS technology allows an extremely low power-down, or *standby*, state to be achieved, with current demand measured in microamps. The device mentioned above has a quoted standby current of 20 microamps, though during normal operation this rises to over 100 milliamps. The very low power-down requirement means that a small battery, such as a 3.6 volt Ni-Cd battery, can be used to retain data, which is then no longer volatile.

Example 7.1: Design a stand-alone semiconductor memory unit with a storage capacity of 16 Kbytes, giving details of all necessary control inputs. The block address selection is to be adjustable over the range 0 through 64K.

We will make use of two of the Hitachi HM6264 CMOS 8K x 8 bit RAM chips as the basis of the design. Each 8 Kbyte RAM has two chip select lines, $\overline{CS1}$ and CS2. $\overline{CS1}$ must be low and CS2 must be high for the chip to be selected. The input and output data circuits operate in parallel, giving an eight-bit bidirectional port. When not selected, this port is set to its high-impedance state. The write enable line, $\overline{WE}$, must be taken low to write into the RAM, and so the following input signals are required: address lines A_0-A_{15}, read/write control $R/\overline{W}$, and a timing waveform. Selection of the memory chips to give two contiguous blocks of 8K can be controlled by address bit A_{13}, figure 7.7. When A_{13} is at zero, CS2 on chip 1 is high; chip 2 is not selected and its port is in the high-impedance state.

The available address space with 16 address bits is divided into four blocks of 16K by means of address bits A_{15} and A_{14}. Selection of a particular block is achieved by connecting point A to the appropriate decoder (74LS139) output. Timing is controlled through the low-active $\overline{\text{Enable}}$ input to this decoder. The selected output goes low only when this input is low, and the RAM chips, through $\overline{CS1}$, can be maintained in their high-impedance state until the address line signals have settled down.

The addressable range determined by the number of address bits available is known as the *memory map*, and address bits A_{15} and A_{14} define the position of the 16K block on this map. Thus

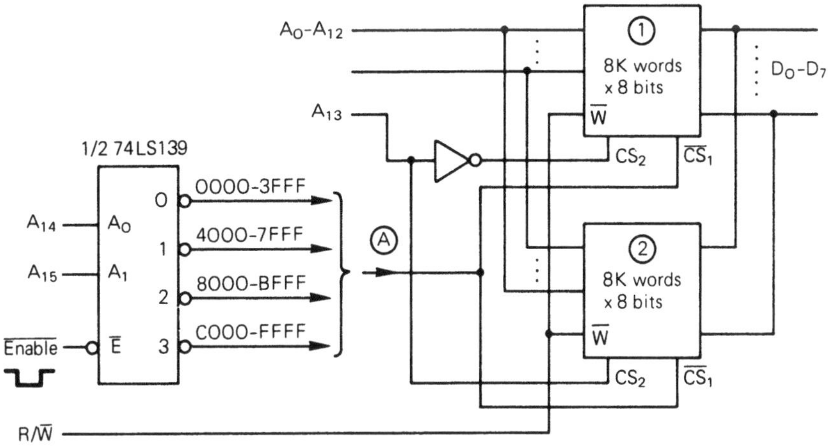

Figure 7.7 Use of chip select inputs to control two 2K words × 8-bit memory devices

	A_{15}	A_{14}	A_{13}	A_{12}	...	A_0	
	0	0	0	0	...	0	RAM chip 1
0000-	0	0	0	1	...	1	address 0000-1FFF
3FFF	0	0	1	0	...	0	RAM chip 2
	0	0	1	1	...	1	address 2000-3FFF
	0	1	0	0	...	0	RAM chip 1
4000-	0	1	0	1	...	1	address 4000-5FFF
7FFF	0	1	1	0	...	0	RAM chip 2
	0	1	1	1	...	1	address 6000-7FFF
	1	0	0	0	...	0	RAM chip 1
8000-	1	0	0	1	...	1	address 8000-9FFF
BFFF	1	0	1	0	...	0	RAM chip 2
	1	0	1	1	...	1	address A000-BFFF
	1	1	0	0	...	0	RAM chip 1
C000-	1	1	0	1	...	1	address C000-DFFF
FFFF	1	1	1	0	...	0	RAM chip 2
	1	1	1	1	...	1	address E000-FFFF

Note that when the chips are not selected, the power consumption drops from around 200 mW to a mere 0.1 mW; only the memory array is supplied with power and, being constructed in CMOS circuitry, it requires very little sustaining current.

For memory arrays of 16K and greater, *dynamic* RAMs are widely used. A typical dynamic memory cell, consisting of a single transistor and capacitor is shown in figure 7.8. The area occupied by the cell is considerably smaller than that required by the bistable cell and power demands are also less.

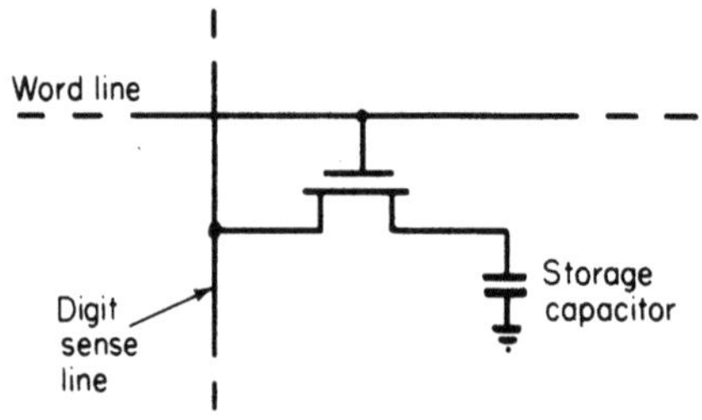

Figure 7.8 A single-transistor dynamic memory cell

To write into the memory, a word line is selected and data on the digit/sense line is stored as a charge on the capacitor. The term dynamic is used because of the need to refresh the stored data as the charge leaks away. At present these devices require such refreshing every two milliseconds. This is not such a stringent demand when one considers that more than four thousand memory accesses can be made during this period. To read the information, the word line is again selected and an amplifier senses the charge on the capacitor. This charge is very small; in the 16K RAM for example, a typical value for the capacitor is 0.04 pF!

The capacitance associated with the digit/sense line may be as much as ten to twenty times greater than the storage capacitor, and, because of charge distribution between this capacitor and the line during sensing, the voltage to be sensed falls to around 100 to 200 mV from a value of, typically, 2 volts, though this will depend on how much charge has already leaked away. The sense amplifier must be able to detect this small voltage reliably, and to achieve this a dynamic flipflop arrangement is used. The flipflop is held in a quasi-stable state with the two sides clamped to the same level until sensing is required. The clamping is then released, with one side of the flipflop connected to a dummy reference signal and the other to the digit/sense line. The minute difference in signal level between the two sides causes the flipflop to settle to the appropriate state. The regenerative switching action automatically replenishes the charge on the line, thus making it possible to refresh the stored data during a normal memory operation.

In one important application, refreshing can be achieved without any interruption in the use of the memory. This application is in video displays, in which the storage demands can be very great, and the large capacity single-cell dynamic RAM memory chips are ideal. To store a high-resolution picture of 1000 by 1000 picture dots, or *pixels*, with a 16-colour facility, for example, would require a memory of around 20 Kbytes. However, as a cathode ray screen needs refreshing continuously to maintain the picture, it is not difficult to ensure that the repetitive reading of the picture memory for screen refreshing is arranged in such a way that each dynamic RAM cell is read, and automatically refreshed, within the required time.

If an application does not allow the refreshing of the memory in this way, then a special refresh cycle must be included. Dynamic memory, DRAM, chips are now available which include refresh circuitry 'on-board', and to the user they behave as *static* RAMs, the refreshing being transparent. Note that the term 'static' is used to describe any RAM cell which does not need periodic refreshing to retain its data.

The large capacity memory arrays made possible by the dynamic cell lead to a major packing problem. Consider for example the Texas Instruments TMS4416/64kx one-bit array, which requires sixteen address lines merely to select a single bit whereas industry standards demand that an 18-pin package be used for the total array. To overcome the pin-limitation problem the address lines are multiplexed. That is, the memory address selection is organized as 256 rows by 256 columns and the row and column addresses, eight bits each, are applied in succession to the same set of address pins. Gating signals are used to strobe the appropriate set of signals, in time sequence, into the chip. Figure 7.9 illustrates this timing sequence, assuming latching on the negative-going edge of the row and column select waveforms.

Figure 7.9b shows one technique for generating the $\overline{\text{RAS}}$ and $\overline{\text{CAS}}$ signals in their proper relationship to a 14-bit address suitable for a 16K DRAM. The circuit assumes that the address line signals, A_0-A_{13}, are stable at the required values before the chip select signal, CS, is generated and that the $\overline{\text{Enable}}$ signal to the chips is present. $\overline{\text{RAS}}$ is developed by inverting CS via T1. Until CS appears the multiplexers, 74LS157, have the select input, S, low, and so address bits A_0-A_6 are selected and appear on the DRAM address pins to be clocked in on the negative-going edge of $\overline{\text{RAS}}$. The changeover on the multiplexers does not occur until after the negative edge of $\overline{\text{RAS}}$ because of the delaying action of inverter T2. When the select input does change, after a short delay the multiplexers switch the address bits to A_7-A_{13}, and the input to inverter T3 is taken high. Now $\overline{\text{CAS}}$ is derived from one of the multiplexers and the delay in T3 is sufficient to ensure that the new address signals stabilize before the negative-going edge of $\overline{\text{CAS}}$ is generated. This circuit can easily be extended for larger memory chips.

To ease the refreshing problems, the internal circuitry of the DRAM is organized so that the selection of a row alone automatically refreshes all the cells associated with that row. This is achieved by reading all the bit cells connected to the addressed row, using parallel sense amplifiers, and the indicated individual bit is then selected using the decoded column address. In this way a 64K DRAM can be fully refreshed using only 256 row selections and $\overline{\text{RAS}}$ signals. In fact, if data is not required to be output, it is not necessary to provide the $\overline{\text{CAS}}$ signal at all.

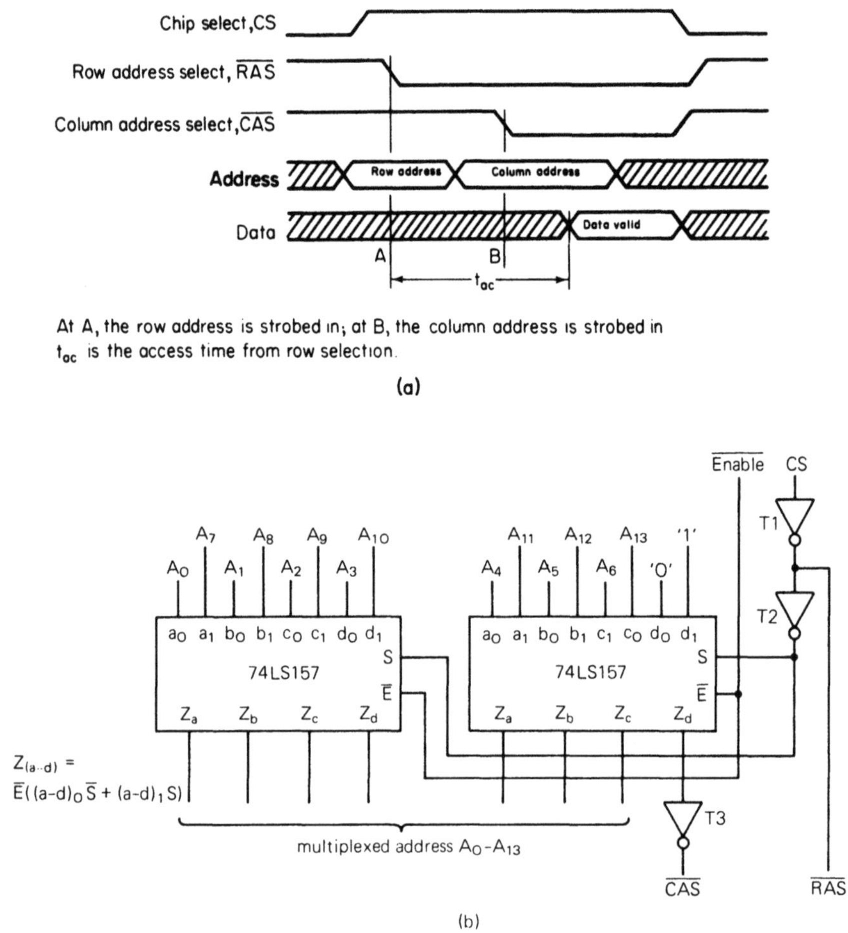

Figure 7.9 (a) Timing diagram for a read operation. (b) Generation of $\overline{RAS}$ and $\overline{CAS}$

Early types of DRAM operated from power supply rails of +5 volts, −5 volts and +12 volts, but more recent designs use a single +5 volts supply, generating the other voltages internally as required.

High-capacity DRAM chips now have cell areas so small that they can be affected by alpha particles emanating from the minute amounts of radioactive substance present in the packaging material, or simply occurring as background radiation. These alpha particles collide with the silicon atoms and create leakage currents which can interfere with the small stored data charges, giving rise to dynamic failures known as *soft errors*. As chip densities increase further and cell sizes reduce, such errors will become a major problem. At the time of writing, the largest-capacity chips commercially available are 256K devices utilizing interconnection line widths of 2.5 microns (micrometres). However, a 16 Mbit DRAM using 0.7 micron technology has recently been announced. Because of the high electrical stress imposed by even five volts, these high-density devices use a 3.3 volt supply. To improve the yield during manufacture, use is made of redundant rows and columns of cells that can be substituted for faulty locations (Goodman and McEliece, 1982).

When designing memory systems using these large capacity memory chips, great care must be taken over the board layout and power distribution. The fast switching speeds of the devices within the chip create large current transients. In order to prevent noise problems efficient decoupling of the power supplies is essential. In fact the same rules apply as with the design of TTL layouts, namely a capacitor of at least 0.1 μF should provide a decoupling path from the power supply rails to ground at every package.

7.2 The read-only memory

The term read-only memory (ROM) is applied to a wide range of devices in which fixed two-state information is stored for later retrieval. Unlike most semiconductor RAM, it is non-volatile. Usually the device is a semiconductor integrated circuit containing a matrix of addressable cells, each cell having been set to either logical one or logical zero. Addressing is basically by 2D selection.

As we shall see later, in a computer system ROM is used to hold the program of instructions and data constants such as 'look up' tables. Other uses for the ROM, which we have already met, are in the generation of codes and the implementation of logic functions.

There are three main types of ROM. These are: mask programmed ROM; programmable ROM (PROM); and erasable PROM (EPROM).

The mask programmed ROM is preprogrammed by the manufacturer during the fabrication process. Both bipolar and MOS transistors may be used as memory cell elements. A MOS transistor can be held permanently off, and effectively disconnected from the matrix, by ensuring that the gate oxide layer is too thick for normal operation. Programming of the MOS memory, therefore, can consist of leaving 'one' cells unchanged during an oxide removal process. The basic memory cell is much simpler and smaller than the RAM cell and larger capacity memories are readily available. Figure 7.10 shows a programmed MOS array.

In the case of a bipolar transistor array, the connection between emitter and load resistor is determined by the program mask. This contact programming is also applied to some types of MOS array, particularly when a fast turn-round time in production is required, since it can be carried out at a later stage in the fabrication than can the 'gate mask' programmed method. Bipolar arrays are inherently faster in information access than the MOS arrays since the on resistance of the selected device is lower. Schottky-clamped transistors are also used to improve the speed but the penalty carried by the bipolar arrays is one of greater power dissipation. Typically, a 1024 bit bipolar memory, organized as 256 words of 4 bits, has an access time of 45 ns and a power dissipation of 425 mW, whereas an 8192 word by 8 bit MOS ROM has an access time of 250 ns

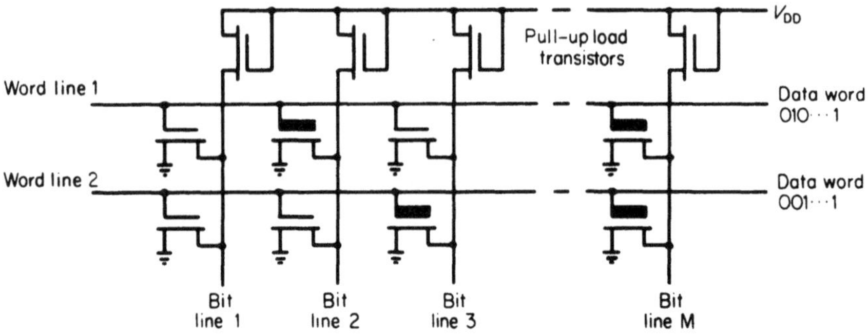

Figure 7.10 A programmed MOS ROM array. The thicker gates indicate the inoperative devices

and a power dissipation of 220 mW maximum with only 35 mW in the standby mode.

The bipolar programmable ROM is manufactured with all bits in the high or logical one state. Programming is carried out by the user and is achieved by blowing a fusible link in each memory cell which is to be set to logical zero. To do this, a current of carefully controlled magnitude and duration is passed through the selected output from the intersection of word and bit lines for the addressed cell (see figure 7.11). The current causes the link to melt, breaking the circuit and changing the state of the cell. Once broken the circuit cannot be remade. Fusible materials in widespread use are nichrome, polysilicon and titanium-tungsten. Nichrome-fused PROMs have proved reliable but necessitate voltages of approximately 20 V during the programming operation. Polysilicon-fused and titanium-tungsten-fused devices do not demand such high voltages, titanium-tungsten links being used on the higher speed devices. It is possible to

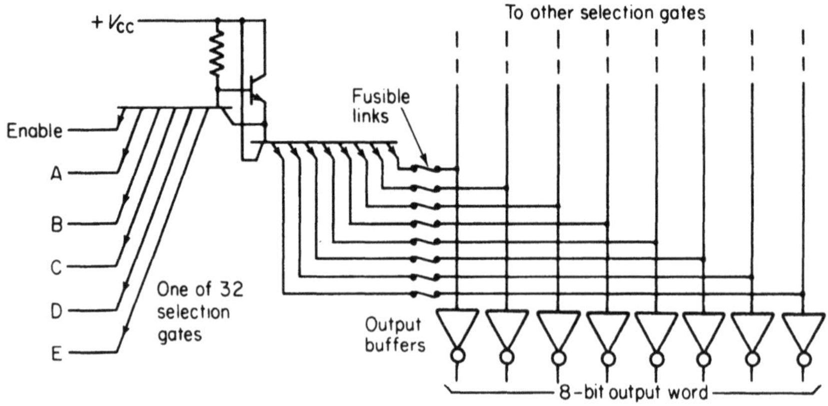

Figure 7.11 A bipolar programmable memory

obtain PROMs and ROMs with pin-for-pin compatibility thus allowing system development and prototyping to be carried out on the more expensive, but user-programmable devices. When the system has been satisfactorily proven, the mask-programmed equivalent can be specified for the production run.

The EPROM is a memory device both erasable and programmable. It offers the user the facility of erasing and reprogramming the memory matrix, and, although much more expensive than the PROM, it is more convenient for program development. The memory cell consists of an MOS transistor with an embedded silicon gate. This gate is electrically isolated within the silicon dioxide insulation and is referred to as a *floating gate* (figure 7.12). This transistor is often called a FAMOS device (*f*loating gate *a*valanche-injection *m*etal–*o*xide *s*emiconductor). Earlier EPROMs consisted of a FAMOS device connected in series with the source of a 'normal' FET in a matrix similar to that in figure 7.10.

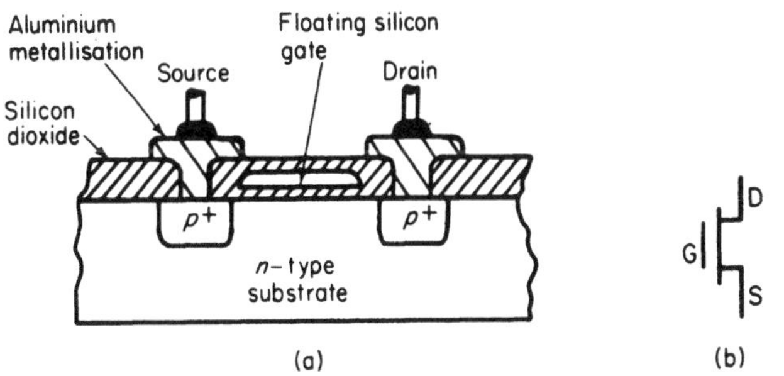

Figure 7.12 The floating-gate memory cell: (a) construction; (b) symbol

A further development, by the Intel Corporation, is the *stacked-gate* memory cell in which the FAMOS device is incorporated into a single FET structure, allowing the word line connection to be made to a second gate positioned above the insulated gate. During programming a relatively high voltage is applied between this gate and the drain, inducing avalanche conditions in which electrons are removed from the floating gate. When the applied voltage is removed the gate retains its charge and it is this charge which creates an inversion layer in the substrate beneath the gate, turning the transistor off, effectively increasing the threshold voltage of the transistor, and so preventing the second gate turning on the transistor when the word line is energized. The presence or absence of a charge on the floating gate determines the information stored in that cell. When reprogramming the memory the whole chip must first be cleared. It is exposed for several minutes to intense ultraviolet radiation of a specific wavelength and the resulting photoelectric current to the substrate discharges all the floating gates. These memory devices are packaged with a transparent lid which should be covered with an opaque label to prevent unintentional erasure, though it is estimated that constant exposure to fluorescent lighting of room-level intensity would not lead to erasure of a typical EPROM in less than three years! At the time of writing memory devices use this structure with capacities up to 64K × 8 bits (512 Kbits) on a single chip, though *wordwide* EPROMs are also available. The Intel 27210 device, for example, is organized as 64K × 16 bits and is currently available in a 40-pin package without the quartz window. Such devices are known as *one-time programmable* since they cannot be erased and reused.

A modified version of the EPROM, known as the *electrically erasable* PROM, EEPROM or E^2PROM is available which can be erased by applying a high voltage pulse of around 30 V to the programming pin.

A more recent development is the *electrically alterable* ROM or EAROM. The data held in a single byte of memory can be selectively changed by application of a voltage pulse. A relatively high voltage is needed, and early devices specified a 21 volt programming pulse. Current devices, however, use a single 5 volt supply and generate the higher voltage internally. A typical device of this sort is the Seeq Technology PQ 2817A, 2K × 8 bit memory. Programming requires around 10 ms for each byte and relies on an *electron tunnelling* mechanism described by Esaki (1958) though predicted as early as 1929 by Gamov and others. They showed that if the barrier region of a semiconductor *pn* junction is made thin enough, there exists a high probability that electrons will pass through at the speed of light. In the EAROM, an oxide layer 2 μm thick is used to insulate the floating gate, and electrons tunnel on to or from the gate in a direction determined by the polarity of an applied electric field.

The main advantage of the EAROM lies with the ability to perform in-circuit modifications of non-volatile programs, which would otherwise necessitate the use of magnetic disc or tape (see section 7.6). This advantage increases as faster programming speeds and lower programming voltages are introduced.

7.3 Bubble memory devices (BMD)

The magnetic bubble memory is a storage device which bridges the capacity/data-retrieval-time gap between magnetic and semiconductor devices on one side and the electromechanical magnetic tape and disc units on the other. Improvements in bubble materials, circuit processing and design now allow bubble technology, to be used in applications requiring one to a hundred megabits and retrieval times less than 5 ms (Twaddell, 1979).

Bubble memories can be viewed as solid-state integrated analogs of rotating electromechanical memories such as discs, drums and tapes. In both rotating and integrated versions information is stored in the form of magnetized regions. These regions, in the integrated versions, are cylindrical domains in a thin layer of magnetic material, with magnetization opposite to that of the surrounding area. The storage material can be either a magnetic garnet, yttrium-iron, grown epitaxially on a non-magnetic garnet substrate, such as gadolinium-gallium, or an amorphous metallic magnetic layer sputtered on to a substrate such as glass.

The magnetic material exhibits uniaxial anisotropy with the easy axis perpendicular to the surface. With no external magnetic field, the magnetic domains are arranged such that the sample is magnetically neutral, half 'pointing up' and half 'pointing down'. As a small external magnetic field is applied, the domains whose polarity is opposite to the field shrink. If the external field is increased further, the domains shrink into bubbles. Any further increase in applied field will cause the bubbles to reverse polarity and collapse. In practice the external field is maintained by a permanent magnet. Typical bubble size is 6 μm, and a 16 Kbit bubble memory can be constructed on a film area of 25 square millimetres.

The bubbles are free to move throughout the film and can be viewed as tiny magnets afloat in a magnetic field 'sea' of opposite polarity, as shown in figure 7.13. They move through the film in much the same way as holes move through a semiconductor. The bubbles must be constrained to move in an orderly way and the most common method is to propagate the bubbles along Permalloy T-bar or chevron patterns (figure 7.14). The patterns are evaporated on to the surface of the film, usually separated by an insulating film of silicon dioxide. To move the bubble a rotating magnetic field is applied to the structure and the bubbles are forced to move along in response to magnetic poles produced in the Permalloy. The substrate chip is placed at the centre of orthogonal coils which, when supplied with a two-phase clock signal, serve to produce the rotating magnetic field.

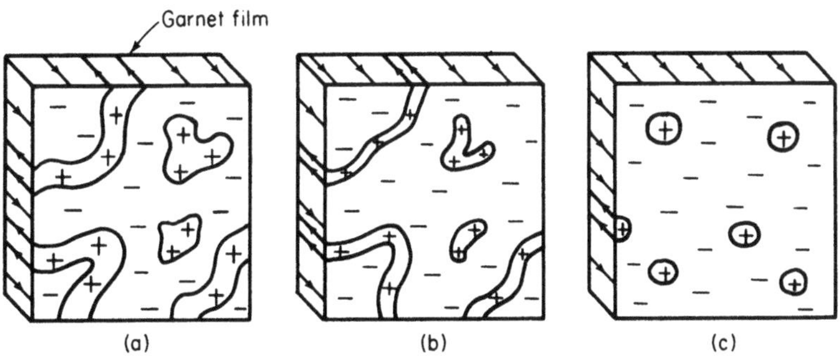

Figure 7.13 Formation of magnetic bubbles in film: (a) no external field; (b) small external field; (c) large external field

Detection of the bubbles is achieved by utilising the magneto-resistive effect which is present in the Permalloy material. The bubble is first elongated into a strip by means of a chevron pattern. Passage of the strip-bubble over an interconnected chevron is possible provided the interconnections do not form a continuous strip. A current is passed through such a chevron and any change in resistance is sensed as a small voltage change when the bubble passes over (figure 7.15).

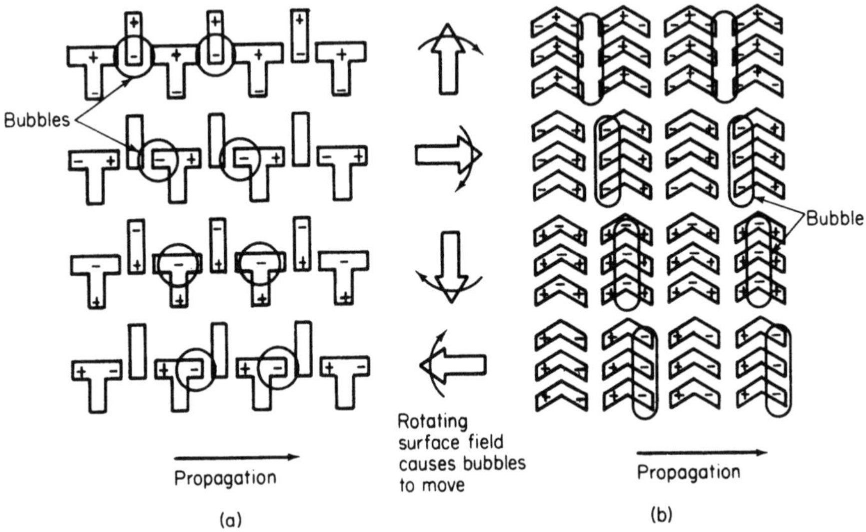

Figure 7.14 Bubble propagation patterns: (a) T-bar; (b) chevron

Bubbles are generated by passing a large amplitude current pulse through a conductor (figure 7.16) to produce a localised magnetic field intense enough to reverse the garnet's magnetisation and thus directly create a bubble. The bubble, after generation, is propagated down the chevron track and on to the propagating patterns. Typically 150 mA is required to generate a bubble.

The bubble memory is basically a shift register and the presence of a bubble at a sensing point signifies a logical one. Trains of bubbles propagate serially along the Permalloy patterns in response to the rotating field. In the absence of

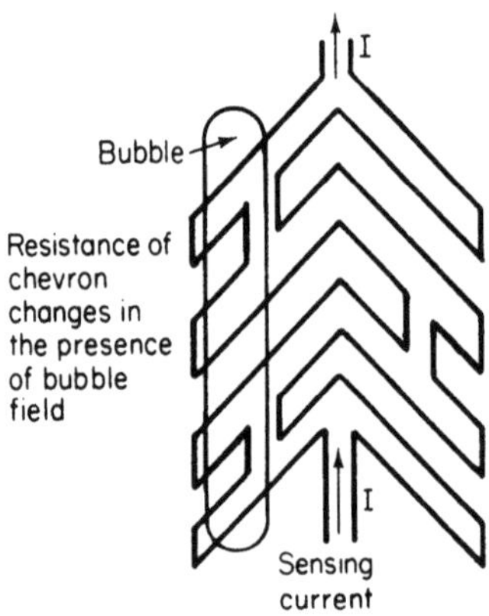

Figure 7.15 Detection of a bubble

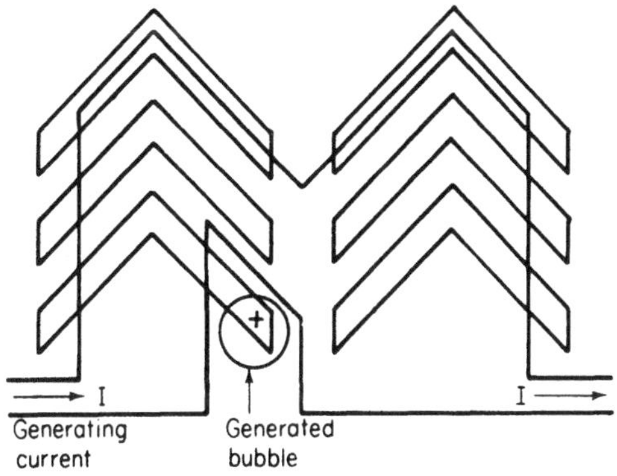

Figure 7.16 Bubble generation

this applied field the bubbles remain stationary, held by the bias field supplied by the permanent magnet. Start/stop operation is a very advantageous feature. Stopping the shifting process completely, or restarting it from a standstill, takes only 10-20 μs; thus after each memory access the bubble chip can be stopped to conserve power. Similarly, during data transfer, start/stop operation can give variable transfer rates and so simplify data buffering requirements.

In order to reduce the access time of the serial bubble memory the memory chip is organized as a series of minor loops, each with a transfer gate into a major line for reading and writing (figure 7.17). Bubbles circulate around the lines and when selected are transferred to and from the major line. Typically for a 64 Kbit

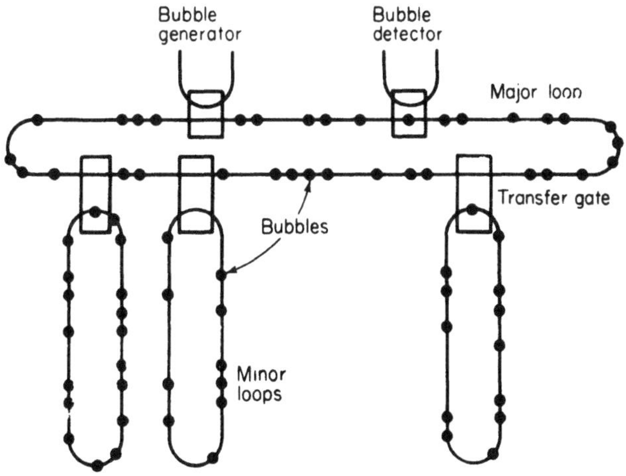

Figure 7.17 Major and minor loop structure. Minor loop data selection is effected by its address-controlled transfer gate

memory chip there might be 128 such minor loops, each containing a maximum of 512 bubbles.

Bubble memories are slower than semiconductor memories but their non-volatility, low power consumption and extremely small size make them competitive with rotating magnetic memories.

7.4 Charge coupled devices (CCD)

Charge coupled devices (Millman, 1979), though volatile, are to a certain extent competitive with bubble memories in mass storage systems. Since their invention in 1970 they have evolved rapidly and commercial devices with capacities in excess of 64 Kbits are readily available.

The basic element is a shift register structure, shown in figure 7.18, in which an input voltage, V_{in}, to the first electrode controls the injection of a charge of minority carriers into the *p*-type silicon near the diffusion. This packet of charge in the vicinity of the first electrode, and subsequent packets, can be moved from electrode to electrode along the structure by means of a three-

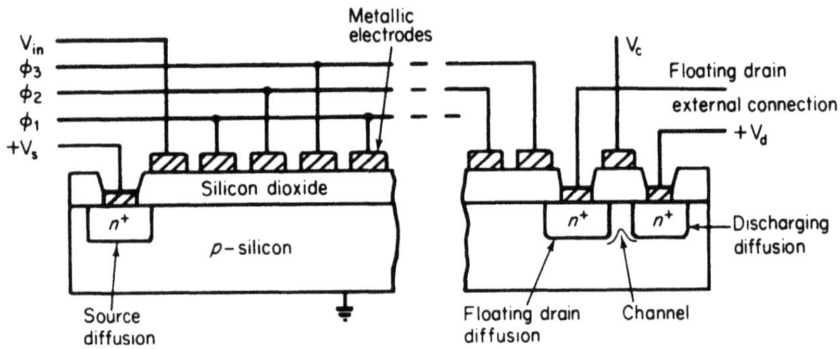

Figure 7.18 Construction of charge coupled device

phase clock ϕ_1, ϕ_2, ϕ_3. At the end of the shift register a floating drain receives the packets of charge which are then discharged through a channel created by the pulse V_c. Before this happens the charge on the floating drain either controls the gate of another device to provide an external signal or is used to inject charge to a subsequent channel.

The CCD shift register has an extremely simple structure and, owing to the low circuit component requirement, can potentially achieve higher bit densities than semiconductor RAM. The two main effects which limit the size of the device are transfer efficiency and storage time. Each time a packet of charge is moved, a fraction of it is lost as a result of charge trapping in the silicon. However, losses are only in the order of one part in 10^4, and registers with lengths of thousands of bits can be constructed before refresh amplifiers are needed to restore the charge level. Storage time refers to the maximum time for which the charge packet can remain in the structure before leakage currents reduce it to too low a level. This time is temperature dependent, but within normal operating ranges, storage times of 2–10 ms can be obtained. This determines the minimum clocking rate, the upper rate being dependent on charge mobility.

As with bubble memories, the internal organization of a typical CCD memory is a parallel arrangement of smaller registers giving a wider data path with a corresponding reduction in access time (Panigrahi, 1977).

7.5 The associative memory

An *associative memory* or *content addressed memory* (CAM) (Foster, 1976) is a memory in which location selection is controlled not by an address word but by a subset of the desired data word itself. This subset is referred to as the *match word*. The use of such a memory is primarily in data sorting and list manipulation since the advantage of this type of selection is that a computer does not need an address code to keep track of words stored. For example, an associative memory could contain the entire record of all vehicles registered in a given period. Each memory word per vehicle could contain the make, the colour, the owner's name, and the registration number. From such a memory the computer could list the owners of all two-year old green Ford cars with a registration number containing the letter A, say, by comparing the match information with the appropriate part of every word in the memory. Associative memories often carry out data searches using criteria other than equality. Typical are those of 'greater than', 'less than', and 'between limits'.

The most commonly used technique for locating the desired word in an associative memory is the *bit search* technique, where bits n of all words are compared with match bit n, then bits $(n + 1)$ with match bit $(n + 1)$, and so on until all the match bits have been compared. Only those words which compare correctly with all match bits are output. Selection by bit search rather than by word search is fast but makes heavy demands on hardware.

The basic requirements for an associative memory are a storage element with a non-destructive read-out facility, a high speed of access, and the ability to perform comparisons with selection. These properties can readily be provided by semiconductor devices and integrated circuit CAM arrays are available, though the cost of such memories is necessarily greater than that of the conventional RAM.

7.6 Magnetic surface recording

Magnetic surface recording is widely used as a low-cost-per-bit, high-volume, high-density storage medium. The surface, a thin layer of magnetic square-hysteresis-loop material, is moved at high speed past record and sensing heads, as shown in figure 7.19. The information to be stored, encoded in the form of current pulses, causes flux changes in the gap between the pole pieces of the write head and magnetises the surface in the vicinity of the gap. The information is retrieved by a read head which senses the flux changes occurring as the surface passes under the head.

Probably the earliest commercial mass storage system using surface recording was the magnetic drum. Drums are still produced but applications are limited to areas where reliability and ruggedness outweigh poor utilisation of space. Their

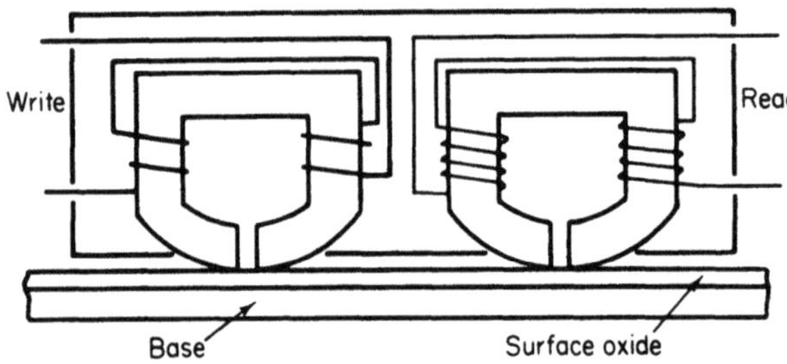

Figure 7.19 Magnetic surface read and write heads

place has been taken by the disc store. Discs can be classified as rigid or flexible (the floppy disc). The rigid disc, with either a plated nickel-cobalt or magnetic iron oxide surface, is capable of storing hundreds of megabytes on a 250 mm (10 inch) diameter surface. To record at very high bit densities, and to obtain good signal strength from the sense heads when reading, it is essential that the heads are positioned as close to the magnetic surfaces as possible but do not *plough* into the surfaces. Early systems used rigidly mounted heads at a distance from the surface defined by the mechanical tolerances. To accommodate unavoidable variations in surface thickness, typical separation was 0.25 mm. Modern systems use the 'flying head' principle in which the head is held a constant distance from the surface by an air cushion, self-generated aerodynamically or supplied from air jets built into the head structure. This method reduces head-to-surface separation by a factor of ten compared with the rigid mounting. The *separation loss* is directly related to the distance separating the head and surface and is a function of the wavelength of the recorded signal.

Disc storage systems are available in a wide range of sizes from single-disc memories to the multiple stacks used in large computer systems, capable of holding in excess of a billion words. At the other end of the scale floppy discs, or diskettes, are available for the smaller system and are used extensively in low-cost microprocessor-based systems where the higher price of the more sophisticated, and more reliable, rigid disc system would not be appropriate.

Basically, the floppy disc consists of a flexible magnetic disc, usually Mylar-based, coated with a magnetic oxide. The disc is 200 mm (8 inches) in diameter, or 130 mm ($5\frac{1}{4}$ inches) for the mini-floppy, and is permanently held in a plastic envelope for protection. This envelope has a slot in it, allowing the read/write heads to come into contact with the oxide layer when inserted into the drive unit (figure 7.20). By running in contact with the surface, bit density is increased but the head-to-surface wear is correspondingly increased as well. The wear is minimized by maintaining contact only during data transfers.

Microfloppy systems use a disc of 90 mm (3.5 inches) diameter, housed in a rigid plastic case which incorporates a dust shutter. Three inch (75 mm) discs are also common.

With both the rigid and floppy discs, data is recorded on to and read from concentric tracks, each of which is allocated a track address. The heads are permanently positioned over the correct track or are moved there under servo control. Each track is divided into *sectors* and positional information is provided from either pre-programmed control tracks, *soft* sectoring, or, in the case of some floppy discs, from photoelectrically sensed holes punched into the disc at appropriate intervals, *hard* sectoring. Typically a floppy disc is spun at 360 revolutions per minute and 77 concentric recording tracks are available for use, though *double-density* discs with 154 tracks per side and a capacity of 2.3 Mbytes are available.

At present, microfloppy discs offer a storage capacity of 0.5 to 1 Mbytes, though this is likely to increase rapidly.

A particular design of rigid disc memory with enhanced reliability, developed initially by IBM and known as the *Winchester* technology, has come into prominence. The main technological advance was in the head design which has a very low mass and is lightly loaded. The head starts and stops in contact with the specially lubricated magnetic surface. The entire mechanism, including the disc, is maintained in a sealed, contamination-free environment. Early Winchester systems, still the most popular, used 130 mm ($5\frac{1}{4}$ inch) diameter discs with a head-to-surface separation of less than 0.5 μm (20×10^{-6} inch). More recently, 200 mm (8 inch) and 350 mm (14 inch) discs have been introduced with bit densities of 8.5 Kbits per track inch (330 bits per millimetre) and storage capacities from 5 to 16 Mbits. Allowing the head to rest in contact with the surface obviates the need for a complex head loading mechanism, reducing weight and enabling the head to fly closer to the surface. This in turn gives a greater signal output, adding to the enhanced reliability of the system. Using thin film technology, recent advances in head design have given bit packing densities of 10 Kbits per inch (400 bits per millimetre).

In magnetic tape recording the magnetic oxide is deposited about 10 μm thick on a polyester base, and is allowed to ride directly in contact with the heads which are stacked together within the width of the tape, normally 12 mm (0.5 inch). The tape moves only when information is to be transferred, unlike

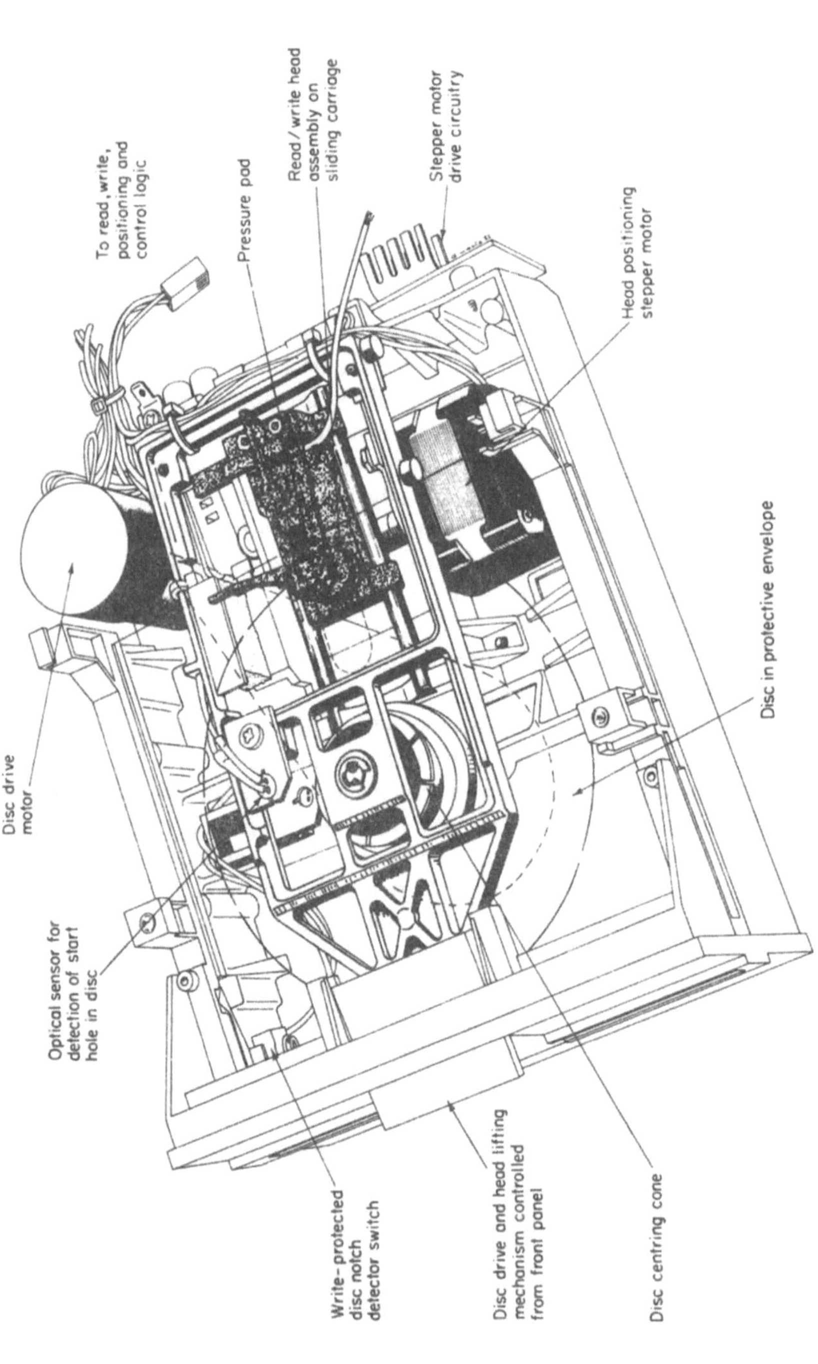

Figure 7.20 Essential features of a floppy disc unit

discs which often run at constant speed throughout, and tape wear is very small. This method of operation poses considerable mechanical problems, however, since the operating speed is 3.8 metres per second (150 inches per second) and in order to minimize tape wastage and time delays, the tape is necessarily subjected to high accelerating and decelerating forces. The tape is driven by pinch wheels and, as the mechanical parts have a high inertia, a reservoir of tape is maintained to give the spools sufficient time to get up speed without restricting the tape motion. The bidirectional spool motors are controlled by servo systems acting on signals from upper and lower limit photocells to maintain the reservoir levels (figure 7.21). An alternative approach is to use mechanical tensioning with

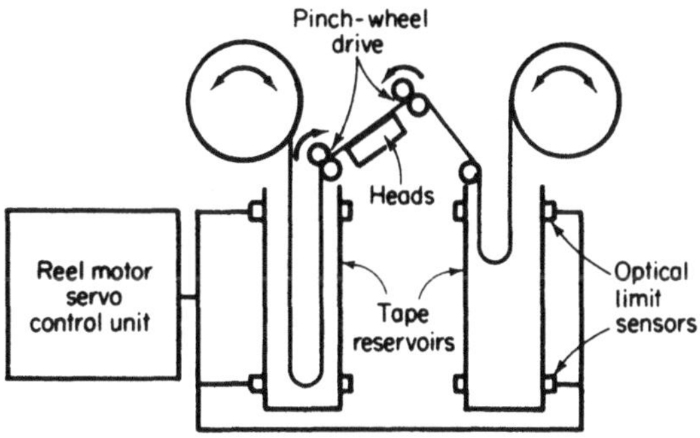

Figure 7.21 Magnetic tape station

the reservoir of tape provided by a system of lightly sprung pulley wheels. Small personal computer systems often adopt special coding arrangements which allow the use of commercial audio cassette recorders operating at 48 millimetres per second (1.875 inches per second).

Because of the inertia limitations and the inherent disadvantage of single dimension addressing it is usual to arrange that information is transferred in blocks to and from a random access memory as it is required.

7.7 Current pulse encoding

Several techniques are in use for coding the current pulses to the recording heads. The earliest and most straightforward method is the return-to-zero (RZ). Referring to figure 7.23 we see that the current pulse in one direction indicates a one to be stored and in the opposite direction a zero to be stored.

In an effort to reduce the number of flux reversals needed, non-return-to-zero (NRZ) systems have been developed. Basic NRZ operation holds the current at its previous level if the information does not change. A more commonly used method known as NRZ1 always changes the current level when a one is to be stored but not when a zero is to be stored.

Two other techniques which achieved popularity are 'double frequency codes' using phase modulation (Manchester method) and frequency modulation. Phase modulation involves a change of current phase in a positive direction for a one to be stored and in a negative direction for a zero to be stored. Frequency modulation, as the name implies, uses two frequencies to indicate the two logic levels zero and one. It is usual to have the upper frequency twice the lower.

A more recent scheme is that of delay modulation or modified frequency modulation (MFM). This uses an adaptive code in which the current level is changed in the middle of a bit period when a one is to be stored. For a zero to be stored the current is not changed unless followed by another zero, in which case the current is changed at the end of the bit period.

The main problems associated with magnetic surface recording are those of pulse crowding and skew. *Pulse crowding* occurs when the spatial separation between successive flux changes is insufficient and interference occurs, imposing an upper limit to bit density. This is particularly apparent towards the centre of a magnetic storage disc as the linear velocity decreases, but with aerodynamically supported 'flying' heads this effect is partially offset automatically by the head riding nearer the surface.

Skew is caused mainly by head-gap misalignment, though in magnetic tape systems additional skew problems are introduced by the flexible nature of the medium, causing jitter at high speed. The time synchronization between signals from tracks operating simultaneously creates difficulties which may be overcome by making all tracks *self-timing*.

Phase, frequency, and delay modulation give rise to regular flux changes, and from these it is possible to extract a timing signal, though the necessity for two flux reversals per bit period in phase and frequency modulation is a limitation not present in delay modulation. Circuits for the generation of these codes are illustrated in figure 7.22.

RZ encoding is seldom used because of the need to pre-erase the surface and because the frequency of flux reversals is high (as can be seen in figure 7.23), leading to pulse crowding at relatively low bit densities. NRZ encoding has a superior performance as far as pulse crowding is concerned but suffers from two other disadvantages: self-timing is not possible, and the misreading of a single bit will cause continued misreading until the bit value changes.

NRZ1 encoding overcomes the misreading difficulty, though pulse crowding occurs earlier. It should be noted that, while a single track cannot be self-timing, an output is obtained for every '1' read and, by including a character odd-parity track, every character encountered must contain at least a single '1', which can then be used for timing. This is only possible, of course, if skew is negligible.

With the improvements that have taken place in flying head design skew can be virtually eliminated by the much closer proximity of the head to the surface, and NRZ1 encoding, with its superior bandwidth performance, retains its popularity over the more complicated self-timing modulation methods.

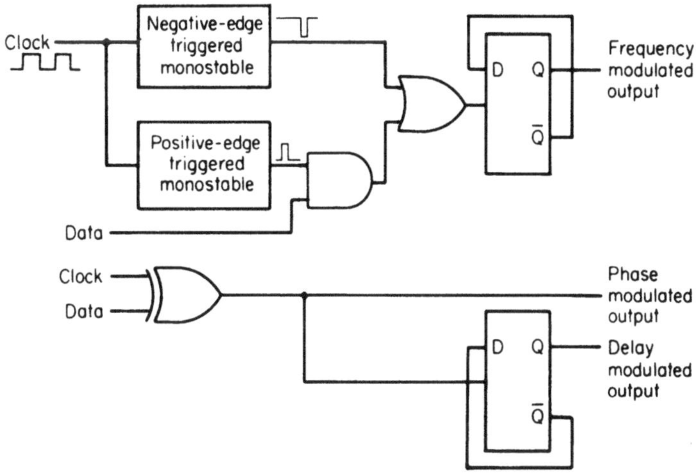

Figure 7.22 Generation of frequency, pulse and delay modulation codes

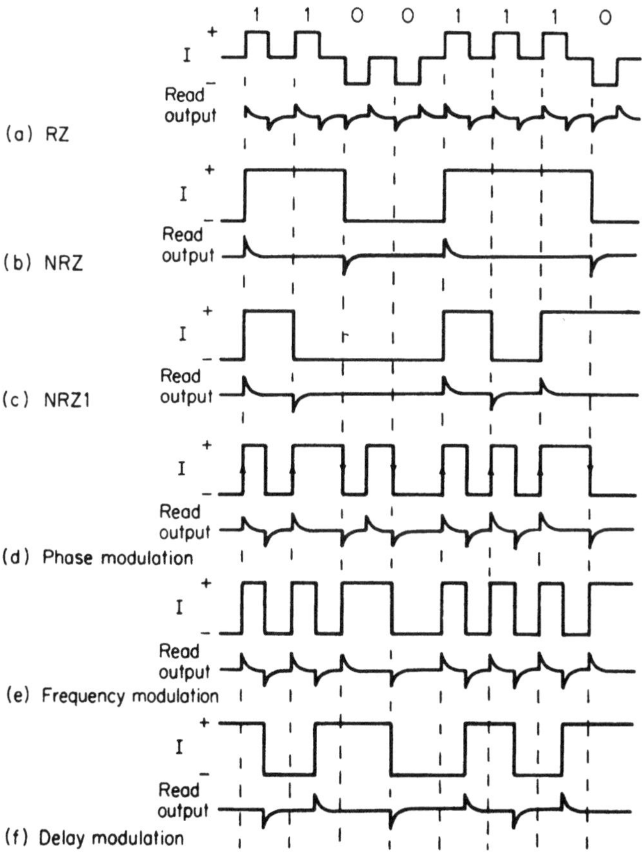

Figure 7.23 Current pulse encoding methods

Problems

7.1 Describe the operation of the basic storage cell used in static MOS and bipolar memories. Explain the addressing and controlling arrangements necessary for the correct operation of the memory with the different cell types, and compare these with the arrangements used in a ferrite core memory.

7.2 What properties must a circuit or device have to be considered for use as a digital memory element?
Describe carefully the operation of a dynamic RAM cell.
You are given a number of 16Kbit byte-wide memory chips. That is, they are organized so that the output is eight bits in parallel. Internally, however, the memory array is square. Show how you would expect the internal decoding to be carried out, and what external control lines you would expect to find.

7.3 Describe with the aid of diagrams the logical construction and method of operation of a bipolar semiconductor memory when arranged as (a) a random access memory, and (b) a read-only memory.

7.4 The effective address necessary to select a word from memory is made up of the word address plus any chip select signals required. Show, using block diagrams, how multiple memory chips may be used (a) to extend the memory word length, and (b) to extend the memory word capacity.

References

Bannister, B. R. and Whitehead, D. G. (1983). *Fundamentals of Modern Digital Systems*, 1st edn, Macmillan, London

Esaki, L. (1958). New phenomenon in narrow germanium p-n junctions, *Phys. Rev.*, **109**, No. 2, 603

Ford, C. (1982). Core memory technology, *New Electronics*, **15**, No. 19 (5 October), 54-6

Foster, C. C. (1976). *Content Addressable Parallel Processors*, Van Nostrand Reinhold, New York

Gilligan, T. J. (1966). $2\frac{1}{2}$ D high-speed memory systems – past, present and future, *IEEE Trans.*, **EC-15**, No. 4, 475-85

Goodman, R. M. F. and McEliece, R. J. (1982). Lifetime analyses of error-control semiconductor RAM systems, *IEEE Proc.*, **129**, Pt E, No. 3 (May)

Millman, J. (1979). *Microelectronics*, McGraw-Hill, New York

Panigrahi, G. (1977). The implications of electronic serial memories, *IEEE Computer*, **10**, No. 7 (July), 18-25

Rajchman, J. A. (1952). Static magnetic matrix memory and switching circuits, *RCA Review*, **13**, 183-201

Twaddell, W. (1979). Technology update; magnetic bubble memories, *Electrical Design News*, **24**, No. 15 (20 August), 80-88

8 Programmable Devices and Systems

The invention of the transistor was announced in 1948 and the first commercial equipment using transistors was available in the early 1950s. Development of improved production techniques, especially the planar process, led to the first tentative integration of circuits about a decade later, and after small scale and medium scale integration, large scale integration was introduced in hand-held calculators and similar equipment in the late 1960s. The first microprocessor, announced in 1971, was a 'spin-off' from the calculator work and since then there has been a rapid development in these very large scale integrated circuits.

The cost of designing, setting up and manufacturing an integrated circuit is very high and is only feasible economically if a large number of each circuit can be sold. Greater flexibility is necessary to maintain the very large markets needed as integration density increases, and this flexibility is achieved by adopting many of the ideas developed in the digital computer field over the last thirty to forty years. Modern digital systems have advanced by combining traditional logic design with programming techniques, at the same time developing circuit integration methods until the equivalent of about a million transistor devices can now be fabricated on a single silicon chip. These powerful chips allow us to build, at low cost and in a very compact form, a wide range of systems from a simple dedicated process sequencer to a complete computer. By the use of programmable devices, and in particular microprocessor circuits, many previously hard-wired logic systems can be replaced and extended in capability by the addition of decision-making logic, providing the ability to adapt the response of the system to the demands of a particular situation. In this chapter, therefore, we look at the fundamental operation of microprocessors and other programmable devices.

8.1 Programmability

In order to appreciate the ideas behind the development of all large scale integrated circuits we must return to the simple logic gate and reconsider our view of its operation.

Let us consider first the AND gate. Figure 8.1a shows the truth table for the two-input AND gate in conventional form. We can, however, allocate one of the inputs to act as an ENABLE input leaving only one logic input and we see that, although the circuitry operates in exactly the same way as before, it is now a *controlled* gate (figure 8.1b). If the gate is not enabled the output is at logical zero regardless of the value at the input.

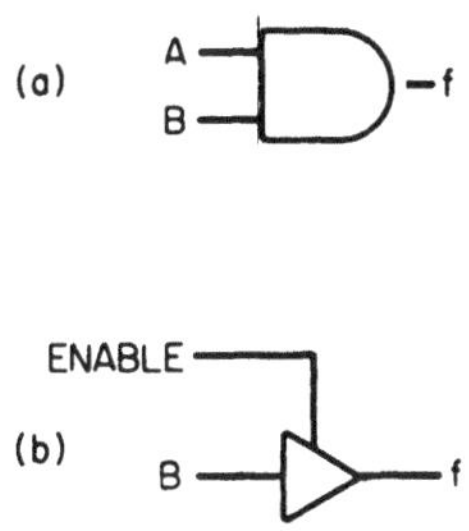

A	B	f
0	0	0
0	1	0
1	0	0
1	1	1

ENABLE	B	f
0	0	0
0	1	0
1	0	0
1	1	1

Figure 8.1 A controlled gate

As a second example, consider the exclusive-OR gate of figure 8.2a with its conventional truth table. Again we consider one of the inputs as a CONTROL input, the value of which controls the action of the gate. It is seen from the table of figure 8.2b that when the control signal is taken to logical one the gate output is the inverse of the input at B; but when the control signal is taken to logical zero the output is the non-inverted input at B. Thus the gate is now more accurately described as a controlled inverter.

It is not necessary to restrict the number of control inputs to one. The circuit of figure 8.3, for example, takes our previous example a stage further and uses

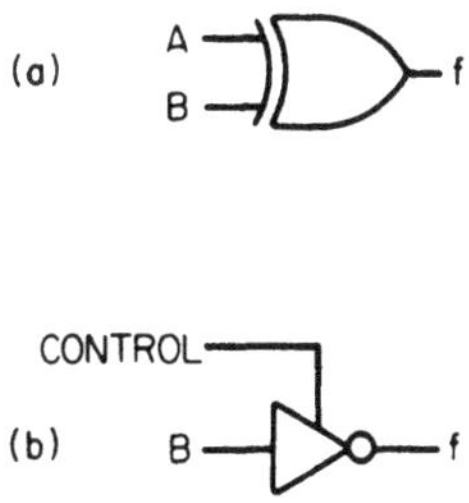

A	B	f
0	0	0
0	1	1
1	0	1
1	1	0

CONTROL	B	f
0	0	0
0	1	1
1	0	1
1	1	0

Figure 8.2 A controlled inverter

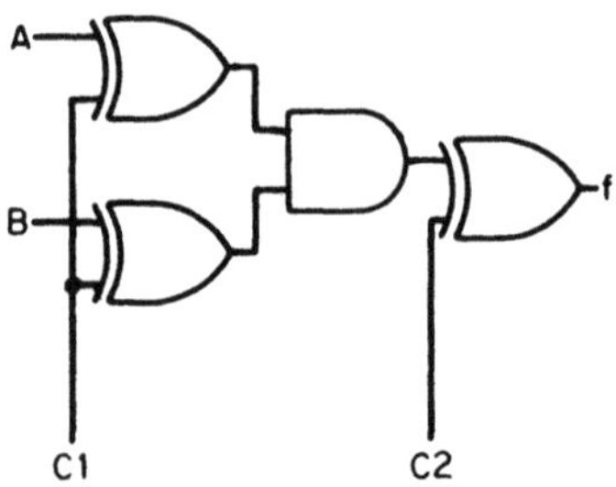

C1	C2	A	B	f	
0	0	0	0	0	AND
		0	1	0	
		1	0	0	
		1	1	1	
0	1	0	0	1	NAND
		0	1	1	
		1	0	1	
		1	1	0	
1	0	0	0	1	NOR
		0	1	0	
		1	0	0	
		1	1	0	
1	1	0	0	0	OR
		0	1	1	
		1	0	1	
		1	1	1	

Figure 8.3 A multi-function gate

two control signals. The operation of the circuit on the two logic, or data, inputs is determined by which of the four possible combinations of the control signals, C1 and C2, is present at any time. Thus by applying the correct control bit values the circuit can be made to act as a two-input AND, NAND, NOR or OR gate.

The enable signals used on LSI circuits, as explained in chapter 3, and the selection of a particular set of inputs on a multiplexer by use of a selection code, are other examples of our ability to vary the action, or mode of operation, of the logic circuit by means of externally applied signals. All these examples illustrate, admittedly on a limited scale, what is known as *programmability*. In the most general sense we can represent a programmable circuit as having a set of inputs carrying input data to be operated upon, a set of inputs carrying control signals specifying what the operation is to be, and a set of outputs for the resultant data, as shown in figure 8.4. In theory we could design and build a

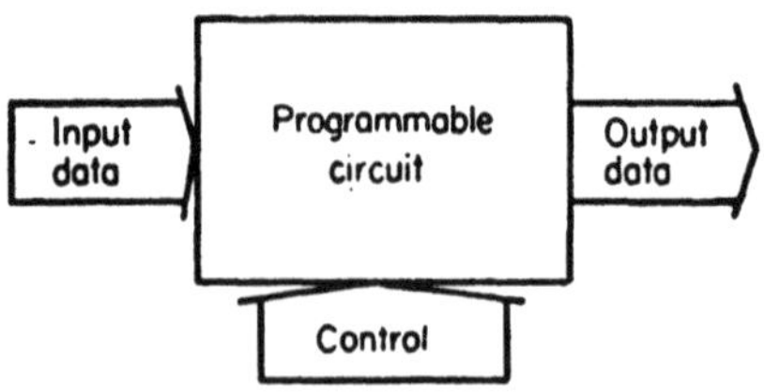

Figure 8.4 The generalized programmable circuit

completely general-purpose circuit, so that the application of the correct control signal pattern, or control word, would program the circuit to carry out a specific operation. In practice, of course, the number of control signals necessary, and therefore the number of control word bits, would be prohibitively large and it is not possible to include all the operations we may think desirable.

Many programmable circuits are available which are limited to a group of specific tasks such as the control of a digital communications receiver and transmitter, or the control of a video display unit or a floppy disc unit. These and more general-purpose devices are designed by extending the idea of time sequencing. Until this point we have considered only parallel operation on the input data but, as we have seen throughout digital system design, parallel and serial operation are interchangeable. Any set of operations which can be carried out simultaneously in parallel can alternatively be carried out, with the same results, as a time-separated sequence of individual operations.

Returning to our general programmable circuit of figure 8.4, we can arrive at a desired end result by carrying out a sequence of operations using a selection of short control words from a limited set available within the circuit rather than using one excessively long control word. In moving to the sequential mode of operation we must introduce memory elements, or registers, to act as temporary storage for the results of intermediate operations as the sequence proceeds. We have, in fact, arrived at the method of working used in the central processing unit, CPU, of the digital computer, and it is for this reason that general-purpose integrated circuits designed in this way are called *microprocessors.*

8.2 The programmable logic controller

The type of operation described above is also found in the programmable logic controller, PLC, which is extensively used in the control of industrial processes. Many of the problems in process control can be broken down into a sequence of comparisons and subsequent simple operations. For example, is the temperature less than that specified? If so, turn on the heater. Are both switch 1 and switch 2 on? If so, start the agitator motor; and so on.

Early programmable logic controllers used relays and small scale integrated circuits, but these have now been replaced by large scale integrated programmable devices which have the structure shown in figure 8.5. The central processor unit is in effect a one-bit microprocessor in that it operates on one data bit at a time. It is provided with a succession of control words, or *instructions*, from the read-only memory, and it is the sequence of instructions known as the *program*, decided upon by the designer of the control system, which determines how the unit is to operate. The instructions are stored in sequence in the memory, and each time the counter is incremented it indicates the location of the next successive instruction. The instruction is taken to the central processor unit where it is decoded and acted upon. Each instruction provides the information necessary to

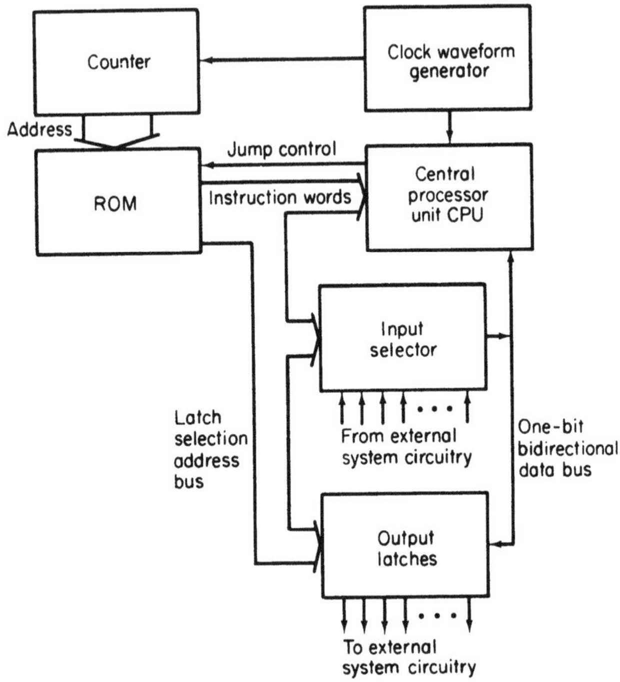

Figure 8.5 The programmable logic controller, PLC (MC14500B Industrial Control Unit, Motorola Semiconductors Ltd)

define the operation required to select the appropriate input or output connection. In most cases the range of operations possible is limited to no more than sixteen, requiring four bits in the instruction word.

These operations are of three main types: logical comparisons between an incoming data bit and a bit previously stored in a one-bit register; input and output data bit selection; and jumps. Jump instructions allow the memory counter setting to be altered so that alternative program sequences can be selected from the memory if the operating conditions require it. The remaining bits in the instruction word in most cases act as the address of the input selector or output latch which is to be used in carrying out the specified operation. For example, an instruction 10110110, 'OUT 6', would indicate that the data bit, 0 or 1, previously stored in an internal one-bit register, is to be transferred to the output latch number six.

The internal sequencing of the controller is based on an alternating cycle, accurately defined by the clocking waveform, in which a *fetch* cycle is followed by an *execute* cycle, which in turn is followed by the next fetch cycle, and so on until an instruction word is reached which halts the sequence. A fetch cycle involves reading the next instruction word from the memory and decoding it in the central processor unit to determine the operation required. The following execute cycle carries out that operation. The complete cycle of fetch and execute is often referred to as the *instruction* cycle.

The programmable logic controller is ideally suited to tasks requiring only decision and subsequent operation but, where calculation or data processing of any sort is involved, the one-bit operation is so severe a limitation that a multi-bit microprocessor would invariably be used.

8.3 The microcomputer

As we have seen, a microprocessor cannot operate in isolation but relies on other circuits to provide memory and input and output registers or *ports*. The complete system is called a microcomputer, and the number of variations in the way the sections of the microcomputer are packaged is very large. Variants arise from the compromises decided on by the design team in arriving at the final product and the applications for which the microprocessor is intended. Thus some microcomputers are complete in one integrated circuit package, with processor and control circuits plus a limited amount of memory and input/output capability. Other integrated circuits contain a more powerful processor designed to be used with external memory and input/output circuits to make up a larger system. Yet other integrated circuits are designed in bit-slice form so that several slices must be operated in parallel, giving the designers a high degree of flexibility.

The complete microcomputer system makes use of external, or peripheral, equipment for the inputting, outputting and bulk storage of data and programs. Industrial control sytems can be very complex but everyday personal systems will normally use a simple keyboard for data inputting, a video display unit, VDU, or small television set, for data outputting and a floppy disc or cassette unit for bulk memory.

The microcomputer, like all digital computers, can carry out only one, relatively simple, operation at a time, though it can work through several hundred thousand operations each second. As with the one-bit controller, each control word, or command, specifying a particular operation is known as an instruction and the complete sequence of instructions necessary to carry out a given task is called the program.

The number of connections to the processor is limited by the physical size of the package and this limits the number of input and output connections which can be provided. Many microprocessors use an 8-bit wide data bus, and so are said to be 8-bit microprocessors. Earlier types used four bits but, increasingly, 16-bit and 32-bit microprocessors have become common and 64-bit microprocessors are forecast.

The basic sections of the computer are shown in figure 8.6. The input/output unit is the means whereby external circuitry gains access to the computer. The data bus interconnects all sections of the system but data on this network is necessarily transient in nature, being synchronized to internal clocking signals.

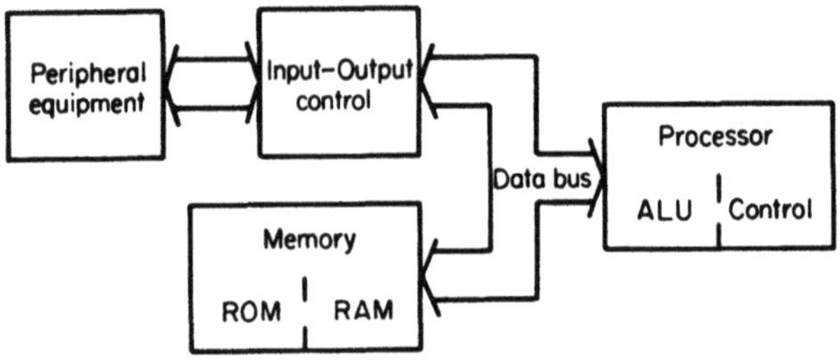

Figure 8.6 Basic sections of a computer

The input/output unit must be capable of latching relevant data words from the bus and presenting them in a suitable form to the external devices. Also, as an input, it must accept signals from external sources and by means of the system control signals transmit the data on to the bus. Most major semiconductor manufacturers make specialized integrated circuits for this purpose and, in order to provide the flexibility necessary for the very wide range of input/output requirements, these devices are programmable.

As a typical example, the Motorola Peripheral Interface Adaptor (PIA) is designed to communicate with an 8-bit data bus and has two 8-bit ports, each output capable of sinking sufficient current to drive two standard TTL gate inputs. When used with a wider data bus, such as a 16-bit, we merely double up, with one PIA handling the low byte and the other the high byte. Within the PIA the eight bits of each port can be programmed independently to act as inputs or outputs and the two ports themselves are selected by means of register addressing pins. Internally, each port has associated with it a data register, a data direction register, DDR, and a control register. The data direction register must be programmed so that each bit at logical one or zero determines the action of the corresponding port bits, either input or output. Because of pin limitations, since the device is packaged in a 40-pin dual-in-line package, access to the data direction register has to be via the same data bus connections. Access to the data register or the data direction register is therefore determined by the contents of the control register which in turn is accessed through this data bus, but only when appropriate control signals are applied to the register select inputs. The programming procedure is fairly complex, but the devices are intended for microprocessor use where the programming, once laid down in the microprocessor program, can be carried out automatically when the system is first switched on. Figure 8.7 shows the data paths through a PIA unit.

The arithmetic and logic unit, ALU, carries out all the arithmetic and logic operations on the parallel data presented to it, and as directed by the instructions of the program. The instructions are held in the memory and each instruction as it is required is taken from the memory and routed to the decoding circuits within the processor. The control section ensures that each instruction is fetched from

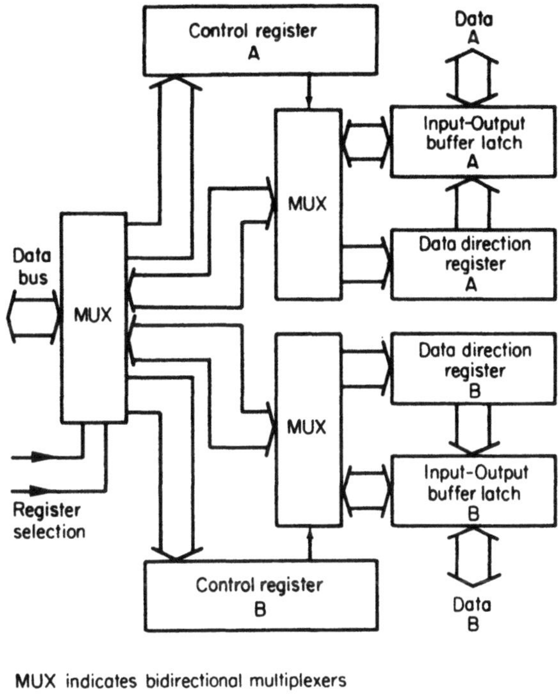

Figure 8.7 The data paths through a peripheral interface unit

the memory at the correct time and that it is decoded and executed correctly before the next instruction is fetched. It is necessary to have a program sequence to carry out any operation, even to bring another program into the memory, and since random access semiconductor memory is generally volatile, so that its contents are lost when the system is switched off, it is usual to have at least some of the program, and in many cases all the program, in non-volatile read-only memory.

Data taken from the memory, or from external peripheral devices via the input ports of the input/output unit, is transferred to the arithmetic and logic unit for processing. New data words resulting from the operations are returned to the memory until required for subsequent operations, or are outputted to external circuits via the output ports. The data memory is invariably random access memory, since both reading and writing are necessary. In order to reduce the need to move data to and from the memory, the processor section itself is often provided with a small high-speed extension to the memory in the form of a set of *working registers*, so that many operations on data are possible without having to use the main memory.

We have seen in chapter 5 that a register consists of an array of flipflops with controlling gates allowing data to be written in and held for as long as is necessary, and to be read out as and when required. Certain microprocessors with integral memory use the low-valued locations in RAM as the working registers (the MCS51 family of devices manufactured by Intel is one example). Other microprocessors have taken the idea still further and allow any block of the available random access memory to be used as the set of working registers (Texas Instruments TMS 9900 is one such). The first location in the block currently in use is specified by the value held in the *workspace pointer*, WP, register, and instructions are provided to change the pointer under program control. Depending on the manufacturer's design philosophy, the working registers are generally two, four, eight or sixteen in number.

The processor contains several additional registers, which are included for specific purposes to ensure the smooth running of the overall system, but such control registers are not in general directly available to the user.

8.4 Bus transfers

By far the most common internal operation of the computer is the transfer of data from one register to another. Both the source register and the destination register can be a working register, a control register, an input or output port register, or a memory location, depending on the task involved. Data is gated in parallel on to the data bus from any selected register, by enabling the three-state output gates, and gated into any other selected register by operating the appropriate register input gates. Each line of the data bus links the corresponding input and output gates of all the registers involved in the transfers (figure 8.8).

In most cases the number of bits in the data word of a microprocessor does not give sufficient combinations to allow the addressing of a large enough memory, and it is usual to provide a wider bus for addresses. Most 8-bit microprocessors, for example, have a 16-bit address bus which allows addressing of 65 536 locations in memory; 16-bit microprocessors use from 20 to 32 bits, allowing addressing of a hugh amount of memory, from 1 megabyte up to 4 gigabytes. Since addresses and data need not be carried on the bus at the same time, it is sometimes arranged that part of a single bus is shared between data and addresses by time multiplexing, as in the Intel 8085A 8-bit processor and most 16-bit and 32-bit processors.

The lines carrying the various gate control and timing signals, which must be passed between the processor's registers, ports and memories to ensure the correct operation of the data transfers, are often referred to as a separate *control bus*.

We have seen that the processor contains several elements including an array of working registers to act as temporary memories, additional registers for use in the running of the processor, an arithmetic and logic unit, and a control section

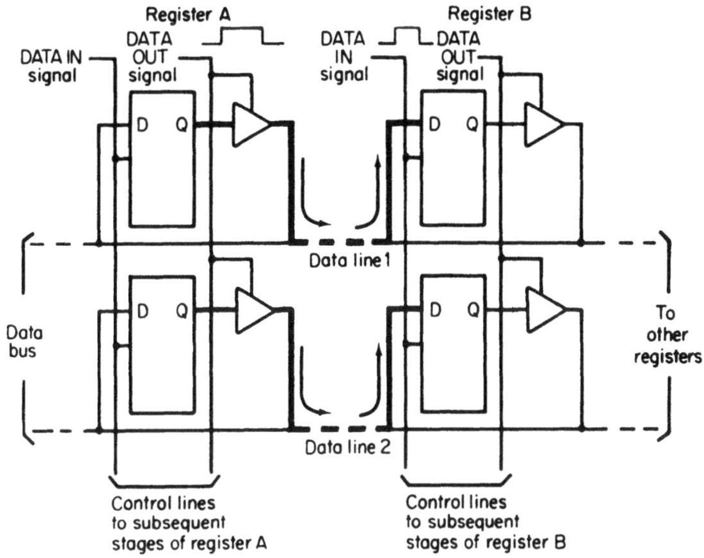

Figure 8.8 Single-bus operation. Register A DATA OUT signal enables the output gates: data passes on to the bus and into register B, selected by the Register B DATA IN signal

in which the instructions are decoded and the signals necessary to carry out the indicated operations are generated. The sequence of signals required for each instruction is itself a small program, known as a *microprogram*. In most processors the individual signals, or *microsteps*, are not available to the user, who works only with the set of microprograms, the *instruction set*, built into the processor at the design stage.

The processor elements referred to are shown in figure 8.9 together with the

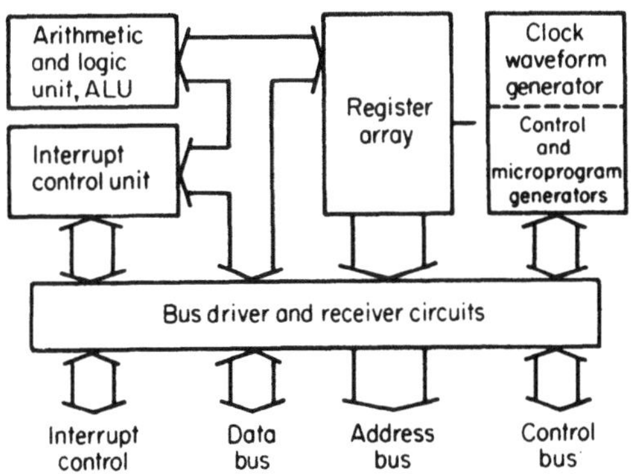

Figure 8.9 The processor

three-state bus driver and receiver buffer circuits and a fifth section, the interrupt control unit, which we shall consider later. The elements of figure 8.9 are shown in more detail in figure 8.10 including the internal bus arrangements normally used in an eight-bit microprocessor. An arrow on to the bus indicates that data can be gated on to the bus under the control of the appropriate signal generated in the microprogram circuits. Similarly, arrows from the bus indicate that data may be gated into the register by the appropriate control signal. Note that figure

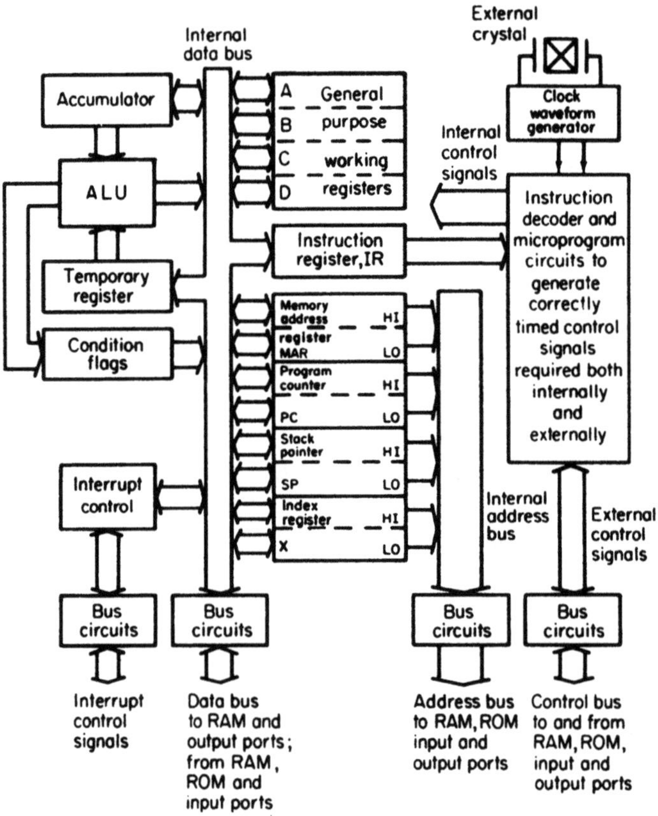

Figure 8.10 Detailed structure of the processor shown in terms of an 8-bit microprocessor. 16-bit and 32-bit microprocessors use similar arrangements with larger registers

8.10 is generalized; not all microprocessors will contain all the registers shown. For simplicity we shall illustrate the operation in general in terms of 8-bit microprocessors, but the principles are the same no matter how many bits are dealt with simultaneously.

8.5 The register array

The register array contains both the general-purpose working registers and the control registers, each of which has a specific part to play. In some microprocessors a single accumulator is provided, separate from the general-purpose registers, to deal with addition and other arithmetic operations. The name 'accumulator' is given to a register which is used in conjunction with the arithmetic circuits, providing one of the operands and accepting the resultant. It is the register in which the results of arithmetic operations accumulate. In other microprocessors, any two registers can be selected and the contents added or subtracted as required.

The *Program Counter*, PC, holds the address of the next instruction to be fetched from the memory, and is used in conjunction with the *Instruction Register*, IR. Recall that the sequencing of the processor relies on instruction cycles which are made up of a fetch phase and an execute phase, shown in figure 8.11. On the completion of an instruction a new cycle begins with the address from the program counter being gated on to the memory address bus.

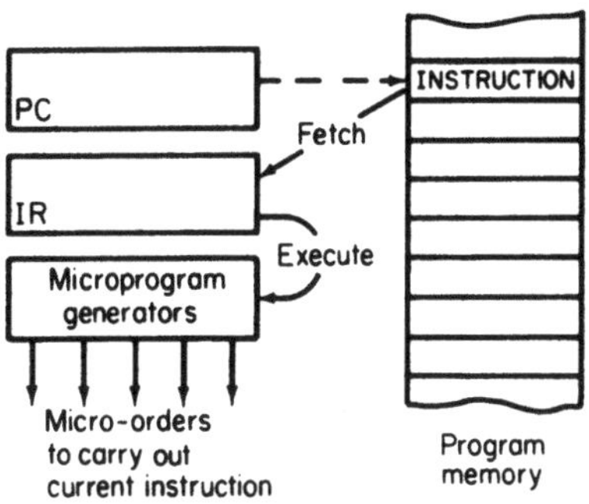

Figure 8.11 Processor FETCH and EXECUTE phases

The data from the selected memory location is returned on the data bus and routed to the instruction register to form the opcode byte of the instruction. As we shall see later, some instructions can be specified completely in one byte, but in most cases one or more further bytes are required. In all cases the instruction register provides the inputs to the instruction decoding circuits in the microprogram-generating section. When the fetch phase is completed the execute phase follows immediately, with the correct sequence of control signals being generated by the microprogram circuits according to the code held in the instruction register. At the end of each fetch operation, and before the execute phase begins, the value in the program counter is incremented by one so that it indicates, or 'points to', the address of the byte or instruction which is next in sequence in the memory. Some instructions can modify the value in the program counter to any value specified in that instruction, so allowing the user to *jump* or *branch* from one program routine to another when certain conditions occur. Special *call* or *jump to subroutine* instructions allow a later return to the original program counter setting if required.

In many cases the processor is required to operate on data held in the memory. The data word is specified by the address of its location in the memory, and the address forms part of the instruction. In order to retrieve or to store the data, the effective address must be assembled or computed from the information provided in the instruction. In eight-bit microprocessors the address is transferred directly to the *memory address register*, MAR, during an early part of the execute phase, so that it can be gated on to the address bus at the appropriate time. In many 16-bit and 32-bit processors, however, the effective address is arrived at only after several processing steps have been carried through. In all cases, the specified data is then transferred on to the data bus.

As it has become possible to provide larger and larger memories at reasonable price, some manufacturers have decided that it is desirable to use a segmented address space rather than a linear address space. This is said to be a better approximation to the way in which programmers use memory in allocating blocks of memory for specific procedures and associated data. We have so far assumed *linear addressing* of a memory range, which means that all possible locations can be selected by the address word. Thus, for example, a 16-megabyte linear address range, as provided in the Motorola M68000 family, necessitates a 24-bit address. In other processors, such as the Intel 8086 family and the Zilog Z8000 family, however, the total address space (1 Mbyte and 8 Mbytes, respectively) is divided into *segments* of 64K bytes. The programmer specifies a *logical address* which refers to a particular segment, and uses a 16-bit *offset* value to address any location within that segment, relative to its starting or *base* address. The current base addresses for the segments are held in a small set of registers, a *register file*, and the appropriate one is used by the memory management unit in converting the logical address used by the programmer to its equivalent physical address which is needed by the memory circuits. This process is illustrated for the Z8000 family in figure 8.12, where the final result is a 24-bit address.

Some microprocessors include other special-purpose registers such as additional memory addressing and refreshing registers (Zilog's Z80A and Z8000 processors are the main examples). The *memory refresh register* allows the processor to refresh dynamic memories automatically and without interfering with the normal operation of the processor.

The operation of the Stack Pointer and Index Register are described later.

8.6 The arithmetic and logic unit, ALU

This unit carries out all the arithmetic and logic functions and makes use of the accumulator or temporary registers to provide the input data to the adder cir-

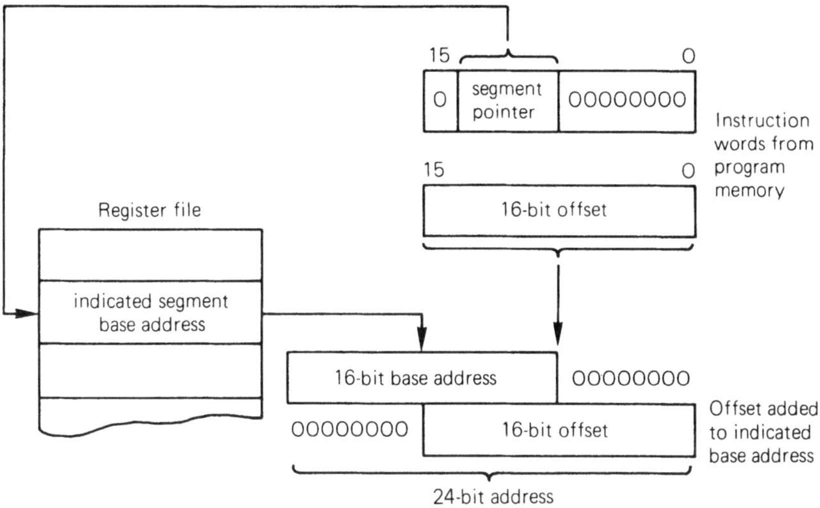

Figure 8.12 Preparation of physical address from logical address in the Z8000 microprocessor

cuits. The two steps needed to carry out a typical addition are shown in figure 8.13. We shall assume the instruction has been decoded as: Add the contents of register B to the contents of the accumulator. The first step is to transfer the contents of register B to the temporary register. The second step connects both the temporary register and the accumulator to the parallel adders of the arithmetic unit, and the resultant is routed back to the accumulator. Thus on the

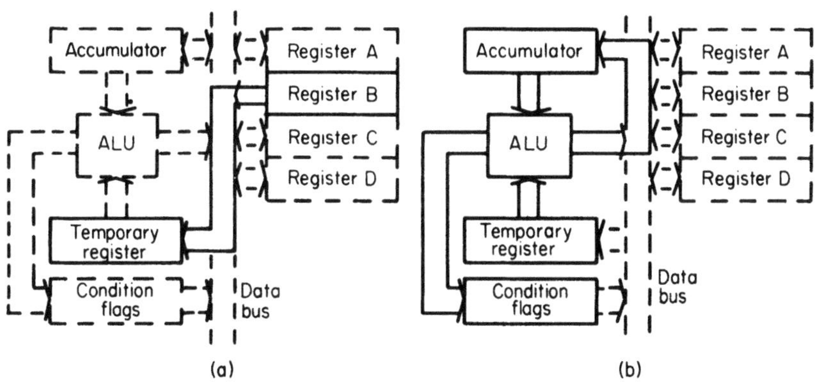

Figure 8.13 Addition of contents of Register B and Accumulator. (a) First step: Register B contents moved to a temporary register. (b) Second step: contents of temporary register and Accumulator added and result returned to Accumulator

completion of the instruction, the accumulator holds the sum of its original value and the value held in the B register. Note that the original value in the accumulator is lost, but the B register value is unchanged. Figure 8.14 shows one stage of an arithmetic unit in which the accumulator uses edge-triggered flipflops. If edge-triggering were not used a second temporary register would be necessary in order to separate the outgoing and incoming data at the accumulator.

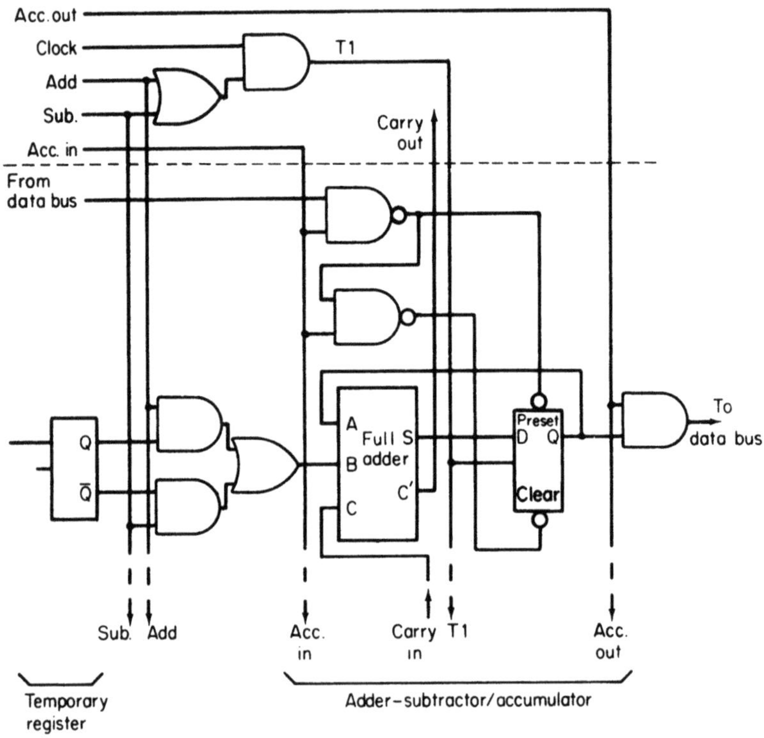

Figure 8.14 One stage of an ADDER/SUBTRACTOR unit. CARRY IN at the least significant stage equals one when subtracting, giving two's complement addition

Most microprocessors have the capability of parallel addition in both fixed point binary and binary coded decimal, and perform subtraction by two's complement addition. If the microprocessor does not include multiplication and division instructions in the instruction set, as is the case with most eight-bit processors, then these must be achieved by programming a succession of additions or subtractions.

As a minimum, four logical operations are normally provided in the instruction set. These are AND, OR, exclusive-OR and Invert or Complement. The inversion, or one's complementing, changes each bit of the specified register word to its opposite value, zero to one, and one to zero. The three other functions involve a bit-by-bit comparison between the word held in one register and a specified second word held in another register or taken from the memory.

The logical AND instruction gives a bit-by-bit comparison of the two words and a one results only where that bit position in both words is a one. The instruction is often used to select a particular byte from a word and involves a *mask* which contains ones in those bit positions to be compared. Thus, the instruction 'Logical AND (A) and (N)' would require the microprocessor to read the mask (N) from the memory and compare it with the register A contents.

For example, suppose we wish to select the lower four bits from the 8-bit register A;

Content of register A	01001101
Correct mask from location N	00001111
Result of logical AND to register A	00001101

Thus, register A contains the selected four bits.

The logical OR instruction is used to insert bits into a word. The comparison in this case results in a one in the bit positions corresponding to ones in either word. Thus

Content of register A	00000110
Bits to be inserted, from memory	00110000
Results of logical OR to register A	00110110

When a mask is used with the exclusive-OR instruction the selected bits are inverted but the others are not affected. The bit comparison in this case gives a one only if the two bits compared are not equivalent, which can be used in comparing words for equality. The resultant is zero only if the two words are identical. Any mismatched bits give a one in that bit position of the resultant word. In a few microcomputers the comparison instructions are extended to include searching of stored data. In a masked equality search, for instance, the search is continued until a word is found that matches a given word.

The second step of our addition example, figure 8.13b, uses a register marked *condition flags*, whose purpose has not yet been explained. The condition flags are flipflops, which together make up the *condition code*, CC, or *program status word*, PSW, register. The register indicates the conditions resulting from the last arithmetic or logical operation, and a few other operations. The flags are used in conjunction with instructions which allow alternative operations dependent on a flag setting, normally jump or branch type instructions. The most common flags are: carry, C; overflow, V; zero, Z; negative, N, or sign, S; parity, P; and half-carry, H, or auxiliary carry, AC. Occasionally other flags such as an interrupt enable, I, flag are also included for use with the interrupt control unit. It is not always obvious which flags are affected by a particular instruction, and it is important to check in the manufacturer's manuals before making use of them.

The carry flag is set only if the last operation resulted in a carry, or borrow, from the most significant digit (msd) position. It is used in multiple additions or subtractions where a carry bit is to be transferred from one operation to the next. Certain rotate and shift operations also make use of this flag.

The overflow flag is set if the result of the last operation was too large, either positive or negative, for the register word length. Overflow, which is only meaningful in signed arithmetic operations, is indicated during addition if the two operands have the same sign and the resultant has the opposite sign. In subtraction, overflow occurs if the two operands have different signs and the resultant has the sign of the subtrahend. This is summarized in figure 8.15 which also includes a simple overflow detection circuit.

(a)

A	B	Sign of result	Setting of O/F flag
+	+	+	0
+	+	−	1
+	−	+	0
+	−	−	0
−	+	+	0
−	+	−	0
−	−	+	1
−	−	−	0

Addition (A+B)

A	B	Sign of result	Setting of O/F flag
+	+	+	0
+	+	−	0
+	−	+	0
+	−	−	1
−	+	+	1
−	+	−	0
−	−	+	0
−	−	−	0

Subtraction (A−B)

(b)

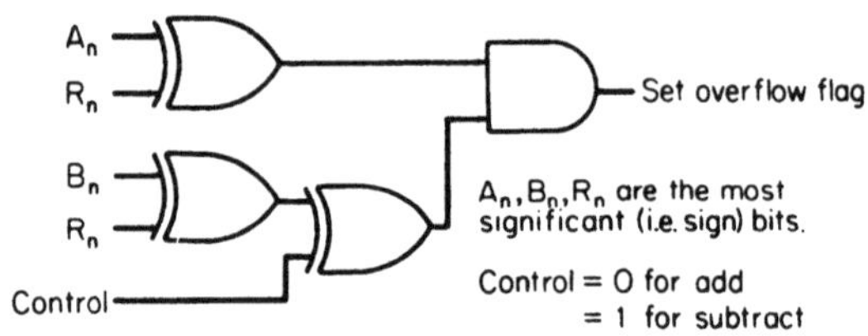

Figure 8.15 Detection of overflow: (a) conditions for overflow; (b) detection circuit

The zero flag is set if the result of the operation was zero; that is the flag is set to '1' if the result was zero, and '0' if the result was non-zero. This flag is often used in comparisons and search operations.

The negative or sign flag is an indication of the value of the most significant bit, or sign bit, of the result of an operation. The flag is set if the sign bit is '1', indicating a negative number.

The parity flag is useful in checking data words for error and, when set, indicates even parity in the word resulting from the last operation. It is set if the modulo-2 sum of the bits in the word is zero.

When two 8-bit words are added or subtracted the half-carry or auxiliary carry flag is set if a carry occurs between the lower four bits and the upper four bits of the resultant. Its principal use is in conjunction with the decimal adjust instruction which is provided in many microprocessors to facilitate addition in binary coded decimal form. In BCD addition the 8-bit word is considered as two 4-bit BCD numerals. As we saw in an earlier chapter, when adding BCD numerals we must adjust the result if it gives one of the six invalid codes or if a carry is generated. The code is invalid if it is greater than nine. In either case, the correct value is found by adding six.

We can illustrate the use of the condition flags by considering the addition of two 8-bit words as follows

Operand A	1000 1001
Operand B	0111 1001
Result	1 0000 0010

Since there is a carry from the msd,	carry flag, C = 1
there is no overflow,	overflow flag, V = 0
the result is non-zero,	zero flag, Z = 0
the sign bit is zero,	sign flag, S = 0
there is an odd number of '1's in the 8-bit resultant,	parity flag, P = 0
there is a carry from the lower to the upper 4 bits,	half-carry flag, H = 1

Thus the condition code register will contain

C	V	Z	S	P	H
1	0	0	0	0	1

The operands in our addition may be interpreted in several ways. They could, for example, represent unsigned binary operands: the decimal sum is then 137 + 121 = 258 which is correctly shown as a carry to the next most significant bit position, 256, plus the value 2.

1000 1001	137
0111 1001	121
1 0000 0010	258

Alternatively, the operands may be considered as signed binary operands so that the sum represented is $-119 + 121 = +2$

1 0001001	−119
0 1111001	+121
0 0000010	+ 2

Finally, the operand may represent BCD numbers, in which case the sum shown is wrong since $89 + 79 \neq 102$. We must look more closely at this last case, and see how a decimal adjustment is made to correct the result. The decimal adjust instruction adds six to either four-bit code if a carry was generated or the code is greater than nine. The six is added first to the lower four bits and may result in a carry to the upper four bits. On adding the operands in our example, both the carry and the half-carry flags are set, so when the decimal adjust instruction is used six is added to both four-bit codes, as follows

First result	1 0000 0010
Lower code valid but	0110
H flag set so add 6	0 1000
Upper code still valid but	0110
C flag set so add 6	1 0110 1000
	(1) (6) (8)

The result is now correct, giving 89 + 79 = 168.

Several other facilities are normally provided in the arithmetic and logic unit, such as incrementing and decrementing of specified registers. Incrementing is achieved by adding zero to the register value while forcing a carry in at the least significant bit. Similarly, decrementing is carried out by subtracting zero with a borrow at the least significant bit.

Compare and *Test* instructions allow two operands to be compared as if subtracting, and the condition flags set appropriately. In some cases bit test instructions are included in which the two operands are compared by a logical AND function of each bit position. Occasionally an instruction is included to allow the two sets of four bits in an 8-bit word to be interchanged. The half-byte is sometimes known as a nibble (or nybble)! In 16-bit processors the equivalent operation, where provided, swaps 8-bit bytes within a register.

Shifting right and left are important operations of the arithmetic unit. Shifting left one place multiplies a number by two; shifting right divides it by two. Certain difficulties arise with the sign digit when we shift right, and it is necessary to differentiate between two types of shift. If the word 11010 is shifted right one place, we get 01101. Interpreting this as 26 divided by 2 the indicated answer is correct, but interpreting 11010 as a signed number in two's complement we get −6, which when divided by 2 is −3. Minus three is 11101 not 01101, and we see that the correct result is obtained, after shifting, only if we duplicate the sign digit in the space left when the bits are shifted right. Duplication of the sign digit is automatic in *arithmetic shift right* instructions whereas *logical shift right* treats the word purely as a pattern of ones and zeros without any numerical significance. This distinction is not necessary on shifting left since a change of sign occurs only if the original number was already greater than half the maximum number capability. On shifting left, or multiplying by two, the number range is exceeded and the overflow check circuits are available to cover this situation. The carry flag is normally set to the value of the bit shifted out.

Rotate instructions, as the name implies, are shift instructions in which the bit shifted out is re-entered at the other end of the word. Rotate instructions can include the carry bit or not, and the various possibilities are illustrated in figure

8.16a-g. Rotate and shift instructions are widely used in scaling, normalization and multiplication sequences when carrying out arithmetic; in merging, separating and matching character patterns when processing data; and in converting between parallel and serial data form.

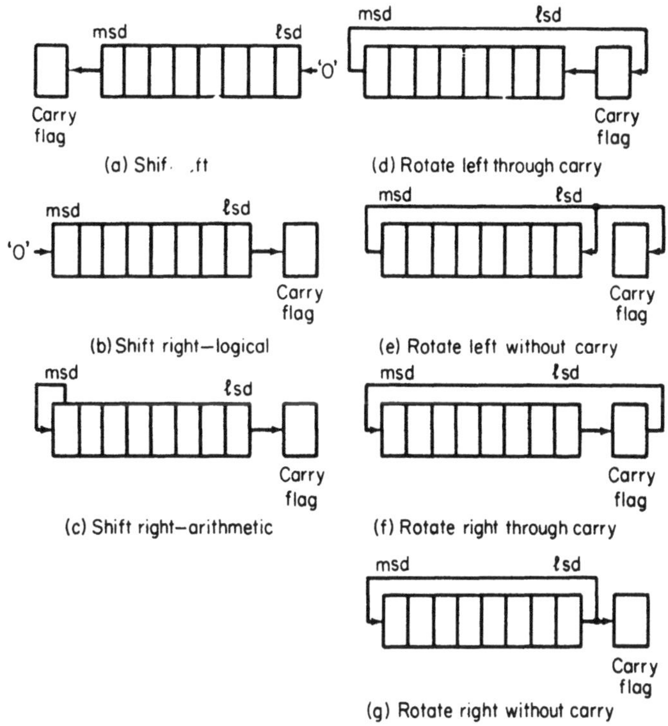

Figure 8.16 Shift and rotate operations

8.7 The control section

The control section is responsible for the generation of the accurately timed sequences of signals required in fetching, decoding and executing instructions.

The operation of the control unit is synchronized to a basic clock waveform, usually from a crystal-controlled oscillator, and the period of the clock cycles, the *clocktime*, is the smallest discrete unit of time recognized by the microprocessor. The clock frequency dictates the fundamental processing speed of the microprocessor, and clock frequencies are generally in the range 1-16 MHz. Because of the dynamic nature of the memory circuits used in some microprocessors, it is not permissible to inhibit the clock signal to these devices during operation and a minimum clock frequency is stipulated.

The next largest discrete unit of time utilised by the microprocessor is called a *state*. A state may be several clock cycles long and is the smallest time unit of processing activity within the microprocessor. Each state is, in fact, associated with the execution of a step in the microprogram of the control unit. The sequencing of these micro-operations is clock-synchronous, and hence a state corresponds exactly to one or more clock cycles, as shown in figure 8.17a. Note that some microprocessors use two separate non-overlapping clock signals in place of the single clock waveform and thus a single clock cycle may, in fact, contain two clock signals (figure 8.17b).

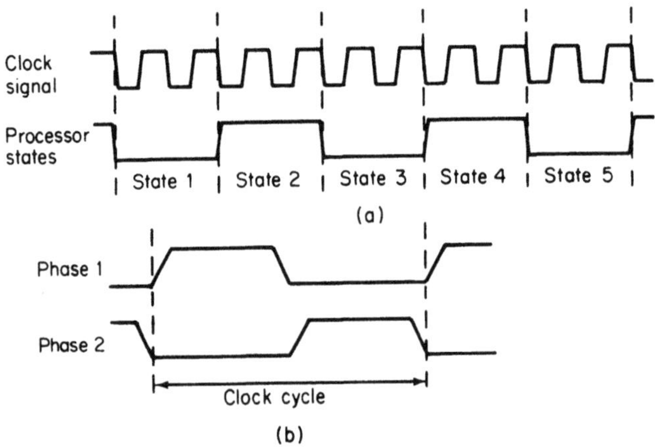

Figure 8.17 Microprocessor timing: (a) processor states; (b) two-phase non-overlapping clock signal

The next largest unit of time is a *machine* cycle. A machine cycle is several states in length, and is characterized by the fact that the first state in a machine cycle always controls a fetch operation, with the instruction opcode byte being retrieved from memory and placed in the instruction register for decoding and subsequent execution.

Finally, the execution of a complete instruction requires up to around seven machine cycles, involving several memory accesses, depending on the number of bytes in the instruction. Each instruction therefore has clearly defined fetch and execute phases. For example figure 8.18 shows the breakdown of a one-machine-cycle instruction: 'move the contents of register 0 to 1'. The processor states and their relationship to the instruction are internal phenomena. However, for the processor to interface correctly to external systems, control and synchronizing signals must be provided by the processor. The signals are used to control the sequence of data words and addresses across the microcomputer buses. A typical set of such control signals, C_1, C_2, C_3, is shown in figure 8.19. In addition to these control signals, others may be present to enable external devices to

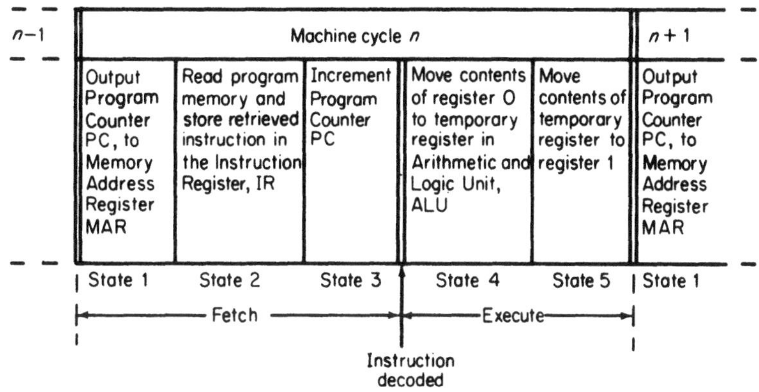

Figure 8.18 Example of processor phases

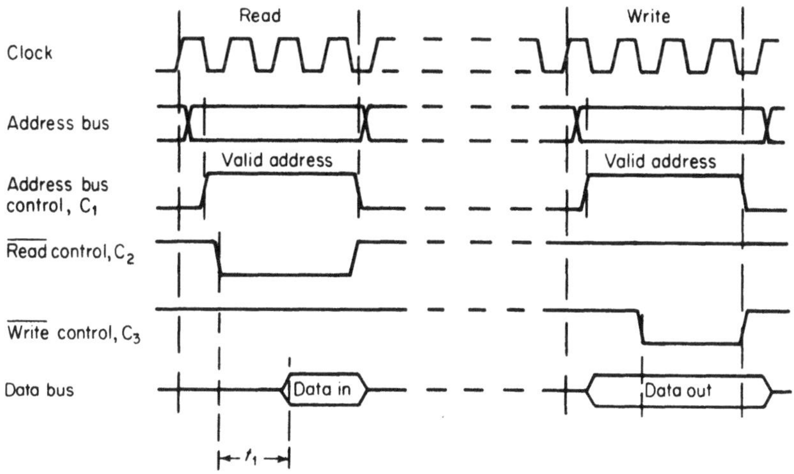

Figure 8.19 Control signals from the microprocessor: C_1 indicates that the address bus data is a valid address; C_2 and C_3 indicate the setting of the data bus (read or write), and ensure that the memory acts in the correct mode. Time t_1 is the access time of the memory

control the processor, by means of interrupt signals for example. Also, a reset signal is needed to set the processor into the correct state each time the system is switched on.

Some instructions are complete in themselves, requiring only one operand: 'complement the data held in the accumulator', for example. Others require a second operand to be defined. This second operand can be stated explicitly in the instruction or can be indicated in terms of the register or memory location which contains it, as in figure 8.20. In either case the instruction must carry an

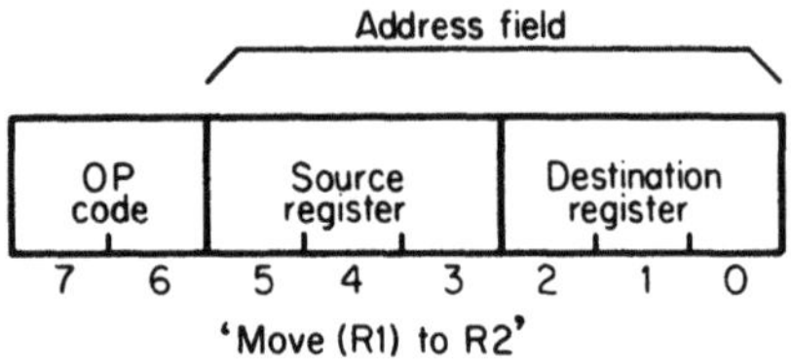

Figure 8.20 Example of a single-byte instruction

extra section or *field* to hold the additional information. In most microprocessors the complete instruction word varies in length and the length is dependent on the instruction type, though it is always a multiple of the byte or word length defined by the width of the data bus. The coding of the first byte indicates to the control circuitry how many more bytes, if any, are needed to complete the instruction. Eight-bit microprocessors use one, two or three and occasionally four bytes in an instruction. Sixteen-bit and 32-bit processors use up to six bytes for most purposes, but certain instructions demand eight or even ten bytes.

The second byte of an instruction, and if necessary the remaining bytes, form the *literal field* and may contain either data or an address defining the location of the data (figure 8.21). In branch instructions, the literal field holds address information to enable a jump to be made.

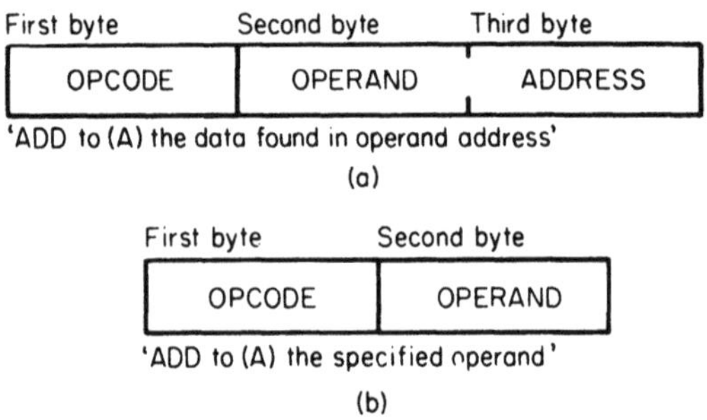

Figure 8.21 Examples of multi-byte instructions: (a) three-byte instruction; (b) two-byte instruction

Many different methods are available for defining operands, either directly or via an address. We can divide them into eight categories, or *addressing modes*, though the number provided in any one microprocessor depends on the design philosophy of the manufacturer.

The addressing mode to be used in a particular instruction is decided by the user in writing the program, and the choice of mode is normally made so as to keep the instructions short but without sacrificing flexibility and efficiency. A short instruction requires less memory space, less time in the fetching, and is probably simpler, which in itself is often a benefit to the user since mistakes are less likely. However, some operations, and especially those involving the manipulation of tables or arrays of data, are dealt with much more efficiently by the use of the more complex addressing modes. The *opcode byte* contains within its code an indication of the addressing mode which is to be used and in many cases more than one addressing method may be used in a single instruction.

The eight main addressing modes are

Implied
Immediate
Direct
Register direct
Indirect
Register indirect
Relative
Indexed

Implied, or *inherent*, addressing is used in all instructions in which the opcode itself indicates the accumulator or register which is to provide the first operand. For example, the instruction 'Add 6' actually means 'Add six to the value held in the accumulator', but the fact that the accumulator holds the first operand is only implied in the instruction as written. Furthermore, the instruction implies that the resultant must be left in the accumulator. Instructions making use of the stack, such as 'Push', also rely on implied addressing.

In the *immediate* mode the instruction contains the operand itself. Our previous example, 'Add 6', specifies the second operand in immediate mode and in practice the byte or bytes immediately following the opcode byte in the memory contains the required operand. The example in figure 8.21b is another example of immediate addressing. This mode of addressing is widely used in loading constants or initial values into registers or memory locations. Where the register acts as a memory pointer, and can, therefore, hold a large number of bits, it may be loaded by using extended immediate addressing so that the full address is held as a succession of bytes immediately following the opcode byte. The order of the bytes is important in indicating the high-order and low-order bytes. Manufacturers differ in their choice of standard as to which is which and care must be taken to avoid confusion.

Direct, or sometimes *absolute*, addressing is the name given to the mode in which the operand is specified by means of its address in the memory. In 8-bit processors, where the address is sixteen bits long, two bytes are necessary, and a longer instruction results. In many cases, therefore, direct addressing is provided in two forms which are known as *short direct* mode and *extended direct* mode, or just *direct* and *extended* mode. Short direct mode makes use of the fact that for much of the time many programs operate in only a limited area of the total memory space available, and it is not necessary to specify the complete address in each instruction. The memory address register, MAR, holds the total address as a 16-bit word and, if we restrict the number of locations which can be addressed by this mode to 256, we need only specify in the instruction the lower eight bits of the address. The total address is formed by combining the specified lower byte with the upper byte already held in the memory address register.

Thus short direct addressing affects only the lower byte of the memory address register, extended direct addressing affects both bytes (figure 8.22). From the programming point of view, short direct addressing leads us to the idea of the total memory being made up of sections, which have become known as *pages*. Each page holds as many words as are addressable using the short direct mode, which in our case is 256, and the address of the word within the page is held in the lower byte. The additional addressing bits, which we specify only in extended direct mode, give the page number and this is held in the upper byte of the memory address register.

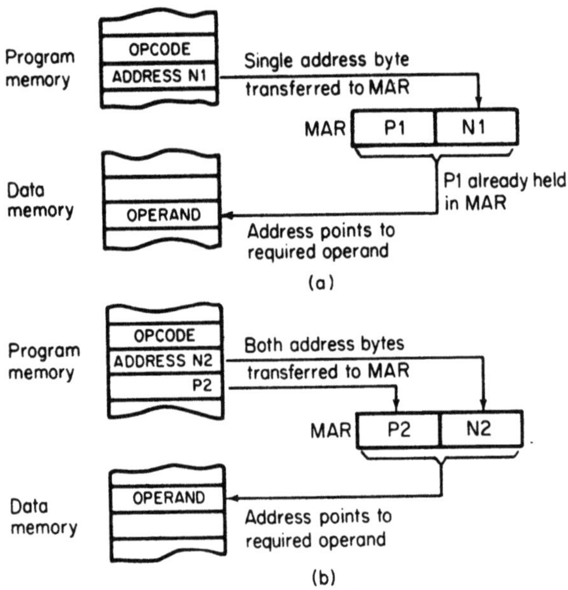

Figure 8.22 Direct addressing modes: (a) short direct addressing (current page); (b) extended direct addressing

The short direct mode allows addressing of any word on the page, and it is in fact *current page* addressing since it uses only the page number already held in the memory address register. Current page addressing is sometimes augmented or replaced by an alternative short direct form, in which the lower byte of address is used alone. This mode, known as *page-zero* addressing, makes the upper address byte appear as all zeros and allows fast access to the low-address section of memory, which can be set aside for important data.

Direct addressing is also used in many branch instructions and examples are given in figure 8.23. The address may be in short or extended mode, though some microprocessors allow conditional branch instructions only on the current page, making an unconditional branch instruction necessary in order to jump to a new page.

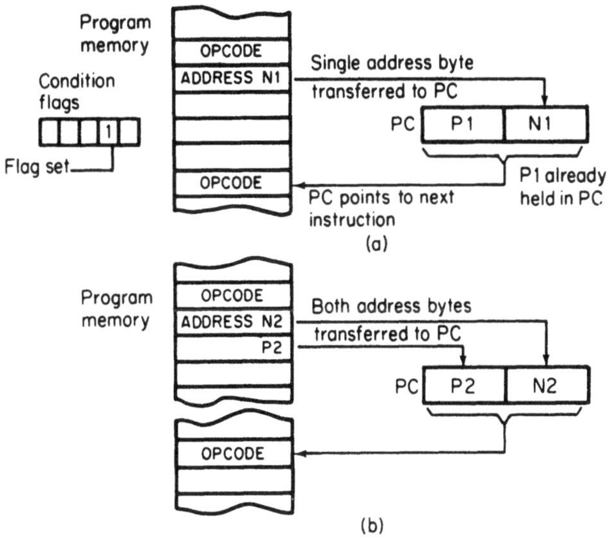

Figure 8.23 Direct addressing in branch instructions: (a) current page, conditional jump; (b) unlimited unconditional jump

The number of variations on the direct addressing mode is very large and includes many different allocations of bits to page and word numbering, as well as provision of page number registers. The *restart* instruction provided on some eight-bit microprocessors is an example of even more restricted short direct addressing. By using page-zero addressing and only a few address bits, this single-byte instruction can provide the opcode and sufficient address information to specify a call to one of eight blocks of addresses on the zero-page very rapidly. In 16-bit and 32-bit microprocessors direct addressing is also provided in two forms, using either a short, 16-bit, address or a long, 32-bit, address contained in two or four bytes of the instruction.

Since the working registers operate in exactly the same way as high-speed memory locations, they may also be addressed directly. *Register direct addressing* is widely used for two main reasons: the limited number of bits necessary to specify the register can be included in the single opcode byte, and it is fast since data need not be fetched from memory. This is particularly useful when several operations are to be carried out on the same data word, otherwise the necessity of having to fetch new data from memory offsets the gain in speed achieved by the register direct addressing. As we saw in figure 8.20, some instructions address two registers directly.

Indirect addressing means that the address contained in the instruction is not the address of the operand but the address of the memory location which holds the address of the operand. The processor must go to the location specified and read out the second address which can then be interpreted as the direct address required by the instruction. Indirect addressing is much slower than direct addressing, because of the extra memory accesses, but is much more flexible. It is intended for use in cases where data must be moved around in the memory, or where a standard subroutine is to be used to operate on different sets of data which are positioned at different places in the memory so that the processor must be directed to the current data location.

As with direct addressing so with indirect addressing; a working register rather than a memory location may be used to hold the address, as shown in figure 8.24. This is known as *register indirect addressing* and has the same advantages as register direct addressing. In addition, register indirect addressing can be used in

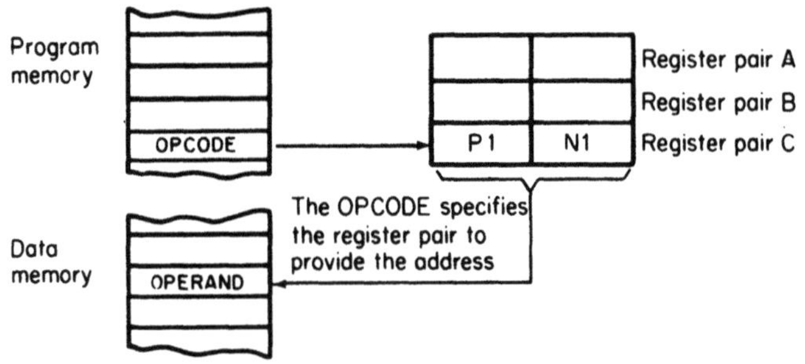

Figure 8.24 Register indirect addressing

any system whereas memory indirect addressing must have the use of random access memory. For this reason, although register indirect addressing is very widely used, memory indirect addressing is not so commonly provided. Many variants and developments of register indirect addressing are used. For example, where scratchpad registers are an integral part of the microprocessor, register indirect addressing is used to specify the particular register required by referring to the scratchpad pointer register. In other microprocessors automatic incrementing or decrementing of the selected register has been included so that operations on blocks of data are possible.

Both *relative addressing* and *indexed addressing* require the processor to add some value to the given address to form the effective address. As we have seen with segmented addressing, this address is in fact made up of two parts: the *base* address and the *displacement* address, which must be added together to arrive at the required address. The displacement address is normally limited to a single byte or word so that the speed of operation is high, and a signed value is used so that the displacement can be positive or negative. The great value of this mode of addressing lies in the ease with which fast jumps to nearby locations in memory can be made and entries in data tables can be selected. In addition, in both cases, the program instructions are inherently *relocatable*; that is, since the addresses are relative to a base address the same program block can operate, unmodified, in different parts of the memory merely by ensuring that the base address is correctly adjusted.

Although similar in action, relative and indexed addressing use different registers to provide the base address. Relative addressing takes the base address from the program counter, PC, and its operation is illustrated in figure 8.25. Note that if at the time the instruction is carried out the program counter has already been incremented, the final address is relative to the next instruction location. Thus, for example, although an 8-bit displacement address would give a nominal displacement specifying the address, the displacement is, in fact, up to $+(127 + n)$ or $-(128 - n)$, where n is the number of bytes in the instruction.

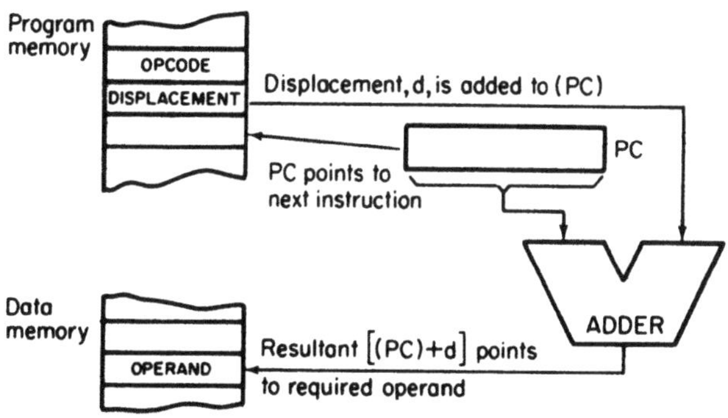

Figure 8.25 Relative addressing

Indexed addressing takes its base address from a special register, the *index register*, which is one of the registers included on figure 8.10 but so far unaccounted for. Some microprocessors include more than one index register, and many allow general working registers to be used as index registers. In such cases the opcode byte must carry an indication of which index register is to be used. Figure 8.26 illustrates the general principle.

As a simple example of the indexing principle let us consider a system in which data is taken in from a keyboard. The keyboard circuitry is such that when a key is pressed, an 8-bit code is generated giving in effect a 'grid reference'

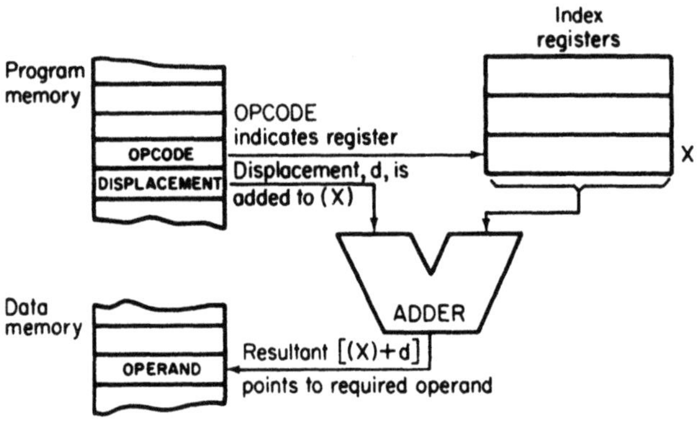

Figure 8.26 Indexed addressing

of the pressed key, as in figure 8.27a. In order to convert the grid reference to the corresponding character code we hold in the program memory a table of codes, T1 to T*n*, and use the grid reference to index the address contained in the instruction. The address in the instruction, until indexed, points to the first entry in the table of codes.

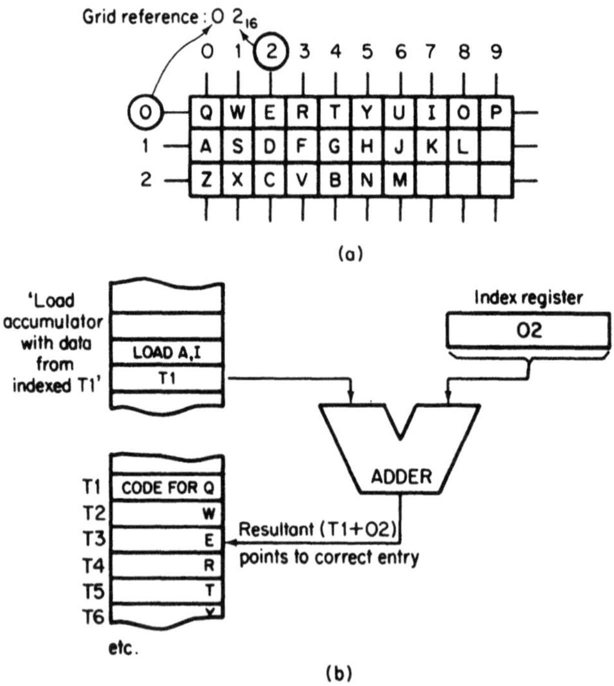

Figure 8.27 Keyboard encoding

Suppose the pressed key is 'E' so that the grid reference is 02_{16}. This value is transferred to the index register and the standard instruction 'Load accumulator with data from location T1, indexed' is used. The pointer to the table, which is initially T1, is increased by the value 02 held in the index register, so that the code selected is in fact the code for 'E', as required (figure 8.27b).

8.8 The interrupt control unit

All microprocessors have the capability of responding to an externally generated *interrupt* signal, to break into the normal flow of instructions in an orderly manner and to transfer control to a predetermined program sequence.

Without an interrupt capability all operations must be initiated by the processor itself under program control. Thus if, for example, some program sequence is to be carried out when a particular switch is operated, the program must contain a subroutine which is used at frequent intervals to check whether or not the switch is yet pressed. This is time-consuming in operation and wasteful on memory, which must be used to hold the subroutine. Where several switches or other peripheral devices are to be dealt with, an extended *polling* sequence is required in which each device is interrogated in turn to determine whether it requires servicing. The interrupt method relieves the microprocessor of such tasks and allows it to respond almost immediately to an external signal which is generated when the key is operated. It does, however, require the provision of extra hardware in the form of an interrupt control unit. As a direct result of the randomness of the interrupt signals inclusion of interrupt handling subroutines makes debugging of systems more complex, since the programs become less well-structured, and it should be noted that, in a system using high external data transfer rates, the processor can be required to spend so much time in dealin with the interrupts that little time remains for directly useful work. In such cases polling offers definite advantages.

In a system using interrupts, signals from external devices can occur at any time rather than being synchronized to the microprocessor cycles as in a polling system. In order to deal with these asynchronous signals it is arranged that the interrupt signal sets a flipflop and it is only at the end of the instruction cycle that the flipflop is checked and the appropriate interrupt handling routine called up. By this means the external signal can interrupt the program sequence at the end of the current instruction. There are times, however, when it is necessary to prevent interrupts, such as during system initialization or when some very important section of program is being carried out. The instruction set therefore includes instructions which allow the user to *disable interrupts* and subsequently to *enable interrupts* (*clear interrupt disable*). This is achieved by setting and resetting an internal flag flipflop, and the state of the flag is sometimes indicated in the condition codes. In many cases the acceptance of the interrupt automatically disables further interrupts and it is necessary to include an *enable interrupt* instruction at a convenient point in the routine. Where more than one interrupt input is provided, it is possible for the user to disable them selectively, at will, by setting the correct masking or 'ANDing' pattern of ones and zeros in a register which is usually known as the *interrupt mask register*. Special instructions called *set interrupt mask* and *clear interrupt mask* are provided to allow this to be done.

The initial sequence of events in response to an interrupt signal is fairly common to all microprocessors. Figure 8.28 illustrates a typical operating sequence for the microprocessor. At the end of each instruction the interrupt flag is checked and if found to have been set the control unit uses a call or restart instruction to transfer operation to the *interrupt routine* program sequence. The use of a call type of instruction ensures that the program counter value at the time of the interrupt is not lost, but is retained until the interrupt handling routine is completed. The interrupt handling routine will in most cases need to make use of some or all of the processor's working registers. The data already held in the registers may be conserved by pushing the contents on to a *stack* from which they can be retrieved at the end of the routine. Alternatively a duplicate bank of registers is brought into use by means of a *switch* (or *exchange*) *register banks* instruction. If more than one interrupt is to be allowed at a time,

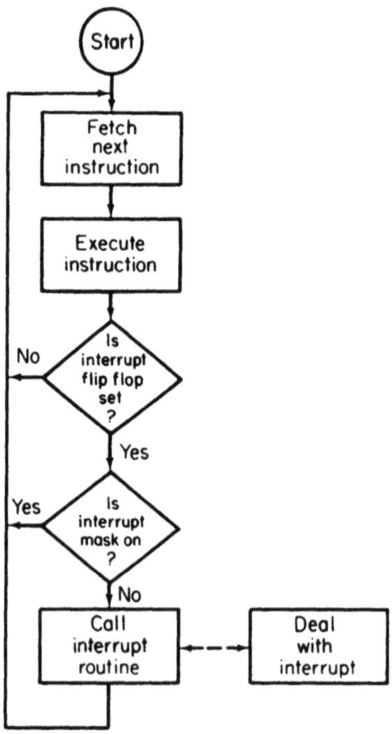

Figure 8.28 Typical microprocessor sequence

the second approach requires multiple register banks and these are readily provided by microprocessors using different sections of memory as the registers, with an adjustable workspace pointer register indicating which section is in use at any time.

The stack is a section of the random access memory which is standard in every respect except for the addressing arrangements. Instead of using a register, which would normally hold the address of the memory location required, the stack takes the address from a counter which is incremented or decremented automatically each time the memory is used. The counter acts as the *stack pointer*, SP, and instructions are provided to allow the stack pointer to be set to any desired starting value. In some cases the stack is made up of a limited number of registers in the processor provided for the purpose, but in most cases a section of the random access memory must be set aside for stack usage. This gives greater flexibility to the user but it is then necessary to ensure that proper control is retained over stack operations and that they do not cause other data in the memory to be overwritten. On writing into the stack, the data is stored in the location specified by the stack pointer, which is then decremented. On reading from the stack, the stack pointer is incremented and the data read from the location then indicated (figure 8.29a-d). By this means data words are retrieved in reverse order to that in which they were stored, so these memories are also referred to as *last-in first-out*, LIFO, or *nesting* memories.

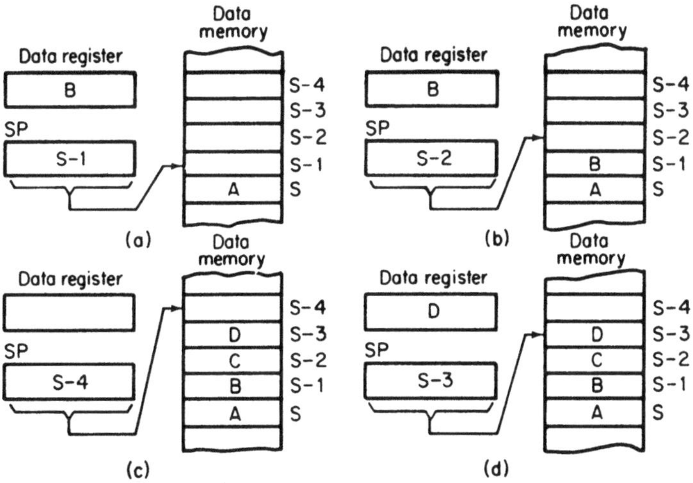

Figure 8.29 Operation of the stack: (a) before writing; (b) after writing; (c) before reading; (d) after reading

The stack pointer operation described above leads to the stack pointer moving to lower address locations as nesting proceeds, but some devices operate in the opposite way. Nesting down is often preferred since the low-address locations of the available memory space can then be used for conventional memory operations and the high-address locations for stack operations. Writing into the stack has become known as *pushing* on to the stack; reading is known as *popping* or *pulling* from the stack. The stack may be used in many ways in addition to storing data temporarily while the registers are required for use by some other section of program. In particular, it is always used to hold the contents of the program counter, so retaining the return address, whenever a subroutine call is invoked. Two or more stacks may be used in a single memory but extreme care is necessary to avoid overwriting of data.

In some microprocessors with only a few registers the data from all registers is pushed on to the stack automatically as part of the interrupt call instruction, and is recovered from the stack by use of a *return from interrupt* instruction. In other microprocessors, registers or register pairs must be pushed on to the stack individually and recovered by popping them off the stack again before using the return instruction. This allows more flexibility in writing interrupt routines to ensure fast response times.

The way in which the address to be used in conjunction with the interrupt call instruction is provided varies considerably among microprocessors and leads to differences in the amount of external circuitry necessary for correct operation. In the simplest case the microprocessor has a single interrupt input, and an *interrupt request* signal on this input causes the processor to carry out a call instruction using a specific address to link it to the interrupt handling routine. If more than one device is able to interrupt it is necessary for the routine to determine which of the devices generated the interrupt signal. In addition, since more than one device may interrupt at any one time, a *priority* must be assigned for the handling of the interrupts, and this is dependent on the system requirements.

The response to the interrupt signal may be controlled by program (software), by additional circuitry (hardware) or by a combination of the two. In the software approach, a polling method is used to check which device generated the interrupt signal. In the simplest case, when an interrupt occurs the microprocessor takes in the *interrupt status word* from a dedicated input port receiving the interrupt signals and the position of the 'one' in the word indicates which device requires servicing. Priorities between competing devices are resolved by arranging that the program checks the bits corresponding to the higher-priority devices first. The hardware methods involve the provision by the interrupting device of an identifying code or address which can direct the microprocessor automatically to the appropriate handling routine. This is called *vectored interrupt*.

One popular vectoring method using both hardware and software is the *daisy chain method*. The signal from the interrupting device is taken to the single interrupt request input of the microprocessor. The processor responds by sending an acknowledgement in the form of a signal which is routed to the first device in the daisy chain. Note that this device therefore always has the top priority. The interrupt flipflop remains reset in each circuit that did not originate the interrupt, and the acknowledgement signal is passed on to the next circuit down the chain until, eventually, the circuit is reached in which the interrupt flipflop is set and the acknowledgement signal can propagate no further. This signal is now used to enable the output latch array which gates a predetermined recognition code on to the data bus. The code identifies the interrupting device and provides the necessary information for the microprocessor to branch to the appropriate handling routine.

Hardware vectoring methods are used in those microprocessors having several interrupt inputs, each of which automatically restarts the program at its predetermined location in memory. In such cases priorities between the inputs are fixed. Sixteen-bit and 32-bit microprocessors have designated sections of memory to hold tables of vectors which are used to provide the restart addresses for particular interrupts. In general, each entry in the table consists of four bytes: in processors using segmented addressing, two bytes are used to form the new program counter value and two to act as a segment indicator or program status word, PSW; in those using linear addressing the four bytes are all used to form the new program counter value.

As an alternative to multiple interrupt inputs, the handling of interrupts is sometimes delegated to a specialized external circuit, so that the processor is relieved of the need to resolve priority clashes. The reader is referred to the more specialized texts listed at the end of this chapter for a further discussion of these methods.

8.9 Bit-slice devices

We have seen in earlier sections of this chapter how computers and other programmable devices are made up of simpler subsystems, such as arithmetic and logic units, multiplexers, and register arrays, all fabricated on the same chip. The concept of microprogramming enables these separately identifiable parts to operate together as a system, whose functions are controlled by a combination of a fixed internal microprogram and the user-supplied, external program instructions.

As it became possible in the mid 1970s to increase the number of bipolar gates on a single chip beyond the five hundred mark, an alternative design approach evolved, in which the elements making up the microprocessor, including the microprogram generator, were fabricated as programmable slices of the complete circuit, generally in 2-bit or 4-bit wide slices. One of the earliest slices available was the SN74181, which was a TTL 4-bit arithmetic and logic unit. To construct a larger unit it was necessary to cascade several parts, the advantage being that it was just as easy to build a 16-bit or 24-bit unit as to build an 8-bit unit. Bipolar technology is used to give high speed, and some bit-slice parts, such as the Motorola 10800 series, are fabricated in emitter-coupled logic. Cascading four of the 10800 arithmetic units gives a 16-bit capability, and with carry-look-ahead provided by a Motorola 10179 chip, the total delay through the adder is only 41 nanoseconds. Probably the most popular bit-slice family is the 2900 series, developed by Advanced Micro Devices, AMD, and based on low-power Schottky TTL circuitry, each part being a four-bit slice.

The most important applications of bit-slice devices are in special-purpose machines which need, for example, high-speed multiplication and division, or the complex processing found in graphic display systems. Image analysis processors commonly use the bit-slice approach. In all these cases the flexibility of design and the ability to specify one's own instruction set can be of great benefit in maximizing the speed of operation.

Problems

8.1 Explain the purpose and use of the condition flags commonly provided in a microprocessor. How does the decimal adjust instruction make use of some of these flags? Illustrate your answer by indicating the steps needed to carry out the addition

$$29_{10} + 65_{10}$$

8.2 A stack memory is often used in a microcomputer. When using standard RAM, how is the stack addressing organized? A certain microprocessor has no stack control routines built in; devise a software push-on-to-stack algorithm, in the form of a flow diagram, to make use of up to eight consecutive locations in RAM.

8.3 Why is the data bus in most microprocessors bidirectional, but the address bus is unidirectional?

8.4 The A register (accumulator) in a certain microprocessor holds the value 72_{16}, and the B register holds the value $C3_{16}$. The following condition flags are provided: carry, C; sign, S; zero, Z; overflow, V; half-carry, H; and parity, P.

What values would the registers hold and what would be the settings of the flags after each of the following operations?

(i) Add (B) to (A)	(iv) Exclusive-OR (A) and (B)
(ii) AND (A) and (B)	(v) Rotate (A) right without carry
(iii) OR (A) and (B)	(vi) Exclusive-OR (A) with itself

What are the corresponding values and flag settings if the value in register A is first one's-complemented?

8.5 Explain clearly the sequence of operations which occurs within a microprocessor in response (1) to a subroutine call, and (2) to an interrupt demand.

The top of the stack contains 27_{16} and the next byte down is 36_{16}. The stack pointer contains $3F62_{16}$, and a call subroutine instruction is located at memory address $10B2_{16}$. The call subroutine instruction is three bytes long and refers to location 2073_{16}. What are the contents of the program counter, PC, the stack pointer, SP, and the stack (a) after the call instruction is executed, (b) after return from the subroutine?

Further reading

AMD (1978). *Build an AM2900 Microcomputer*, Advanced Micro Devices Inc., California

Bacon, J. (1986). *The Motorola MC68000*, Prentice-Hall, Englewood Cliffs, New Jersey

Bannister, B. R. and Whitehead, D. G. (1986). *A Tutorial Guide to Transducers and Interfacing*, Van Nostrand Reinhold, New York

Givone, D. D. and Roesser, R. P. (1980). *Microprocessors/Microcomputers – an Introduction*, McGraw-Hill, New York

Intel (1982). *The 8086 Family Users Manual*, Intel Corp., Santa Clara

Intel (1984). *Microcomputer Handbook*, Intel Corp., Santa Clara

Leventhal, L. A. (1978). *Introduction to Microprocessors; Software, Hardware, Programming*, Prentice-Hall, Englewood Cliffs, New Jersey

Lipovski, G. J. (1980). *Microcomputer Interfacing*, Lexington Books, Lexington, Massachusetts

Motorola (1977). *M10800 4-bit Slice Microprocessing Course*, Motorola Semiconductors Inc., Phoenix, Arizona

Motorola (1983). *16-bit Microprocessors Data Manual*, Motorola Semiconductors Inc., Phoenix, Arizona

Nichols, K. G. and Zaluska, E. J. (1982). *Theory and Practice of Microprocessors*, Edward Arnold, London

Osborne, A. and Kane, G. (1981). *Osborne 4- and 8-bit Microprocessor Handbook*, McGraw-Hill, New York

Osborne, A. and Kane, G. (1981). *Osborne 16-bit Microprocessor Handbook*, McGraw-Hill, New York

Skinner, T. P. (1985). *An Introduction to Assembly Language Programming for the 8086 Family*, John Wiley, New York

Wakerley, J. F. (1981). *Microcomputer Architecture and Programming*, John Wiley, New York

Zilog (1983). *Microprocessor Applications Reference Book*, Vol. 2, Zilog Inc., California

9 Data Transmission and Conversion

An important part of any system is the means by which it communicates with its surroundings. In the case of the digital systems we have discussed in this book, the interconnecting circuits, or *interface*, may deal with both serial or parallel data, and signals in both analog and digital form. The aim of this chapter is to explain the various methods and devices used to effect this data transfer.

Digital signals are transferred between devices in either a *synchronous* or an *asynchronous* manner. By synchronous we imply that the receiver is made aware of precisely the instant in time that data will appear. To achieve this synchronization a common timing signal must be present at both transmitter and receiver. By contrast, asynchronous data transfer can take place without the need to convey additional timing information. If the data is to be transmitted in one direction only from, say, a computer to a printer, or from a keyboard to a display, the operation is termed *simplex*. *Duplex* operation refers to two-way data transmission, for example data transmission from a terminal to a computer, with an echo back from the computer to the terminal. If the two-way transmission is simultaneous then the term *full-duplex* is used whereas *half-duplex* refers to the two-way transmission, one way at a time.

9.1 Synchronous transfer

The simplest form of synchronous data transfer is that which, as we have seen, takes place between registers within a clocked digital system such as a computer. Here, individual devices communicate via a common parallel data highway. For example, in response to certain control signals a computer memory unit will output a parallel data word on to the data highway within a time period defined by a clock signal, $\phi 2$. During this period the central processor unit will gate data from the highway into its internal registers. The data transfer is therefore synchronized to the particular time slot referred to as $\phi 2$.

The term 'data highway' may be a grandiose term for the connections between adjacent devices on an integrated circuit chip, or it may refer to a parallel set of wires or printed connections on a backplane, interconnecting a number of circuit boards. In many cases, address highways are also required so that one device may be uniquely selected. In such cases more than one control or clock signal may be required. Naturally, if the transmitting and receiving devices are positioned physically some distance apart, care must be taken to ensure that time delays over the timing and data paths are equalized.

Synchronous *serial* data transmission, in which only a single line is used to transmit the data, demands that the listening device must be able to determine where one data word finishes and the next begins: it must be able to *frame* a character from among a string of continuous data bits. To do this a special predetermined synchronizing character is transmitted at the beginning of the data stream or message. Considering figure 9.1, when the bit stream which has entered the receiving shift register corresponds to the synchronizing code a match signal

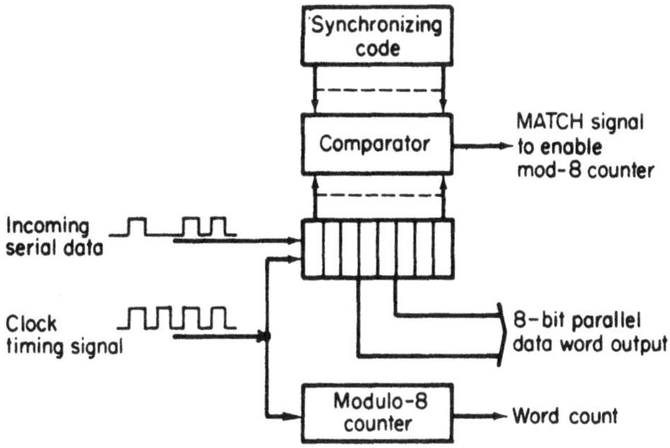

Figure 9.1 A serial synchronous receiver

is generated and the modulo-8 word counter enabled. Thereafter, the receiver accepts every successive eight-bit group as a single word. To reduce the possibility of random noise appearing as a synchronizing code, it is common to require the recognition of two consecutive synchronizing codes. Such serial synchronizing devices are available as LSI circuits and can often perform the dual role of transmitter and receiver, the Motorola MC6852 synchronous serial data adapter being one such unit.

9.2 Asynchronous transfer

Perhaps the commonest form of data transfer between independent peripheral devices is that termed asynchronous serial data transfer, which was initially developed for use with electromechanical teleprinter devices. It is used predominantly to transfer intermittent data to and from data terminals such as the video display unit (VDU) or a Teletype machine. Data transmission takes place effectively over a single wire though in practice three wires are required if the communication is to be two-way since *send* and *receive* lines are usually separate and a common return line is required. The format used is shown in figure 9.2. Until the receiver senses that the incoming line voltage has been taken low it

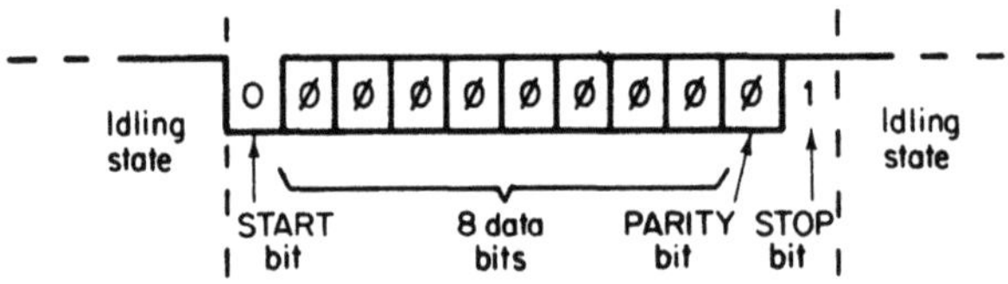

Figure 9.2 Asynchronous serial data format

remains in an *idling* condition. On the first negative-going transition detected the receiver is synchronized to the beginning of the character. In the electromechanical teletypewriters this *START* bit is used to engage an electromechanical clutch to spin a commutating contact once around a segmented disc which carries nine contacts. The data rate of the incoming signal is synchronized to the rotational speed of the commutator so that each segment in turn makes contact with the commutator during the successive bit periods. The result is that the incoming serial data stream is demultiplexed into a series of successive data pulses on separate lines. These data pulses then energize solenoids in an electromechanical decoder. To transmit data from a Teletype machine a reverse process takes place with the same commutator picking off data signals from the segments and converting them into a serial bit stream.

Because of the widespread use of this form of data transfer, LSI serial transmitter/receivers are widely available for use with more modern data terminals. Such a device is known as a *Universal Asynchronous Receiver/Transmitter* (UART). There are many variations (the Intel 8251 USART and the American Microsystems S1883 UART, for example) but the general structure is as shown in figure 9.3. The electromechanical devices were usually limited to a data rate of up to 30 characters per second. Although the data transmission is termed asynchronous, this really refers to the fact that no timing signal need be transmitted along with the data. In practice, timing is obtained from the a.c. mains frequency, this determining the speed of the commutating motors. With the all-electronic systems there is no such speed restriction and data rates up to 500 Kbits per second can be obtained. Timing information is not obtained from the mains but from local oscillators, usually crystal controlled. Naturally both sender and receiver must agree on the data rate to be used! The mechanism of data extraction from the serial bit stream is tolerant of small variations in frequency as the receiver resynchronizes upon receipt of each character.

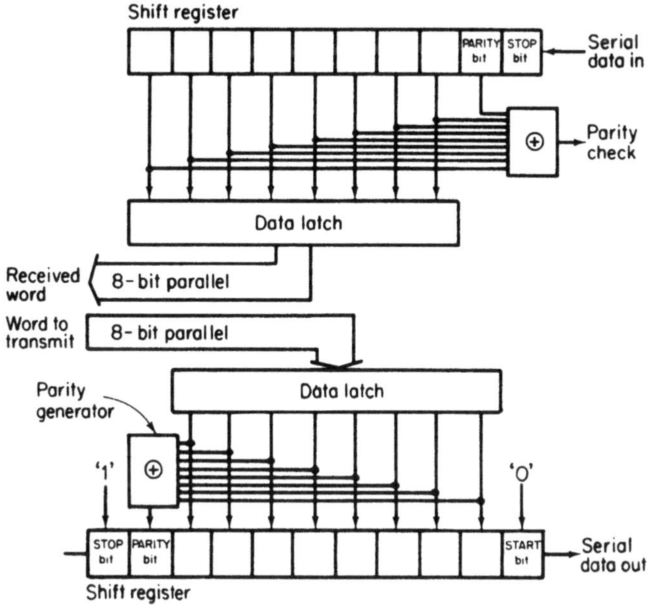

Figure 9.3 The universal asynchronous receiver/transmitter (UART)

In operation, a clock signal must be provided for the UART.The frequency of this timing signal is 16 times the data rate of the signals to be received. When a negative-going edge is detected in the incoming bit stream the next nine clock pulses are counted, during which time the received level is monitored. If the level remains in its low state the detected negative edge is assumed to be the beginning of the 'start' bit of an incoming character. Sixteen further clock pulses are counted and then the signal on the data line at this time is gated into the shift register. Data continues to be fed into the shift register on every succeeding sixteenth clock pulse until ten such data values have been entered. Parity is then checked and, if correct, an 8-bit data word is output. Figure 9.4 shows this sequence. Sampling the data stream in this manner ensures an accurate time

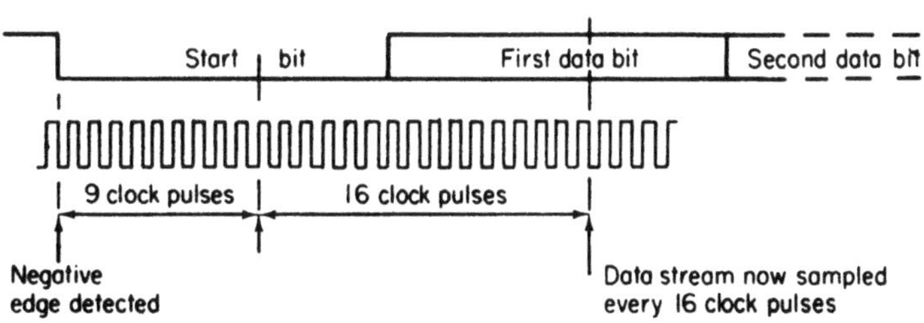

Figure 9.4 Receiver synchronization method

reference to the centre of each data bit even if the transmitter and receiver clocks are not exactly the same. Transmission of data is more straightforward, the parity and start/stop bits being added to the shift register prior to the data being clocked out at the selected rate. The stop bits constitute a delay between successive frames since they are indistinguishable from the idling condition and serve simply to provide a recovery time for the system before the next data word appears. The UART generally contains facilities for selection of odd or even parity and the number of stop bits can usually be adjusted to the requirements of different receiving devices.

If data is to be transmitted between devices operating at different clock rates the use of a *queue*, *elastic buffer* or *first-in first-out* (FIFO) memory can greatly ease transmission problems. This device is basically a register into which data can be entered at one clock rate, and be accessed at a different rate. The circuitry is organized in a similar manner to a shift register in that data entered first is available first at the other end.

FIFO memories find their widest use as peripheral buffer stores: for example, a computer is able to output an entire line of characters to a printer at its own microsecond clock rate and then move on to another task while the printer accesses the data from the FIFO memory at its much lower clock rate. A typical commercial device is the Fairchild 9403 FIFO buffer memory, which is organized in a 24-pin dual-in-line package as 16 words of 4 bits. Facilities are available for expanding this in both a horizontal and vertical direction; that is, a 'wider' word length can be obtained by paralleling several chips and, if desired, more words can be stored by serially increasing the number of chips. FIFO devices are provided with various flags indicating the status of the buffer: full, half full and empty are the usual ones.

9.3 Signal handling

Careful consideration of the medium by which signals are transmitted from one device to another is vitally important if error-free messages are to be received. Over short distances, for example from one circuit board to another, CMOS or low-power Schottky TTL can be used, as shown in figure 9.5a. However, a change in the ground reference of a volt or so due to electrical disturbances can cause severe problems in logic gates with less than one volt of noise margin. A satisfactory technique for removing the effects of unwanted noise is to use a *differential line* or *twisted-pair* (figure 9.5b). The characteristic impedance of these lines is in the order of 100 ohms and, accordingly, we require special drivers, such as the 75172 differential line driver and the corresponding 75173 differential line receiver. A positive and negative-going signal pair is transmitted and received by a voltage comparator having a high common-mode rejection ratio. If a noise signal occurs the wires have identical signals induced in them, the differential, true, signal remaining unaffected.

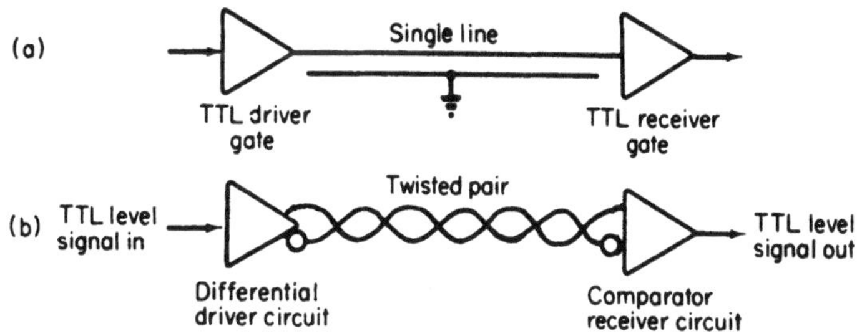

Figure 9.5 Interconnection lines: (a) over a short distance; (b) differential method for longer distances

The '20 mA current loop' is often used to interconnect display units to computers. The presence or absence of a current in an electrical loop signifies logical zero or logical one respectively in such systems. Historically, the 20 mA current loop was used to drive electromechanical machines since solenoids respond more quickly to a current source than to a more conventional voltage source. With the availability of sensitive opto-isolators, currents less than 20 mA are now used and still provide a very reliable transmission system with a high noise immunity and excellent electrical isolation. Figure 9.6 shows a typical interconnection scheme.

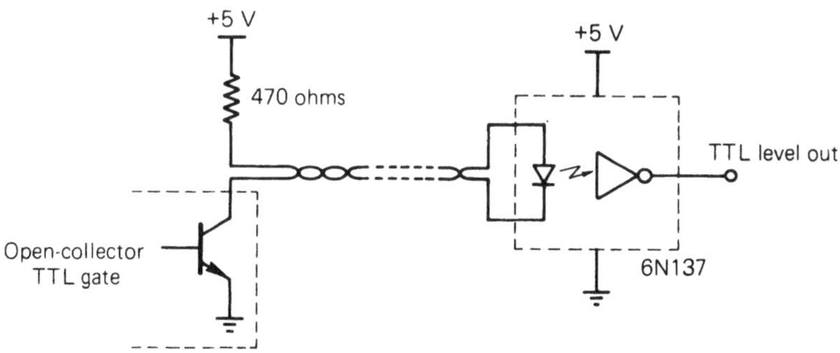

Figure 9.6 A current loop isolation circuit

For the transmission of signals in a high noise environment, optical cables can be used to great effect. Glass fibres, constructed with inner and outer glass layers of differing refractive index can transmit optical signals with very low attenuation, the signals being contained within the inner glass body by total internal reflection. Typically, attenuation values of better than 10 dB/km in the infrared range can be obtained. Although their attenuation is much greater (typically 200 dB/km), cheaper polymer cables are also used, and over shorter distances the lower cost is a dominant factor. Optical cables, with suitable transmitting and receiving semiconductor diode modules, can be used for all applications where high bandwidth, freedom from crosstalk, absence from electrical interference and earth current loops, are at a premium.

Various Standards Committees have put forward codes of practice relating to signal handling in an effort to standardize on a few well-thought-out transmission systems. One such standard, adopted in 1969, was the serial line RS232-C/V.24 interface, developed jointly by the American Electronic Industries Association and the European CCITT. Both versions specify a negative logic system, logical one being a voltage level between -3 and -25 V, and logical zero between $+3$ and $+25$ V. These large signal levels provide good noise immunity. The interface standard also specifies connectors and pin assignments. For a more detailed description refer to appendix C. Associated with this standard are specific transmitter and receiver integrated circuits. The Motorola quad line driver MC1488 and receiver MC1489 are typical and both interface directly with TTL, though extra power supplies are required.

More recently newer standards have emerged, designed to replace the older RS232 as the main specification for the interface between data terminal equipment. Change was necessary as a result of the advances in integrated circuit design since 1969, to permit higher signalling rates (up to 2 Mbits per second), and greater distances between equipment. The cable length is not directly specified since this depends on the type of cable and the signalling rate required. For example, for a twisted pair cable of 25 SWG (0.5 mm diameter) a 1 Mbit per second data rate can be sustained over a 100 m length.

The new standards (Electronic Industries Association) are loosely designated as RS449 but in fact this standard, which defines the signal functions and mechanical characteristics such as pin designations and connectors, specifies the use of transmitters and receivers which are defined by two other standards, RS422 and RS423. These standards define the electrical characteristics of balanced voltage and unbalanced voltage digital interface circuits respectively. Signal voltages either between balanced lines or between signal line and ground are required to be less than 200 mV for the *mark*, or binary one state, and to lie between 2 and 6 V for the *space*, or binary zero state. Integrated circuit drivers and receivers are available, the National Semiconductor DS3691/DS3486 line drivers/receivers being typical examples.

Various standards have been defined to allow an orderly exchange of data in bus-organized systems. The specification for the S100 bus is probably the most widely accepted (IEEE-696). (Details of all the standards mentioned in this chapter can be found listed at the end of the References section.)

A more involved interconnection standard extensively used between digital instruments is the IEEE-488 bus standard. The European equivalent is the IEC 625. This standard was originally developed by Hewlett-Packard as the HP-IB (Hewlett-Packard Interface Bus). Like the previous standards, connectors are specified but unlike the simpler serial data standards the IEEE-488 specifies a *protocol* or communications process between a transmitter (talker) and a receiver (listener) for the transmission of parallel data bytes. The process is a *handshaking* procedure which allows 8-bit data bytes to be transmitted and received asynchronously at the fastest rate possible. The term handshaking refers to the action-response communication which takes place between the transmitter and receiver. Figure 9.7 shows the operation in detail.

The handshake removes the problem of delays from long interconnection paths: data will only be accepted when the negative edge of DAV (figure 9.7) appears at the receiver and will only be changed by the transmitter when it, in turn, receives the negative edge of NRFD.

Common-collector circuits are used in the control signal connections for DAV (data available), NRFD (not ready for data) and NDAC (not data accepted).

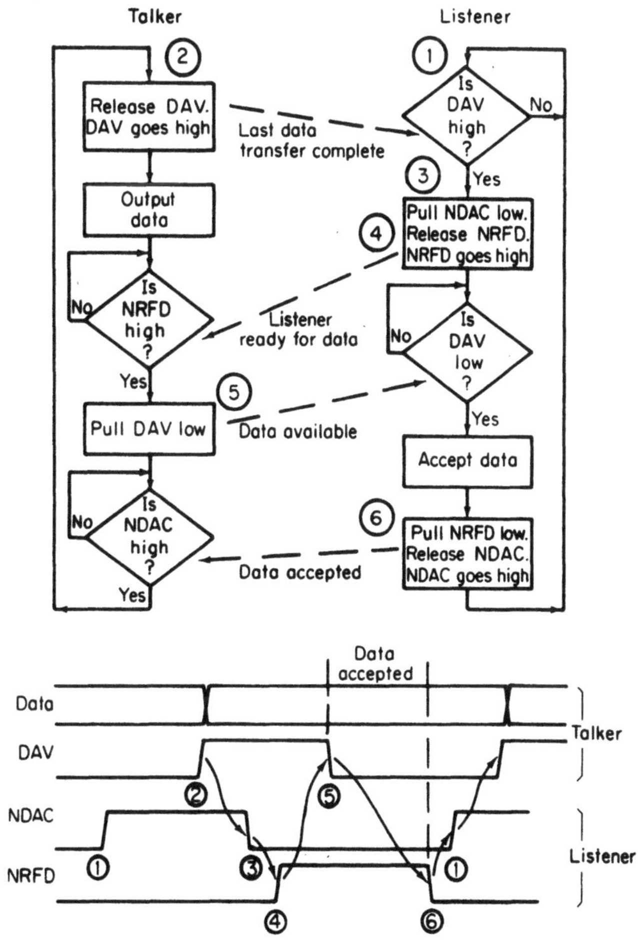

Figure 9.7 Flow chart and timing waveforms of the IEEE-488 Standard communication procedure, showing the 'handshake' principle

Since DAV is in negative logic, the line forms a wired-OR circuit when several devices are to be connected. Any device, by pulling DAV low, can thus become 'bus master'. NRFD and NDAC being in effect in positive logic are wire-ANDed: NRFD is high only if all receivers are ready for data, NDAC is high only when all receivers have accepted data. As with the RS232-C/V.24 interface, integrated circuit devices are available to generate and receive the handshake signals needed for this interface. It should be noted that all communication systems which operate by handshakes must provide some form of 'time-out' circuitry to safeguard against a faulty or non-existent response from the listener. Without such protection a transmitter can 'hang-up', unable to proceed further in its program.

The IEEE-488 standard was developed as an instrumentation bus designed to connect devices having varying operating speeds. Increasingly, communications networks are being required to handle a much higher rate of data flow between devices capable of communicating with each other on an equal basis. For example, several computers may need to share the use of special peripheral devices, such as a printer or a plotter, and the idea of interconnecting equipment within a limited distance, of up to a few kilometres, using a relatively low-cost, high-data-rate transmission medium, has led to the rapid development of local area networks, LAN. Two differing approaches which have been adopted to handle this type of data flow are the broadcast and loop, or ring, systems.

The first important ring system proposed, the Cambridge Ring (Wilkes and Wheeler, 1979), was a loop network, shown in figure 9.8, in which a fixed number of packets of data circulate, unidirectionally, around a series of stations. Each packet is of a fixed length of 38 bits and has the format shown in figure

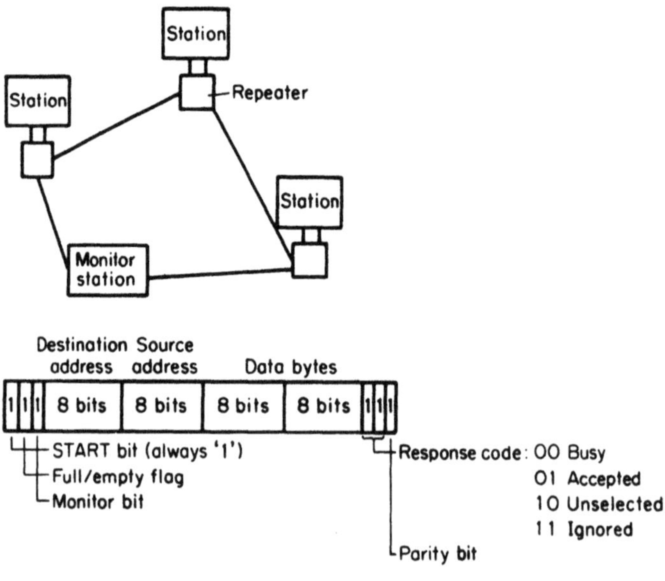

Figure 9.8 The Cambridge Ring and packet format

9.8. The first bit, always set to one, is used for synchronization, there being a gap of several zeros between packets. The second bit indicates whether the packet is *full* or *empty*. If empty, a station wishing to transmit data will generate the appropriate addresses and data, set the full/empty bit to logical one and transmit the packet. Each station constantly monitors the data stream. On detecting its own address the station can accept the incoming data, recoding the response bits appropriately, and pass the packet on around the loop. By coding the destination address with all ones a source can transmit its message to all stations in the loop. When the packet returns to its source station the full/empty bit is cleared and this 'emptied' packet is retransmitted. To avoid stations 'hogging' packets it is usual to prevent a source from 'refilling' the returned packet on that pass.

The data rate of the first-generation Cambridge Rings was 10 Mbits per second. Twisted pairs are used as the interconnecting medium, two pairs being used to transmit a simple self-clocking code: a signal transition on both pairs during a pulse interval denotes a one, a single transition on one or the other pair denotes a nought. The twisted pairs provide good noise immunity to induced noise, and distance between stations of 200 m can be obtained using simple telephone cable quality wire.

The Cambridge Ring is a *slotted ring* in which a finite number of packets circulate. The electrical length of the ring must be an integer number of packet lengths, and a transmitting node inserts its data, two bytes at a time, into packets marked as empty. A second type of ring is the *token passing ring* (IEEE Std 802.5, 1985), using a token package which is passed unidirectionally from node to node around the coaxial cable or twisted pair ring (Bux *et al.*, 1981). When a node wishes to transmit data it must wait until the token is received; it then changes the token bit, figure 9.9, and retransmits the token as a start of package indication, following it with the destination and source addresses, a data string of up to 4099 bytes, and a final tail of six bytes. At the destination node the packet data is copied and the tail is adjusted to indicate receipt. The packet continues round the ring and is removed by the transmitting node on its return. The length of the ring therefore is not critical, as with the Cambridge Ring, and can be any length beyond that necessary to contain all the bits of the token package. In order to ensure that this requirement is satisfied, it is usual to include at least one monitoring node containing a shift register of sufficient length to hold the token. This is called a *latency buffer*.

When the tail of the packet is received by the transmitting node, the token is regenerated and passes on to the next node. By this means, and because the packet length is limited, each node is guaranteed access to the network within a defined time. The ring is continuously monitored to ensure that the token circulates correctly, since operation fails if the token becomes corrupted (or duplicated). In some cases a single monitor station is used, but, where higher reliability is essential, the monitoring function can be duplicated at each node, with a seniority arrangement among nodes to determine which should regenerate the token in the event of failure.

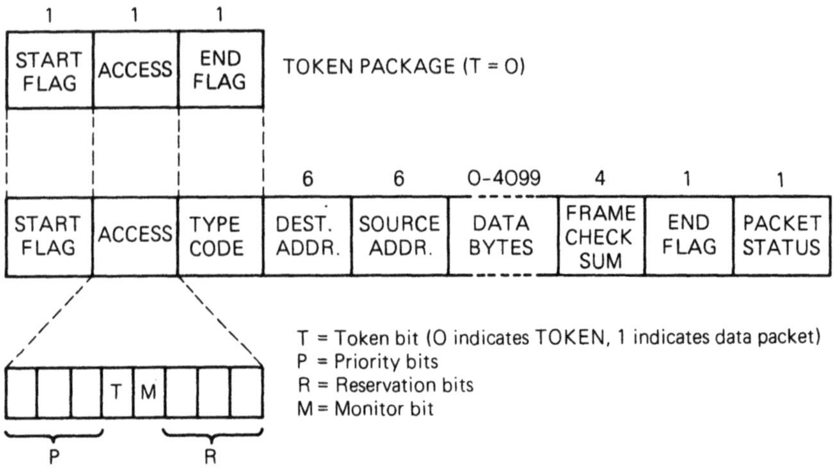

Figure 9.9 The Token Ring packet format

An alternative form of token ring using a broadcast, or bus, arrangement is known as the *token bus* (IEEE Std 802.4, 1985).

Typical of the broadcasting systems is the Ethernet (Xerox, Intel and Digital Equipment Corporation, 1980); see Metcalfe and Boggs (1976). The network consists of a number of access units connected to a coaxial cable (figure 9.10). Each unit can operate as transmitter or receiver and, typically, the separation between station units can be up to 500 m, with data transmission rates of 10 Mbits per second, phase encoded. In operation, stations monitor the line and all receive a transmitted packet. Stations wishing to transmit data do so whenever the line appears to be idle. Unlike the ring system with its circulating

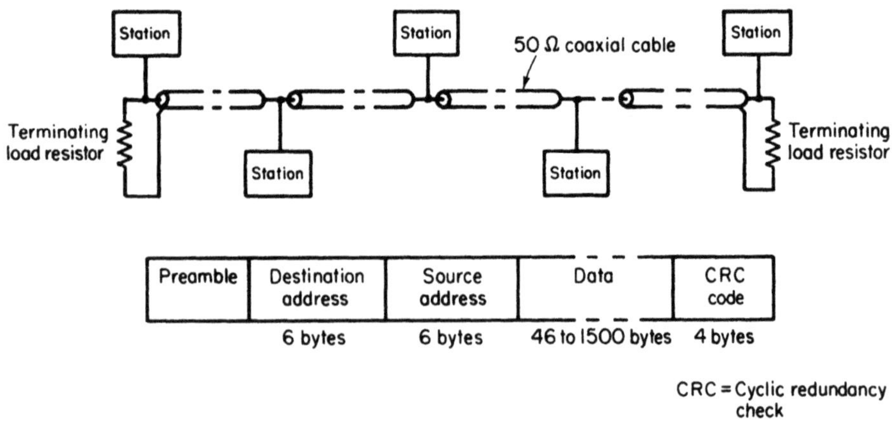

Figure 9.10 The Ethernet system and package format

packet it is probable that conflicts will arise when two or more stations attempt to transmit on to the line at the same time. To deal with this problem each transmitting station monitors the line during transmission. If the data appears to be corrupted, that is another station is simultaneously transmitting, transmission of that message ceases and a 'jamming' signal is sent out, lasting sufficiently long to ensure that all the other stations attemping to transmit become aware of the conflict. Thereupon all transmitting stations disengage from the line for a random time interval. If a collision occurs again stations disengage for a longer random interval. Since a station wishing to transmit will only do so when it detects that the line is idle, collisions can only occur if two stations wish to transmit during the time taken for a signal to travel from one to the other, of the order of microseconds. Such collisions therefore are rare and do not cause a serious problem.

The growing importance of communications networks has led to the establishment of several packet switching protocols. Two important schemes are the High Level Data Link Control (HDLC) and the Synchronous Data Link Control (SDLC). Several semiconductor manufacturers now provide controller circuits to implement the complex protocols required by HDLC/SDLC sytems. Typical are the Motorola MC6854 Advanced Data Link Controller and the Intel 8273 SDLC Protocol Controller. These devices are intended to operate from within a microcomputer and are configured in a similar way to the PIA described in chapter 8.

The HDLC protocol was established by the International Standards Organization (ISO) as the discipline used to implement their X.25 packet switching system (International Telecommunication Union); see Pouzin and Zimmermann (1978). The SDLC is an IBM Corporation communications link protocol. Both protocols are bit-orientated, being designed for serial links, and their use is indicated wherever high integrity data transmission is demanded.

In both systems, a primary control station issues commands and sends or receives data from secondary slave stations. Figure 9.11 shows the SDLC frame structure. The HDLC is similar save that the address field may be any length and the control field 8 or 16 bits wide. The flags delineate the frame information and consist of the bit pattern: 01111110. Various control codes are available such as 'ABORT', this being indicated by a sequence of seven ones in the HDLC protocol and eight ones in the SDLC protocol. In order that differing codes can be presented in this way it is arranged that no other part of any frame can contain more than five consecutive ones. This seemingly impossible restriction on

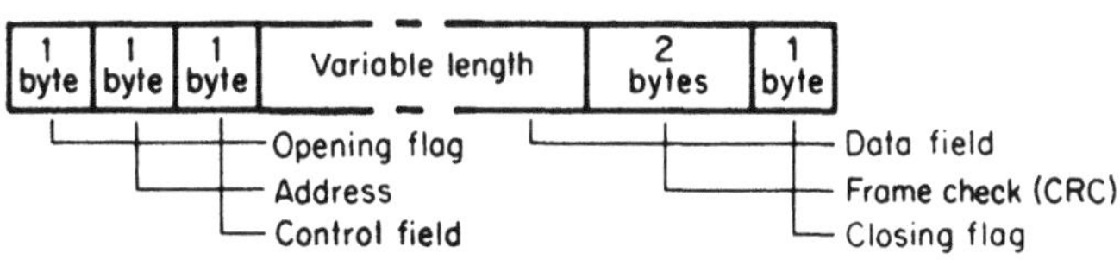

Figure 9.11 Typical frame format for HDLC/SDLC protocols

the data is in fact transparent to the user since, every time five consecutive ones are encountered, the transmitting device automatically inserts zero 'spacers' immediately before transmitting the data, and the receiving device removes them before the data stream is presented to the user. Error checking is carried out automatically using a *cyclic redundancy check* (CRC) and detected errors are communicated to the computer by means of flag setting in a status register.

Cyclic redundancy checking is common to many serial data communications systems. Basically, the data stream which can be considered to be a random number, is made evenly divisible by a known, fixed, number termed the *generator polynomial*. This is achieved by serially dividing the data stream by the known number and then appending the remainder to the end of the data stream to form the frame check bits. Thus the 'coded' data with the appended remainder, when received by the listening station, should be exactly divisible by the same number. If there is no remainder after division then the probability that the data has been received correctly is very high since the polynomial is chosen so that the generated codes (data and remainder) are sufficiently different from one another to make the probability of errors, resulting in the production of another multiple of the polynomial, very low indeed. The divisor number is large; usually 17-bit numbers are used so that the remainder can always be contained within 16 bits, and they are conveniently described in terms of their polynomial. For example, the Motorola 6854 uses the number 10001000000100001_2, usually referred to in its polynomial form as $X^{16} \oplus X^{12} \oplus X^5 \oplus X^0$.

A full explanation of error checking coding techniques is given in Lin and Costello (1983).

9.4 Analog–digital conversion

Digital control systems must be able to accept or transmit *analog* signals as well as digital signals. These signals are usually expressed as voltage levels and, in order that they may be processed digitally, analog-to-digital (A-D) conversion is needed. Conversely, since the output of a digital system may need to be connected to an analog device, digital-to-analog (D-A) conversion is also required.

The task of converting digital signals into their analog equivalent is fairly straightforward. The commonest method is to use a resistor ladder network and figure 9.12 shows a typical system. Digital information is represented by either a voltage V_R or zero being applied to the resistors.

If the ladder is correctly terminated with 2R, the output voltage V_o is given by

$$V_o = \left(\frac{1}{2}B_1 + \frac{1}{4}B_2 + \frac{1}{8}B_3 + \ldots + \frac{1}{2^n}B_n\right) V_R$$

where n is the number of bits and B_n is the value of bit position n, being either 1 or 0.

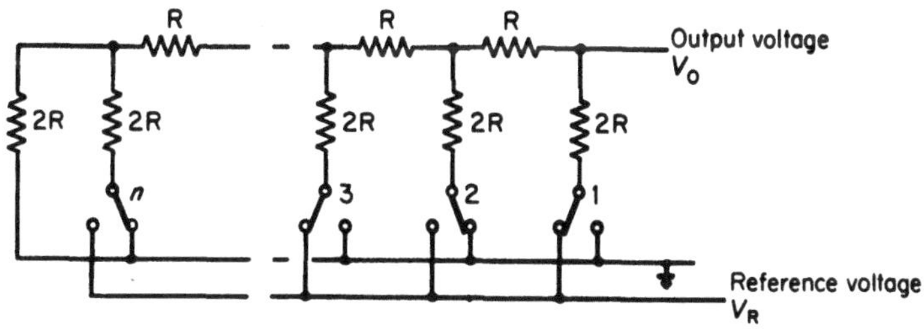

Figure 9.12 Digital-to-analog conversion using a ladder network

The full-scale output is given by

$$V_{\text{ofs}} = (2^n - 1)\,\frac{V_R}{2^n}$$

Note that if the reference voltage supply V_R is of low impedance, then the network presents a constant impedance of R. The great advantage of this type of circuit is that it is not the absolute value of the ladder resistors that is important but their relative values. Ladder networks for D–A converters are readily available constructed from thick-film or thin-film circuits suitable for integrated circuit applications.

One of the simplest A–D converters is the *pulse width modulator*. This relies on the comparison of the unknown signal with an accurately timed ramp signal. Figure 9.13 shows a converter typical of this type. The control unit initiates the

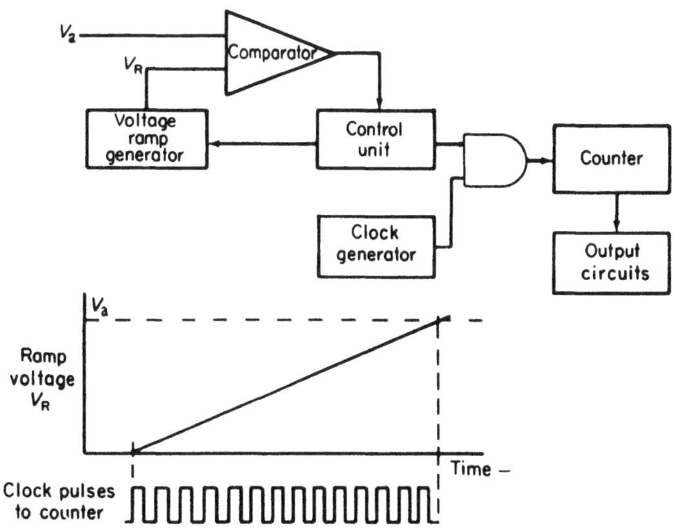

Figure 9.13 Pulse width modulation analog-to-digital conversion

ramp voltage and gates the clock signal into the counter. The output from the comparator is a pulse whose width is proportional to the magnitude of the analog voltage V_a. When the ramp voltage is equal to V_a the clock pulses are inhibited. The count then indicated by the counter is proportional to the applied voltage. This is a very simple A-D converter which relies heavily on the linearity of the ramp voltage and stability of the clock frequency. A more accurate converter which minimizes the effect of variations in ramp voltage is the up-down integrator shown in figure 9.14.

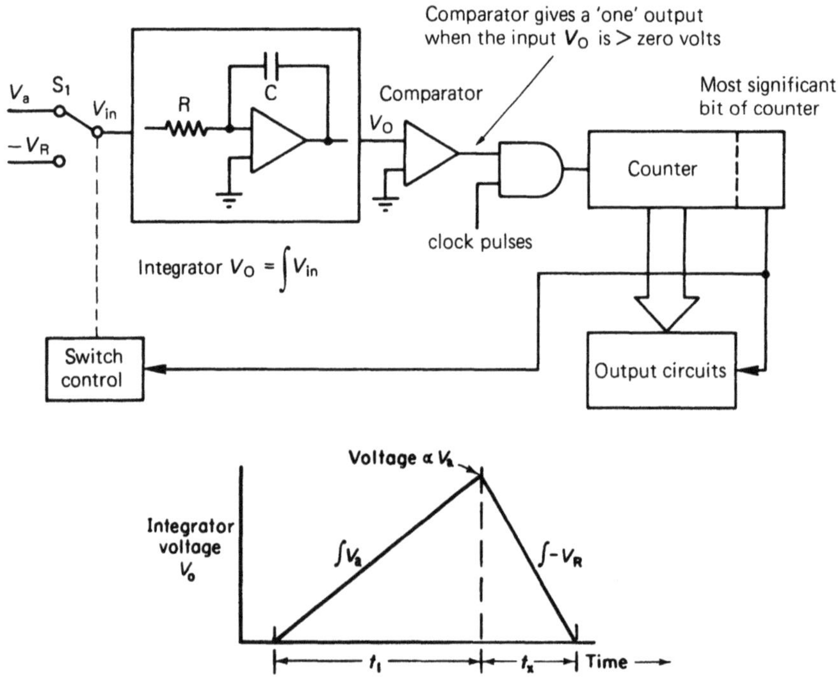

Figure 9.14 The dual-ramp analog-to-digital converter

The operation is as follows. After V_a has been applied, switch S_1 is closed. The output from the integrator is a positive ramp function and the comparator registers an output. This output signal enables the AND gate allowing the counter to count the clock pulses. When the most significant bit of the counter is set the output gates are enabled and the switch drive unit changes the input so that the integrator receives the negative reference voltage. V_o now starts to ramp downwards at a rate defined by V_R. As long as the comparator output is positive the counter continues to count. Now, however, the output gates are open and the count is displayed. As soon as the integrator output goes negative the count is stopped by the comparator.

The time taken to set the most significant bit of the counter is constant and will be termed t_1. Let the time taken for the negative-going ramp to reach zero be t_x.

We have

$$V_{ot_1} = \frac{1}{RC}\int_0^{t_1} V_a \mathrm{d}t = \frac{V_a t_1}{CR}$$

assuming V_o is zero at $t = 0$.

Also

$$V_{ot_x} = \frac{V_R t_x}{CR}$$

Now

$$V_{ot_1} = V_{ot_x}$$

therefore

$$\frac{V_a t_1}{CR} = \frac{V_R t_x}{CR}$$

where

$$V_a = V_R \frac{t_x}{t_1}$$

Since V_R and t_1 are known constants, t_x is proportional to the applied unknown voltage V_a, therefore the count accumulated in the counter during the time t_x is a digital representation of V_a.

This method has the advantage that the accuracy of conversion is independent of clock frequency drift. For increased accuracy it is necessary only to increase the capacity of the counter, though naturally a penalty must be paid. In this case the penalty is increased conversion time.

Another type of A-D converter commonly encountered involves the use of a feedback loop. Figure 9.15 illustrates the principle. In the simplest system the counter starts from zero and is incremented by successive clock pulses. By means of a D-A converter a voltage V_o is generated. When V_o is equal to the applied unknown voltage V_a, the counter is stopped. The count stored in the counter is thus a measure of V_a. The conversion time for this counter is quite long, and a more efficient method known as the *Successive* (or *Logical*) *Approximation* technique involves first setting the most significant bit of the counter. If the resultant V_o is greater than V_a then the next most significant bit is set instead. If the analog output produced is now less than V_a then that bit remains set and the next most significant bit is set also. The process continues until voltages V_o

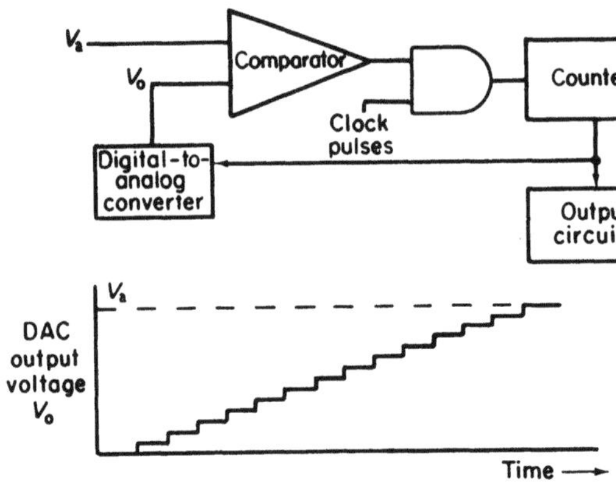

Figure 9.15 A feedback analog-to-digital converter

and V_a are equal (within the limits of resolution). The circuit of the converter is similar to that of figure 9.15 but the counter and the controlling AND gate are replaced by a *successive approximation register*. This register includes control circuitry with a counter which selects each bit of the output word in turn, starting with the most significant bit. At the beginning of a clock period the indicated bit of the register is set to one and the previous bit, if any, is either retained at one or reset to zero, depending on the signal from the comparator. The required value is indicated directly by the comparator since, if the approximated value is too high, the comparator gives a zero indication. Thus the successive approximation register converts the serial train of signals from the comparator into a parallel representation at the register output. A conversion is always completed in the same number of clock periods which is one greater than the number of bits in the output word of the converter. Figure 9.16 shows a typical digital output of such a converter, the final output being 1011.

Both successive approximation and ramp-type A-D converters are readily available in integrated circuit form, quite commonly with signal levels and control signals compatible with microprocessor bus standards for ease of use with microcomputer systems (Aldridge, 1975; Hnatek, 1976). However, A-D converters as described are relatively slow. Current technology limits these A-D

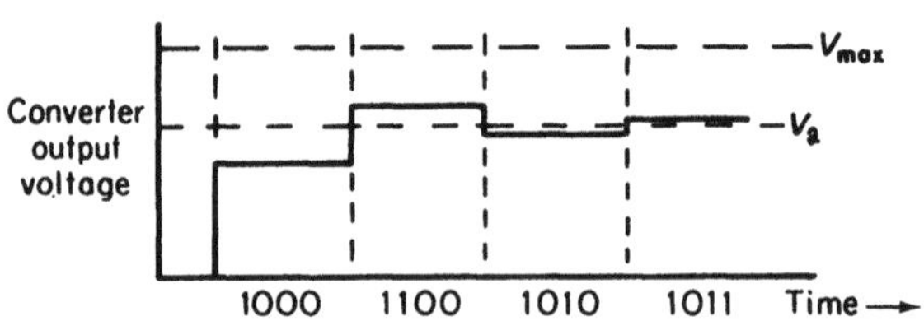

Figure 9.16 Successive approximation analog-to-digital conversion

converters to conversion speeds of approximately 500 ns for an 8-bit resolution and 2 μs at the 12-bit level. A faster technique is that called the *flash* method (Shoreys, 1982). Converters using this technique are able to perform an 8-bit conversion in less than 50 ns. Figure 9.17 shows the basic circuit.

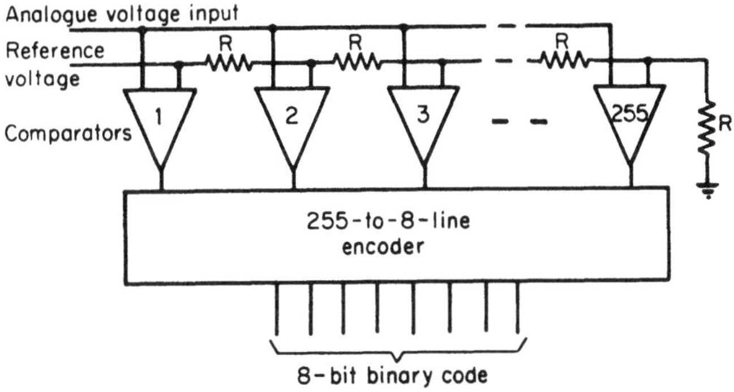

Figure 9.17 The *flash* analog-to-digital converter

A bank of voltage comparators compares the analog voltage with reference signals derived from an internal resistor network. For the 8-bit resolution converter shown, 255 comparators are required. If the analog input voltage is equal to one least-significant-bit, lsb, then only comparator number 255 will be turned on. If it is twice that then both comparators 255 and 254 will switch on. A full-scale input voltage will turn all the comparators on. The comparator outputs are fed into a 255-to-8 decoder circuit which generates a true binary signal. The obvious limitation to this approach is that for an n-bit resolution, $2^n - 1$ comparators are required, and it is only with modern integrated circuit technology that the flash technique has become commercially viable.

It is important that the time taken for a conversion should be as short as possible since a varying input to the converter can introduce considerable errors. In particular, the successive approximation converter can generate substantial errors if the input changes during the weighting process. In order to reduce such errors when rapidly varying signals are being digitized *sample-and-hold*, S/H, circuits are often placed at the input to the analog-to-digital converter. These circuits are capable of *tracking* an analog signal, so that the output accurately follows the input, and then, when conversion is to take place, *holding* the signal so that it remains constant during the conversion time. By this means analog-to-digital converters can operate on samples of signals changing at rates up to the conversion rate of the converter.

The basis of all sample-and-hold circuits is a capacitor which is connected to the analog voltage source by a switch, figure 9.18a. Assuming a slowly changing voltage, when in the *sample* mode, with the switch closed, the capacitor charges to V_0 and the output voltage *tracks*, that is follows, the input voltage. In the

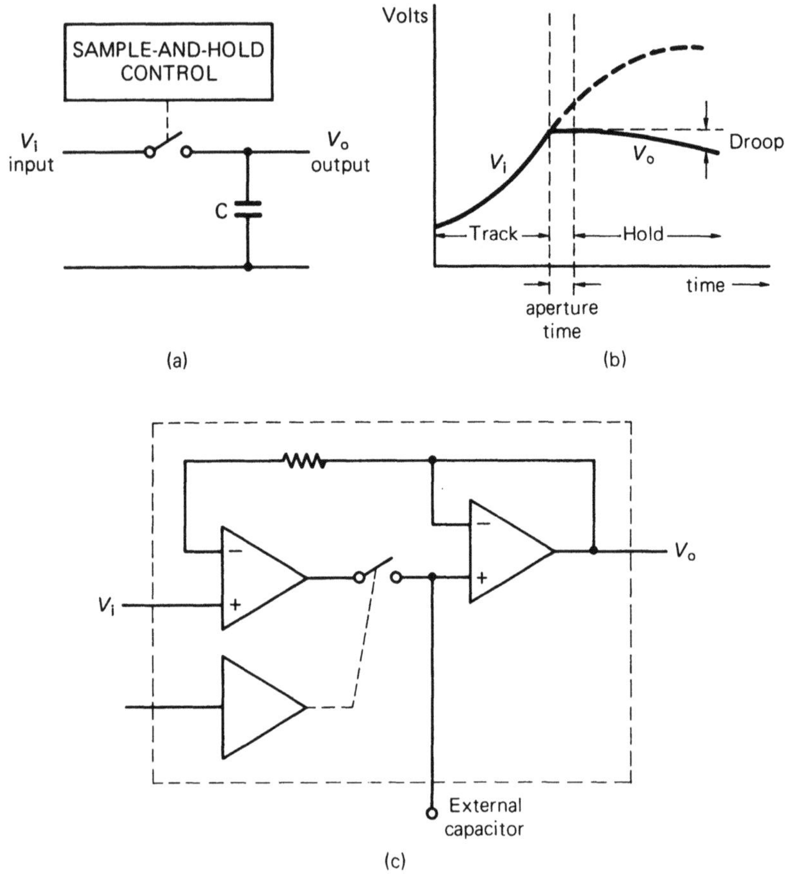

Figure 9.18 Sample-and-hold circuit

hold mode, the switch is opened and the capacitor maintains the voltage at the output. Inevitably, this voltage falls as the capacitor discharges through the load impedance, and the amount by which it falls in a given time is known as the *droop*. The capacitor must always be a low leakage, low dielectric absorption type, using dielectrics such as polypropylene or polycarbonate. During sampling the capacitor must track the changing input voltage accurately, and this demands a very small time constant compared with the large time constant required during hold. These conflicting demands can be resolved by the use of buffer amplifiers at input and output. The time taken to switch from sample to hold mode is known as the *aperture* time, and is quoted in nanoseconds. In practice, the majority of sample-and-hold circuits are monolithic and, rather than designing an individual circuit, the user needs to be able to interpret manufacturers' specifications and select a device most appropriate to the application. A commonly used circuit is the AMD LF398, which can achieve a droop rate as low as 5 mV/minute, though this does depend on the value and quality of the capacitor used. Logic inputs controlling the sample-and-hold sequence are TTL compatible.

The performance of a sample-and-hold circuit is defined by its *slew-rate*, that is the maximum rate of change of signal it can follow accurately, its *droop*, the rate at which the output falls during the hold period, and the *aperture time*, the time taken for the hold command to be implemented. As a general rule, the droop of a given sample-and-hold circuit should not exceed 0.1 of the least significant bit during the conversion time. Sample-and-hold circuits can be obtained with aperture times as short as 20 ps but with droop in the order of 10 μV per ms (Precision Monolithics Inc.).

Problems

9.1 Discuss the problems encountered in the transmission of data from one device to another. In particular, consider the problem of serial asynchronous data transfer and describe, with the aid of a signal timing diagram, a data transfer scheme to enable a number of asynchronous devices to communicate with a central unit.

9.2 Several parallel bus systems have been evolved for the transmission of signals from one device to another. Describe the essential features of a system designed to operate between a central processor and an asynchronous transmitter-receiver. What extensions to the system would be necessary if several data stations were involved?

9.3 Data pulses contain many harmonics; the shorter the pulse the higher the harmonics. There is therefore a limit to the number of pulses which can be carried and resolved by a network and the limit is expressed by $B \geqslant 1/(2T_{\text{min}})$, where B is the network bandwidth and T_{min} is the minimum pulse duration.

(a) A teletypewriter operating in asynchronous mode transmits ASCII characters (with parity) at ten characters per second. Each transmitted character has one start bit and two stop bits. What is the minimum bandwidth required of the system?

(b) The telephone system can handle frequencies up to 4.8 kHz. If a video terminal is to accept data over the telephone network what is the maximum baud rate possible? (One baud equals one bit per second.) Assuming 8-bit characters, what is then the maximum data transfer in characters per second, using: (i) asynchronous operation with one start and one stop bit; (ii) synchronous operation with two SYNC characters every 1000 bytes?

9.4 The UART shown in figure 9.3 is double-buffered in that the received character is transferred from the shift register to a staticizing data latch to extend the time during which the data is available to the microprocessor. If the incoming data is arriving at 9600 bits per second, how long has the microprocessor to respond to the DATA RDY indication and to transfer the character, before an overrun error occurs, that is the next character is written into the latch? What would the time available be if the double-buffering were not used?

9.5 When analog data is to be used in a microprocessor system it is often sensible to make use of the microprocessor in the conversion process. Develop a block diagram for such a system assuming that an 8-bit D–A converter and a voltage comparator are available in addition to the microprocessor system.

9.6 Draw a flow diagram describing the logical sequence in an analog-to-digital conversion for two decade sections of an 8421 BCD successive approximation A–D converter. Discuss the advantages of building a 3-decade BCD successive approximation A–D converter using (a) hard-wired sequential logic, and (b) a single chip microcomputer supported by suitable software.

9.7 The basic considerations in choosing an analog-to-digital converter are the conversion rate, the output precision, the accuracy and cost. What factors affect these considerations, and are there any conflicts which arise?

References

Aldridge, D. (1975). Analog-to-digital conversion techniques with the M6800 microprocessor system, *Application note AN-757*, Motorola Semiconductors Inc., Phoenix, Arizona

Bux, W., Closs, F., Janson, P., Kummesle, K. and Muller, H. R. (1981). A reliable token-ring system for local-area communication, *National Telecomms Conference, New Orleans, November*

Hnatek, E. R. (1976). *A User's Handbook of D/A and A/D Converters*, John Wiley, New York

Lin, S. and Costello, D. J. (Jr) (1983). *Error Control Coding: Fundamentals and Applications*, Prentice-Hall, Englewood Cliffs, New Jersey

Metcalfe, R. M. and Boggs, D. R. (1976). Ethernet; distributed packet switching for local computer networks, *Comm. ACM*, **19**, No. 7, 395-404

Pouzin, L. and Zimmermann, H. (1978). A tutorial on protocols, *Proc. IEEE*, **66**, No. 11, 1346–70

Shoreys, F. (1982) New approaches to high-speed, high-resolution analogue-to-digital conversion, *IEE Electronics and Power*, **28**, No. 2, 175-9

Wilkes, M. V. and Wheeler, D. J. (1979). The Cambridge digital communication ring, *Local Area Communications Symposium*, Mitre Corporation and National Bureau of Standards, Boston, Massachusetts

Xerox Corporation, Intel Corporation and Digital Equipment Corporation (1980). *Ethernet; a Local Area Network, Data Link Layer and Physical Specifications, Version 1.0*, Intel Corporation, Santa Clara, California

Standards

ECMA-RR. *Local area networks: token ring.* European Computer Manufacturers' Association, January 1985

EIA Standard RS232-C. *Interface between Data Terminal Equipment and Data Communications Equipment employing serial binary data interchange.* Electronic Industries Association, Washington D.C., 1969

EIA Standard RS449. *General-purpose 37-position and 9-position interface for Data Terminal Equipment and Data Circuit-Terminating Equipment employing serial binary data interchange.* Electronic Industries Association, Washington D.C., 1977

IEC 625-1. *An interface system for programming measuring apparatus (byte serial, bit parallel).* Publication 625-1, International Electrotechnic Commission, Geneva

IEEE-488-1978. *Standard digital interface for programmable instrumentation.* Institute of Electrical and Electronic Engineers, New York

IEEE-696-1981. *Standard specification for S-100 bus interfacing devices.* Institute of Electrical and Electronic Engineers, New York (previously known as S-100 bus or Altair/IMSAI bus)

IEEE Standard 802.4. *Token-passing bus access method and physical layer specifications*, Institute of Electrical and Electronic Engineers, New York, 1985

IEEE Standard 802.5. *Token ring access method and physical layer specifications.* Institute of Electrical and Electronic Engineers, New York, 1985

Recommendation V.24. *List of definitions for interchange circuits between data-terminal equipment and data circuit-terminating equipment*, in CCITT vol. 8.1, *Data Transmission over the Telephone Network*. International Telecommunication Union, Geneva, 1977

Recommendation X.25. *Interface between data terminal equipment (DTE) and data circuit-terminating equipment (DCE) for terminals operating in the packet mode on public data networks*, in CCITT vol. 8.2, *Public Data Networks.* International Telecommunication Union, Geneva, 1976 (amended 1977)

Appendix A: Logic Symbols

Internationally accepted symbols allow us to represent circuits of different levels of complexity in terms which are readily understood irrespective of the reader's everyday language. Many different levels are needed to cater for the wide range of possible uses, but Kampel (Kampel, 1985) has suggested three main divisions. The first is systems level engineering, requiring pure symbolic logic or conceptual diagrams, describing the system only in block schematic form. The second division is design engineering, in which we are concerned with subsystem interconnections but without defining the precise method of implementation. This requires functional block diagrams. Thirdly we have component level engineering, which is concerned with the physical interconnection of devices, requiring a detailed circuit diagram including pin numbers, device types and positions on a printed circuit board layout, connector details and so on.

Earlier standards have been limited to gate symbols which have been useful only at the third level, and higher levels of abstraction have been prepared, in general, in the form of large block diagrams containing a fair amount of detailed textual description. The most popular example of this gate level type of standard is the American Mil. Spec. Standard, ANSI Y32-14-1973, in which the shape of the symbol denotes the logical operation involved and, in common with most textbooks, we use this standard for its simplicity. More modern methods have replaced Y32-14 with regular rectangular symbols which are more easily drawn on automatic drafting equipment, but there is still a direct equivalence as shown in figure A.1.

The most recent standard, published by the International Electrotechnic Commission in 1983 as IEC Pub. 617-12, 1983, goes beyond simple gate symbols, and allows meaningful symbols to be developed for even very complex circuits, such as VLSI devices, by the use of symbols embedded within symbols and a powerful dependency notation. Many national standards organizations have accepted the IEC standard and have published their own versions. In the UK it is known as British Standard BS3939: section 21, and in the USA as ANSI/IEEE Std 91-1984.

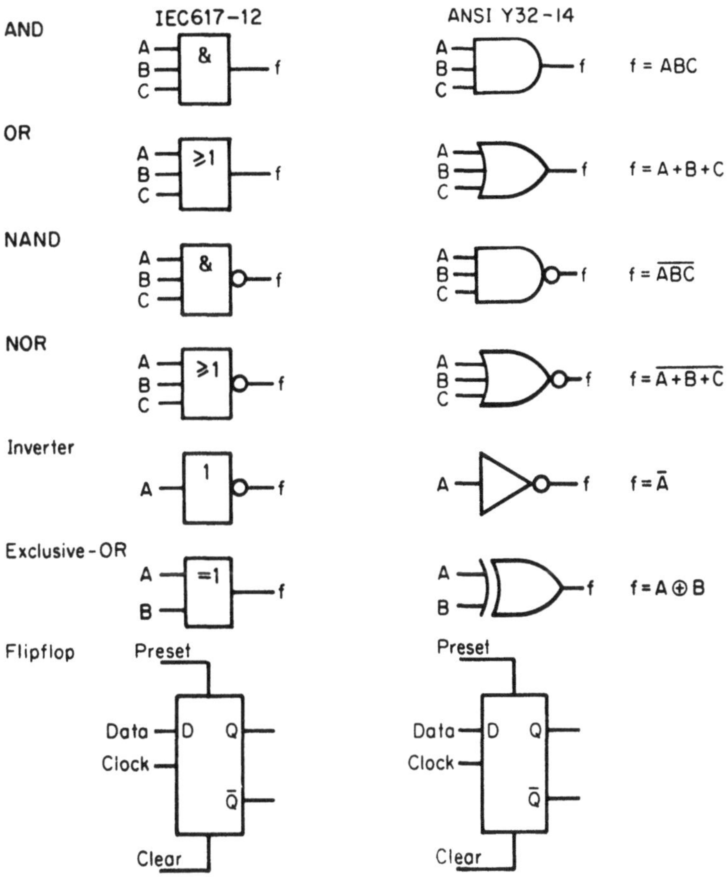

Figure A.1

A.1 An introduction to graphical symbols for diagrams: binary logic elements (IEC Pub. 617-12, 1983)

The outline of each symbol is rectangular, but the height-to-width ratio is not defined. A *general qualifying symbol* is placed centrally at the top of the outline to indicate the logic function. The logic flow in a diagram is conventionally from left to right, and, where appropriate, from top to bottom. Thus, in general, inputs should be shown on the left of the symbol and outputs on the right. Where a flow contrary to the conventional is necessary, arrowheads can be included on interconnecting lines to ensure clarity. Additional *qualifying symbols* relating to individual input and output connections are positioned at the input or output involved, adjacent to the outline. The more common general qualifying symbols and input-output qualifying symbols are shown in figure A.2.

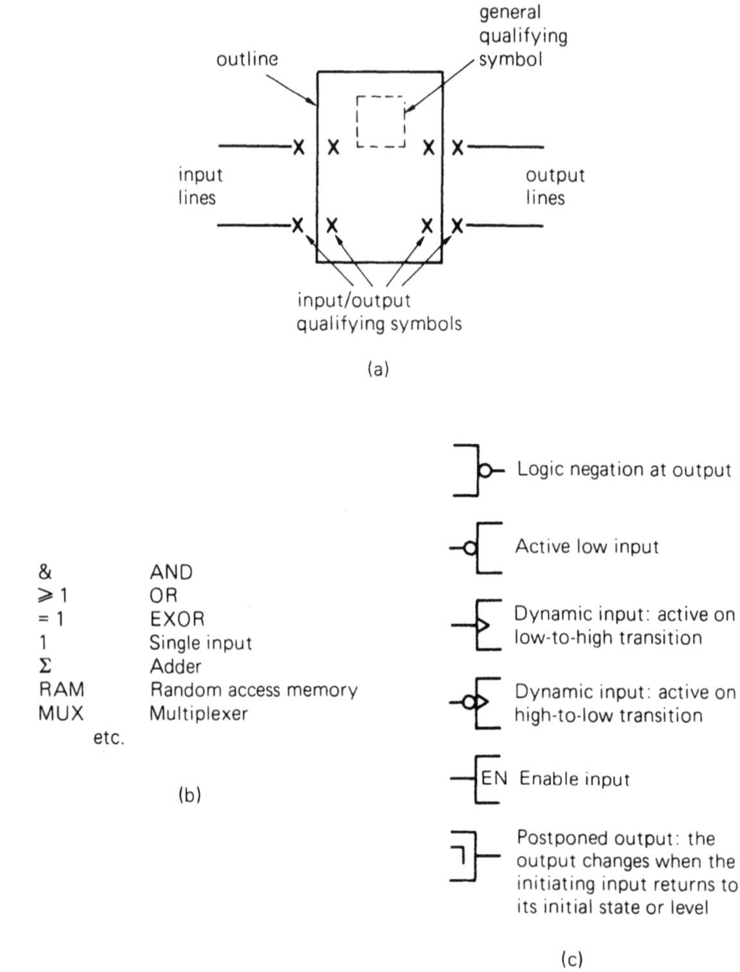

Figure A.2 Standard outline and qualifying symbols: (a) outline details; (b) general qualifying symbols; (c) input–output qualifying symbols. (*Note*: the standard allows positive logic, negative logic or a mixed logic convention; positive logic is used throughout this text unless otherwise stated)

For convenience, symbols may be drawn contiguously, figure A.3a, and no logic interconnection is indicated at joins in the direction of flow. At joins perpendicular to the flow at least one interconnection is indicated. Embedded symbols act in the same way, as, for example, with the multiple-input JK flip-flop of figure A.3b.

Where a circuit has one or more inputs or outputs common to different sections of the circuit, the standard introduces the principle of the common

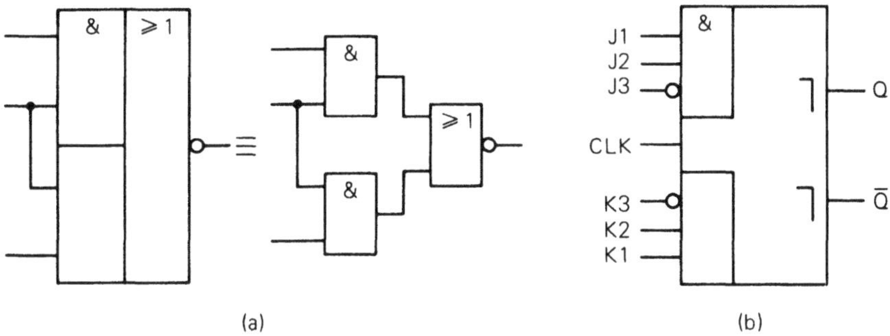

Figure A.3 Contiguous and embedded symbols: (a) continuous representation of AND–OR–INVERT function; (b) multiple-input, master–slave JK flipflop

control block. The specific shape shown at the top of figure A.4a is used to indicate the common control block, and its operation is common to all the elements beneath it.

The dependency notation is a powerful feature not found in previous standards. With large and complex logic circuits the dependency notation allows the interrelationships between inputs and outputs to be clearly defined without

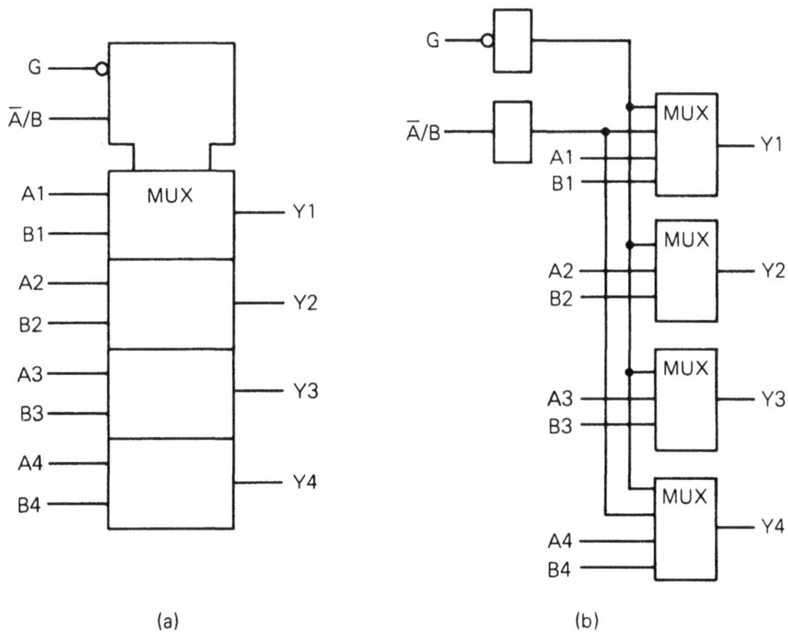

Figure A.4 (a) Use of common control block. (b) Equivalent logic circuit

having to show how they are actually achieved. The principle is simple enough: any input or output affecting other inputs or outputs is labelled with a letter and an identifying number. Any input or output affected by that input or output is labelled with the same number. Certain letters have been reserved for specific relationships; G, for example, indicates an AND relationship and V indicates an OR relationship. The AND–OR-INVERT gate symbol of figure A.3 can now be simplified to the form shown in figure A.5. It should be remembered, of course, that the whole aim of a logic diagram is to present the information as clearly as possible, and the dependency notation is not necessary for straightforward gate interconnections and similar low levels of complexity.

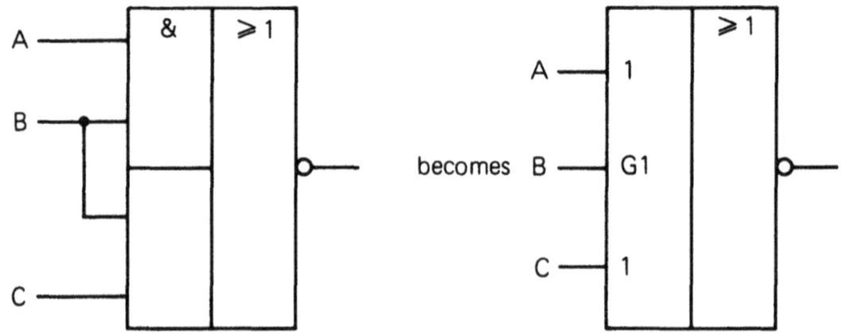

Figure A.5 The dependency notation used in the AND–OR–INVERT function

References

Kampel, I. (1985). *A Practical Introduction to the New Logic Symbols*, Butterworth, London

Mann, F. A. (1985). An explanation of logic symbols – Overview of IEEE Standard 91-1984, in *The TTL Data Book*, Vol. 1, Texas Instruments

Appendix B: Standard Codes

American Standard Code for Information Interchange, ASCII, or ISO code (also known as CCITT Alphabet no. 5).

Bits 6	0	0	0	0	1	1	1	1
Bits 5	0	0	1	1	0	0	1	1
Bits 4	0	1	0	1	0	1	0	1
0 0 0 0	NULL	DLE	SPACE	0		P	@	p
0 0 0 1	SOH	DC1	!	1	A	Q	a	q
0 0 1 0	STX	DC2	"	2	B	R	b	r
0 0 1 1	ETX	DC3	#	3	C	S	c	s
0 1 0 0	EOT	DC4	$	4	D	T	d	t
0 1 0 1	ENQ	NACK	%	5	E	U	e	u
0 1 1 0	ACK	SYNC	&	6	F	V	f	v
0 1 1 1	BELL	ETB	'	7	G	W	g	w
1 0 0 0	BS	CNCL	(	8	H	X	h	x
1 0 0 1	HT	EM	)	9	I	Y	i	y
1 0 1 0	LF	SS	*	:	J	Z	j	z
1 0 1 1	VT	ESC	+	;	K	[	k	{
1 1 0 0	FF	FSR	,	<	L	£	l	|
1 1 0 1	CR	GSR	–	=	M	]	m	}
1 1 1 0	SO	RSR	.	>	N	∧	n	¬
1 1 1 1	SI	USR	/	?	O	←	o	DELETE

3 2 1 0

Bits

Extended binary-coded-decimal interchange code, EBCDIC

Bits 3 2 1 0 \ Bits 7 6 5 4	0000	0001	0010	0011	0100	0101	0110	0111	1000	1001	1010	1011	1100	1101	1110	1111
0000	NULL		DS		SPACE	&	—•—									0
0001			SST				/		a	j			A	J		1
0010			FDS						b	k	s		B	K	S	2
0011		TM							c	l	t		C	L	T	3
0100	PF	RES	BYP	PN					d	m	u		D	M	U	4
0101	HT	NL	LF	RST					e	n	v		E	N	V	5
0110	LC	BS	EOB	UC					f	o	w		F	O	W	6
0111	DELETE	IL	PR	EOT					g	p	x		G	P	X	7
1000		DLE		CNCL					h	q	y		H	Q	Y	8
1001	SOH	DCI		EM					i	r	z		I	R	Z	9
1010	STX	DC2	SM	SS	£	!	∧	:								
1011	ETX	DC3	VT	ESC	.	$	,	#								
1100		DC4	FF	FSR	<	*	%	@								
1101	ENQ	NACK	CR	GSR	(	)	–	'								
1110	ACK	SYNC	SO	RSR	+	;	>	=								
1111	BELL	ETB	SI	USR	|	¬	?	"								ERROR

Control character abbreviations:

ACK	Acknowledge	ETX	End of text	RES	Restore
BELL	Audible signal	FDS	Field separator	RSR	Record separator
BS	Backspace	FF	Form feed	RST	Reader stop
BYP	Bypass	FSR	File separator	SI	Shift in
CNCL	Cancel	GSR	Group separator	SM	Set mode
CR	Carriage return	HT	Horizontal tabulation	SO	Shift out
DC	Device control	IL	Idle	SOH	Start of heading
DLE	Data line escape	LC	Lower case	SS	Start of special sequence
DS	Digit select	LF	Line feed	SST	Significance starter
EM	End of medium	NACK	Negative acknowledge	STX	Start of text
ENQ	Enquiry	NL	New line	SYNC	Synchronous idle
EOB	End of block	NULL	All zero	TM	Tape mark
EOT	End of transmission	PF	Punch off	UC	Upper case
ESC	Escape	PN	Punch on	USR	Unit separator
ETB	End of transmission block	PR	Prefix	VT	Vertical tabulation

Appendix C: The RS232-C Interface

The RS232-C standard defines a logical one data signal as a voltage between −3 and −25 V. A logical zero is defined as a voltage between +3 and +25 V. Four types of signal are defined: data signals, control signals, timing signals and ground signals. Note that in a given interface between equipment all the specified signals may not be required and therefore not present. What must be adhered to, however, is the pin assignment on a 25-pin D-type connector defined as part of the RS232-C specification. Pin assignments are given in the table. The interface is applicable for data signalling rates in the range from zero to a nominal 20 000 bits per second.

Pin number	Signal nomenclature	Signal abbreviation	Signal description	Category
1	AA	–	Protective ground	ground
2	BA	TXD	Transmitted data	data
3	BB	RXD	Received data	data
4	CA	RTS	Request to send	control
5	CB	CTS	Clear to send	control
6	CC	DSR	Data set ready	control
7	AB	–	Signal ground	ground
8	CF	DCD	Received line signal detector	control
9	–	–	–	reserved for test
10	–	–	–	reserved for test
11	–	–	–	unassigned
12	SCF	–	Secondary received line signal detector	control
13	SCB	–	Secondary clear to send	control
14	SBA	–	Secondary transmitted data	data

15	DB	–	Transmission signal element timing	timing
16	SBB	–	Secondary received data	data
17	DD	–	Received signal element timing	timing
18	–	–	–	unassigned
19	SCA	–	Secondary request to send	control
20	CD	DTR	Data terminal ready	control
21	CG	–	Signal quality detector	control
22	CE	–	Ring indicator	control
23	CH/CI	–	Data signal rate selector	control
24	DA	–	Transmit signal element timing	timing
25	–	–	–	unassigned

Solutions to Selected Problems

Chapter 1

1.3 (a) $T = \bar{A}\bar{B}\bar{C}\bar{D} + \bar{A}\bar{B}\bar{C}D + \bar{A}\bar{B}C\bar{D} + \bar{A}\bar{B}CD + \bar{A}B\bar{C}D + A\bar{B}\bar{C}\bar{D} + A\bar{B}C\bar{D}$
$+ AB\bar{C}D + ABC\bar{D}$.
(b) $T = (\bar{A} + \bar{B} + \bar{C} + \bar{D})(\bar{A} + \bar{B} + C + D)(\bar{A} + B + \bar{C} + \bar{D})(\bar{A} + B + C + \bar{D})$
$(A + \bar{B} + \bar{C} + \bar{D})(A + \bar{B} + \bar{C} + D)(A + \bar{B} + C + D)$.

1.4 Mrs Smith went to the Ministry of Secrets.

1.5 $f = (A\bar{B} + \bar{A}B)(A + D)$.

1.6 $(\bar{A} + \bar{B})(\bar{B} + \bar{D})(\bar{A} + \bar{C} + D)$.

1.7 $f = (\bar{A} + B)(\bar{A} + \bar{C})$; $f = \bar{A} + B\bar{C}$.

1.8 The set of prime implicants is $(\bar{B}C)(\bar{A}B\bar{D})(\bar{A}B\bar{C})(\bar{A}C\bar{D})(B\bar{C}D)$.

1.9 The function is fully symmetric.

1.10 Non-equivalence symmetry in AB and equivalence symmetry in $B\bar{C}$ and $A\bar{C}$.

1.12 (a1) $f(ABCD) = \Sigma m(3, 5, 6, 9, 10, 12, 15)$
(a2) $f(ABCDE) = \Sigma m(1, 2, 4, 7, 8, 11, 13, 14, 16, 19, 21, 22, 25, 26, 28, 31)$
(b1) $S^5_2(ABCDE)$
(b2) $S^3_{1,2}$
(c1) $f = ABC$ (c2) $f = A + B + C$ (c3) $f = A \oplus B \oplus C$

1.13 The circuit must generate a function such as
$f = (A + B)(CD + CE + DE) + CDE + AB(C + D + E)$.

1.15 $X = AD(B + C) + A\bar{B}C + B\bar{C}D$; $Y = \bar{A}D + \bar{B}C + B\bar{C}D$; $Z = B\bar{C} + B\bar{D} + \bar{B}CD$.

1.16 $f = (\bar{A} + \bar{B} + \bar{D})C$.

Chapter 2

2.1 (a) 13.375 (b) 38.3125 (c) 427.328125.

2.2 (a) 653.715 (b) 427.900390625.

2.3 (i) 0001011 (ii) 1111011 (–0000101).

Chapter 3

3.1 (i) $NM_1 = 1.0$ V (ii) $NI_0 = 0.6$ $NI_1 = 0.4$.

Chapter 5

5.3 The output code sequence is 0, 1, 2, 3, 4, 8, 9, 10, 11, 12.

5.5 A 3-input AND gate is used to recognize the 000∅ condition and to inject a zero via the 3-input exclusive-OR gate. The sequence is then 0101100100001111

Chapter 6

6.3 There are essential hazards when in state A with inputs 11 and in state C with inputs 00. They are not critical because of the full address decoding used in the ROM.

6.4 The K input of the fourth stage is stuck at '1'. The most likely single fault is a solder short to +5 V on the printed circuit board.

6.10

	Reset	Internal state	Input	Next state	Output
	0	0000	0	0000	000
(0)	1	0000	0	0000	000
	1	0000	1	0001	000
	1	0001	1	0001	000
	1	0001	0	0010	000
(1)	1	0010	0	0010	001
	1	0010	1	0011	001
	1	0011	1	0011	001
	1	0011	0	0100	001
(2)	1	0100	0	0100	010
	1	0100	1	0101	010
	1	0101	1	0101	010
	1	0101	0	0110	010
(3)	1	0110	0	0110	011
	1	0110	1	0111	011
	1	0111	1	0111	011
	1	0111	0	1000	011
(4)	1	1000	0	1000	100
	1	1000	1	1001	100
	1	1001	1	1001	100
	1	1001	0	1010	100
(5)	1	1010	0	1010	101
	1	1010	1	1011	101
	1	1011	1	1011	101
	1	1011	0	1100	101
(6)	1	1100	0	1100	110
	1	1100	1	1101	110
	1	1101	1	1101	110
	1	1101	0	0000	110
	1	1110	0	0000	000

Chapter 8

```
8.1  29  0010   1001
     65  0110   0101
         ----   ----
       0 1000   1110    C = 0, H = 0
                0110    Correction
                ----
            1   0100
     --  ----
     94  1001
```

8.4

	A		C	S	Z	V	H	P	A		C	S	Z	V	H	P
(i)	0011	0101	1	0	0	0	0	1	0101	0000	1	0	0	1	1	1
(ii)	0100	0010	0	0	0	0	0	1	1000	0001	0	1	0	0	0	1
(iii)	1111	0011	0	1	0	0	0	1	1100	1111	0	1	0	0	0	1
(iv)	1011	0001	0	1	0	0	0	1	0100	1110	0	0	0	0	0	1
(v)	0011	1001	0	0	0	0	0	1	1100	0110	0	1	0	0	0	1
(vi)	0000	0000	0	0	1	0	0	1	0000	0000	0	0	1	0	0	1

(B) = 1100 0011 throughout.

8.5 (a) (SP) = 3F60

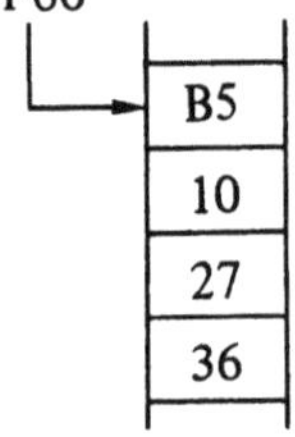

(PC) = 2073

(b) (SP) = 3F62

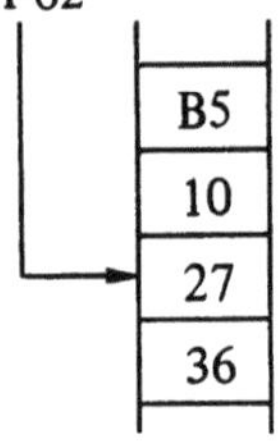

(PC) = 1085

Chapter 9

9.3 (a) Each character has 11 bits giving 110 bits/s. $B \geqslant 110/2 = 55$ Hz.

(b) Baud rate $\leqslant 2 \times 4.8 \times 10^3 = 9600$ baud.

(i) Each character has 10 bits, giving 960 characters per second.

(ii) 1000 bytes require 8016 bits, giving a transfer rate of $9600 \times 1000/8016$, that is 1197 characters per second.

9.4 Each character has 11 bits, so time per character is $11/9600 = 1.15$ ms. The final data bit is latched approximately midway through the ninth bit period of the character, and is followed by the parity and stop bit periods. The time available without double-buffering is, therefore, approximately 2.5 bit periods, or 2.5/9.6 ms; that is 0.26 ms.

Index

MIX
Papier aus verantwortungsvollen Quellen
Paper from responsible sources
FSC® C105338

If you have any concerns about our products,
you can contact us on
ProductSafety@springernature.com

In case Publisher is established outside the EU,
the EU authorized representative is:
Springer Nature Customer Service Center GmbH
Europaplatz 3, 69115 Heidelberg, Germany

Printed by Libri Plureos GmbH
in Hamburg, Germany